Progress in Genomic Medicine
From Research to Clinical Application

Progress in Genomic Medicine
From Research to Clinical Application

Moyra Smith
*Department of Pediatrics and Human Genetics,
College of Health Sciences, University of California,
Irvine, CA, United States*

ACADEMIC PRESS
An imprint of Elsevier

Academic Press is an imprint of Elsevier
125 London Wall, London EC2Y 5AS, United Kingdom
525 B Street, Suite 1650, San Diego, CA 92101, United States
50 Hampshire Street, 5th Floor, Cambridge, MA 02139, United States
The Boulevard, Langford Lane, Kidlington, Oxford OX5 1GB, United Kingdom

Copyright © 2022 Elsevier Inc. All rights reserved.

No part of this publication may be reproduced or transmitted in any form or by any means, electronic or mechanical, including photocopying, recording, or any information storage and retrieval system, without permission in writing from the publisher. Details on how to seek permission, further information about the Publisher's permissions policies and our arrangements with organizations such as the Copyright Clearance Center and the Copyright Licensing Agency, can be found at our website: www.elsevier.com/permissions.

This book and the individual contributions contained in it are protected under copyright by the Publisher (other than as may be noted herein).

Notices
Knowledge and best practice in this field are constantly changing. As new research and experience broaden our understanding, changes in research methods, professional practices, or medical treatment may become necessary.

Practitioners and researchers must always rely on their own experience and knowledge in evaluating and using any information, methods, compounds, or experiments described herein. In using such information or methods they should be mindful of their own safety and the safety of others, including parties for whom they have a professional responsibility.

To the fullest extent of the law, neither the Publisher nor the authors, contributors, or editors, assume any liability for any injury and/or damage to persons or property as a matter of products liability, negligence or otherwise, or from any use or operation of any methods, products, instructions, or ideas contained in the material herein.

British Library Cataloguing-in-Publication Data
A catalogue record for this book is available from the British Library

Library of Congress Cataloging-in-Publication Data
A catalog record for this book is available from the Library of Congress

ISBN: 978-0-323-91547-2

For Information on all Academic Press publications
visit our website at https://www.elsevier.com/books-and-journals

Publisher: Andre G. Wolff
Acquisitions Editor: Peter B. Linsley
Editorial Project Manager: Kristi Anderson
Production Project Manager: Kiruthika Govindaraju
Cover Designer: Christian J. Bilbow

Typeset by MPS Limited, Chennai, India

Dedication

This book is dedicated to Dr. Susan Dyson Bodemer and to Dr. Simon Prinsloo, dear friends and wise counselors since our days in medical school, for their continued friendship and encouragement, even as I chose an unusual path in Medicine.

Contents

Preface xxi
Acknowledgments xxiii
Epigraph xxv

Part I History and Growth of Genetic Medicine

CHAPTER 1   Documentation of units of inheritance and their contribution to phenotype 3
   1.1 Rediscovery of the laws of Mendel 3
   1.2 Genes and genetics 4
   1.3 Nucleic acids 5
   1.4 The structure of DNA 5
   1.5 DNA and chromatin 5
         1.5.1 Consequences of determination of DNA structure 6
         1.5.2 Modifications of DNA sequences 6
   1.6 Applications of studies of chromosomes, genomes, genes, and gene expression to clinical medicine 6
         1.6.1 Chromosome microarray analyses 8
   1.7 Long-read sequencing for detection of genomic variants including structural chromosome abnormalities 9
   1.8 Determination of the significance of structural variants in the genome 10
         1.8.1 Clinical significance of structural genomic variants 11
   1.9 Mosaicism 11
         1.9.1 Chromosomal mosaicism 12
         1.9.2 Mosaicism detection 12
         1.9.3 Mosaicism and genetic diseases 12
   1.10 Germline mutations 13
   1.11 Genetic mosaicism in inborn errors of immunity 13
         References 13
         Further reading 15

CHAPTER 2   Early documentation of inherited disorders through family studies 17
   2.1 The Treasury of Human Inheritance 17
   2.2 Ectrodactyly 17
   2.3 Deafness 18
   2.4 Hemophilia 18

2.5	Achondroplasia	18
2.6	Color blindness	19
2.7	Blue sclerotics and fragility of bone	19
2.8	Hereditary optic atrophy (Leber's disease)	20
2.9	Huntington's chorea	20
2.10	Duchenne muscular dystrophy	22
2.11	Determination of genetic causes of specific diseases and family studies	23
	References	23
	Further reading	25

CHAPTER 3 Discoveries in physiology, biochemistry, protein, and enzyme studies between 1920 and 1970......... 27

References 31
Further reading 32

CHAPTER 4 Early translation of biochemical, metabolic, and genetic discoveries into clinical medicine 33

4.1	ABO	33
4.2	Further information on the ABO blood group system	33
4.3	Secretor status	34
4.4	Mapping of the ABO locus to a chromosome	35
4.5	Rh blood group system	35
4.6	RHD genotyping	36
4.7	Chemical analyses and metabolism incorporating information gathered across the decades	37
	References	38
	Further reading	39

CHAPTER 5 Advances in methods of genome analyses, nucleotide analyses, and implications of variants ... 41

5.1	Introduction	41
5.2	DNA sequencing	41
5.3	Applications of long-read sequencing	44
5.4	Sequence variant interpretation	45
	5.4.1 Sequencing in clinical diagnosis: reinterpretation of data and secondary findings	47
	5.4.2 Long-range sequencing relevance to diagnosis of rare disorders	48

5.5 Additional evidence for digenic or complex inheritance 49
5.6 Variants in nonprotein coding genomic regions 50
5.7 Haplotype phasing 51
    5.7.1 Noninvasive prenatal screening and haplotype phasing .. 51
5.8 Haplotype analysis 51
5.9 Long-range sequencing and identification of structural
    genomic variants leading to disease 53
5.10 Investigations of chromatin structure and genomic function 53
    5.10.1 Chromatin conformation capture 55
5.11 Methylation analyses 55
5.12 Imprinted genomic regions 56
5.13 Genetic disorders where analysis of methylation is important ... 57
    5.13.1 Methylation and cancer 57
    References 58

Part II Clinical Applications of Genomic Medicine

CHAPTER 6 Expansion of use of genome analyses and sequencing in diagnosis of genetic diseases 65

6.1 Measurement toolkit for assessing the clinical utility of
    whole genome sequencing 65
    6.1.1 Next generation sequencing in clinical neurology 66
6.2 One phenotype many genes 68
    6.2.1 Incomplete penetrance 68
6.3 Genome sequencing in pediatric developmental defects 69
6.4 Optical DNA mapping in human genome studies 70
6.5 Transcriptome sequencing 70
6.6 Imprinting 72
6.7 Imprinting disorders 73
6.8 Prader – Willi syndrome and Angelman syndrome 74
6.9 Silver – Russell syndrome 75
6.10 GNAS locus 77
6.11 Epivariations 79
6.12 Multilocus imprinting disorders 80
6.13 Chromosomes genomes and sequence 82
6.14 Structural genomic variants 82
    6.14.1 Population analyses 83
    6.14.2 Dosage sensitivity and haploinsufficiency 83
    6.14.3 ACMG recommendations regarding genomic copy
           number variants analysis and reporting 84

6.15 Assessment of copy number changes in different conditions and at different life stages 84
6.16 Prenatal exome sequence analysis 86
6.17 Deciphering Developmental Disorders Study 87
6.18 Investigations of causes of recurrent miscarriage 88
6.19 Sequencing in prenatal diagnosis: noninvasive prenatal testing 88
References 90
Further reading 93

CHAPTER 7 Improved analyses of regulatory genome, transcriptome and gene function, mutation penetrance, and clinical applications 95

7.1 Introduction 95
    7.1.1 Gene expression 96
7.2 Regulatory genome, gene expression, phenotype, and variability 96
    7.2.1 Long noncoding RNAs (long nonprotein coding RNAs) 96
    7.2.2 Defining functions of specific regulatory elements and their relationship to diseases 97
7.3 Epigenetic factors relevant to gene expression 98
    7.3.1 Disorders of the epigenetic machinery leading to neurodevelopmental disorders 101
    7.3.2 Cohesinopathies 101
7.4 Regulatory circuit: Epimap 102
    7.4.1 Combinations of variants in different genes and impact of phenotype 103
7.5 Genotype phenotype axis 103
7.6 Promoters 104
7.7 Transcription initiation and promoters 105
    7.7.1 Genes with more than one promoter 105
    7.7.2 Ornithine transcarbamylase gene promoters and enhancers 106
7.8 Transcription factors 107
    7.8.1 Transcription elongation and RNA polymerase II 108
7.9 Transcription termination 109
7.10 Polyadenylation 109
7.11 Alternate polyadenylation 109
    7.11.1 Alternate splicing of transcripts 110

7.12	The spliceosome	111
	7.12.1 Generation of microexons	111
7.13	MicroRNAs and posttranscriptional regulation	112
7.14	Translation, ribosomes biogenesis, functions, and defects	112
	7.14.1 Human disorders associated with impaired ribosomal biogenesis or function	112
7.15	Translation of mRNA to proteins and associated defects leading to disease	113
	7.15.1 Aminoacyl tRNA synthases	113
	7.15.2 Noncanonical functions of aminoacyl tRNA synthetases	114
7.16	tRNAs	114
7.17	RNA surveillance	115
	7.17.1 Posttranscriptional control and RNA binding proteins	116
	7.17.2 RNA modifications and regulation of gene expression	116
7.18	Translation	117
7.19	Nonsense-mediated decay	117
	7.19.1 Suppression of nonsense mutations	118
7.20	Nonsense mutations and human disease	119
7.21	Approved RNA targeted therapeutics	119
7.22	Therapy with short inhibitory RNAs	120
7.23	MicroRNAs in therapeutic use	121
7.24	RNA sequencing in diagnosis of genetic diseases	121
7.25	Penetrance of mutations and modified penetrance	122
7.26	Variable penetrance of disease due to polymorphisms in regulatory factors	123
7.27	Penetrance in inherited eye diseases	124
7.28	Primary immunodeficiency and incomplete penetrance	125
	References	126

CHAPTER 8 Standardized phenotype documentation, documentation of genotype phenotype correlations

8.1	Phenotype and clinical genetics	133
8.2	Congenital malformations and syndromes	134
	8.2.1 Inborn errors of development	134
	8.2.2 Twin studies and analysis of gene effects on phenotype	135

8.2.3 Accounting for phenotypic differences in individuals with the same genetic defect 135

8.3 Variable phenotypes associated with specific mitochondrial mutations 136

8.4 Variable genomic abnormalities in individuals with the same phenotype 136

8.5 Standardized phenotype documentation, documentation of genotype phenotype correlations databases 137

8.5.1 Clinical genetics and genomics databases 138

8.6 Phenome-wide association studies 139

8.7 Dysmorphology syndromes with overlapping features due to defect in gene products that function in a specific pathway 139

8.7.1 Gene products involved in the RAS/MAP signal transduction pathway and chromosomal map positions 139

8.8 Phenotypic defects due to defects in sonic hedgehog signaling pathway 140

8.9 Fibroblast growth factor signaling pathway 142

8.10 Fibroblast growth factor receptor defects 143

8.10.1 Mutations reported as pathogenic in achondroplasia multiple submitters 144

8.11 Transforming growth factor beta signaling pathway 144

8.11.1 Phenotypic features Loeys – Dietz syndromes 146

8.11.2 LDS type 4 TGFB2 mutations 146

8.11.3 LDS type 5 TGFB3 mutations 147

8.12 Marfan syndrome 15q21.1 FBN1 147

8.12.1 FBN1 mutations in Marfan syndrome 147

8.13 FBN1 mutations, pathogenic, likely pathogenic, Marfan syndrome multiple submitters, without conflicts identified in Clin Var searches 148

8.14 Connective tissue disorder Ehlers – Danlos syndrome disorders 148

8.14.1 COL3A, vascular EDS, pathogenic/likely pathogenic, mutations, multiple submitters, no conflicts 149

8.14.2 Classic type `9q34.3 COL5A AD collagen type V alpha 1 chain 150

8.14.3 Hypermobile Ehlers – Danlos syndrome 150

8.15 DNA methylation episignatures and phenotypic correlations ... 151

References 152

CHAPTER 9 Expansion of methods of gene editing therapy and analysis of safety and efficacy

- 9.1 Introduction 155
- 9.2 Therapies designed to block nucleotides or RNA derived from a specific gene 155
- 9.3 Oligonucleotide therapies 155
  - 9.3.1 Steric block oligonucleotides 156
  - 9.3.2 RNA inhibition in therapies 156
  - 9.3.3 MicroRNAs as mRNA inhibitors 157
  - 9.3.4 Long noncoding RNAs, small RNAs, endogenous antisense RNAs 157
- 9.4 Delivery challenges in oligonucleotide therapies 158
- 9.5 Splice mutations and diseases 158
- 9.6 Antisense therapies under investigation 159
- 9.7 Genomic data leading to therapeutics 160
  - 9.7.1 Additional RNA modifications to further improve use in therapy 161
- 9.8 Pluripotent stem cells for investigation of disease manifestations and effects of therapies 161
- 9.9 Gene therapy by adding genes 162
- 9.10 Gene therapy 162
  - 9.10.1 Early gene therapy applications 162
- 9.11 Stem cells and importance in gene therapy 164
  - 9.11.1 Hematopoietic stem cells (HSC) and therapies 164
  - 9.11.2 Collection of hematopoietic stem cells for therapy 164
- 9.12 Gene editing 165
  - 9.12.1 Early discoveries 165
- 9.13 Delivery of reagents for editing 166
- 9.14 Preclinical and clinical trials 166
  - 9.14.1 Delivery of agents for gene editing 167
- 9.15 NIH (National Institutes of Health) somatic cell gene editing program 168
- 9.16 Base editing 169
- 9.17 Programmable base editing 169
- 9.18 Prime editing 170
- 9.19 CRISPR-Cas theta 171
- 9.20 RNA editing 171
- 9.21 Gene therapy in specific diseases 171
  - 9.21.1 Eye diseases and retinal degeneration 171
- 9.22 Molecular analyses and therapies relevant to hearing loss 172

        9.22.1 RNA-based therapies in deafness 174
        9.22.2 Efforts to promote hair cell regeneration 174
  9.23 Therapy of cystic fibrosis including genetic approaches 175
        9.23.1 Ongoing problems needing to be addressed in gene therapies 176
        References 176
        Further reading 179

CHAPTER 10 Public health applications of genetics including newborn screening and documentation of gene environment interactions

  10.1 Recessive disorders carrier screening in specific populations .. 181
  10.2 Newborn screening and hemoglobinopathies 181
  10.3 Cystic fibrosis 183
  10.4 Molecular-based therapeutics 184
  10.5 Newborn screening, United States 185
        10.5.1 Aminoacidurias 185
        10.5.2 Endocrine disorders 185
        10.5.3 Fatty acid oxidation defects 186
        10.5.4 Galactosemia 186
  10.6 Methods 187
        10.6.1 Newborn screening for immunodeficiencies 188
        10.6.2 Spinal muscular atrophy 188
        10.6.3 Screening for X-linked disorders 189
  10.7 Newborn screening in other parts of the world 190
  10.8 Expanded carrier screening 190
  10.9 Genetic disorders with high frequency in certain populations 191
        10.9.1 Tay − Sachs disease due to hexosaminidase mutations 191
        10.9.2 Summary of predominant manifestations of disorders with increased incidence in Ashkenazi population and mode of inheritance (AR) autosomal recessive (AD) autosomal dominant 193
        10.9.3 Mutation heterogeneity in Tay 194
  10.10 Genetic disorders with increased frequency in other specific populations 195
        10.10.1 Aspartylglucosaminuria 195
        10.10.2 Familial Mediterranean fever (FMF) 196
  10.11 Porphyrias 196

10.12	Factor V Leiden	196
	10.12.1 APOL1 apolipoprotein L1 mutation in Africans and African Americans	197
10.13	Population-wide screening of adults	197
	10.13.1 Promoting diverse population screening	199
10.14	Hemochromatosis	199
10.15	Other disorders that illustrate the impact of gene environment interactions	200
10.16	Human genetic variation and pathogen sensitivity	200
10.17	Genetic and environmental factors and additional aspects of population screening	200
	10.17.1 Severe visual impairment in children	200
	10.17.2 Iodine deficiency, hypothyroidism	201
	References	201
	Further reading	206

CHAPTER 11 Analysis of variants associated with abnormal drug responses, genetics, and genomics in drug design

11.1	Pharmacokinetics and pharmacodynamics	209
	11.1.1 Important examples of drugs and their binding to receptors	209
11.2	Biotransformation of medicinal compounds	210
11.3	World-wide distribution of genetic polymorphisms in the CYP450 system	210
11.4	Other factors and processes involved in biotransformation of drugs	211
	11.4.1 UDP glucuronyl transferase enzymes and medication biotransformation	212
11.5	Drug responses, variants, genetic and environmental factors	212
	11.5.1 Pharmacogenes and therapeutics	213
	11.5.2 HLA typing: application of DNA sequencing	214
11.6	Gene variants that are disease causing and are also associated with abnormal drug reactions	215
	11.6.1 Glucose-6-phosphate (G6PD) deficiency	215
	11.6.2 Drug-induced hemolytic anemias	216
11.7	Porphyrias	216
11.8	Identifying therapeutic targets and developing therapies	217
11.9	Translation of biomedical observations to treatments and health improvements	218
11.10	Fragment-based drug discovery	219

11.11	Monoclonal antibodies as therapeutic agents	219
11.12	Approaches to target identification and therapeutic design in specific genetic disorders	220
	11.12.1 Neurofibromatosis type 1	220
	11.12.2 Approaches to therapy	221
	11.12.3 Gene-related therapies	221
11.13	Inborn errors of metabolism	222
	11.13.1 Mitochondrial fatty acid oxidation defects and carnitine shuttle disorders	222
	11.13.2 Glycogen storage diseases	222
11.14	Lysosomal storage diseases and enzyme replacement therapy	223
11.15	Ceroid lipofuscinoses	224
	11.15.1 Clinical manifestations	225
11.16	Aminoacidopathies and organic acidemias	226
	11.16.1 Methylmalonic acidemia	226
11.17	Transporter defects	226
	11.17.1 Cystic fibrosis	227
	11.17.2 Wilson's disease	227
11.18	Complexities of mitochondrial diseases and explorations of treatments	228
	11.18.1 Search for treatment of specific mitochondrial disorders	228
11.19	New approaches to cancer therapy	229
	11.19.1 Synthetic lethality in cancer treatment	230
	11.19.2 RAS signaling pathway	231
	References	231
	Further reading	235

CHAPTER 12 Genetic and genomic medicine relevance to cancer prevention, diagnosis, and treatment

12.1	Introduction	237
12.2	Genes with germline mutations predisposing to cancer listed in order of frequency	237
12.3	Gene products with germline mutations that can lead to cancer and function of these products	237
12.4	Specific syndromes that include the presence of tumors	240
	12.4.1 Multiple endocrine neoplasia and associated gene defects	240
12.5	Retinoblastoma and RB1	240

12.6	Germline succinate dehydrogenase gene mutations and cancer predisposition	241
12.7	Hereditary gastrointestinal cancers	242
	12.7.1 Pediatric cancers	242
12.8	Germline mutations and developmental origins of cancer	244
	12.8.1 Medulloblastoma	244
12.9	Osteosarcoma	245
	12.9.1 Autosomal-dominant gene mutations	246
	12.9.2 Syndromic genes associated with osteosarcoma	246
	12.9.3 Developmental origins of pediatric cancers	246
12.10	Genetic alterations in cancers in children, adolescents, and young adults	247
12.11	Adult cancers, driver gene mutations, and passenger gene mutations	248
12.12	DNA damage and repair	250
12.13	Lymphomas and leukemia	251
12.14	Myeloid leukemia	253
12.15	Providing insight into cancer-inducing mechanisms	255
12.16	Genome sequencing in cancer	255
12.17	Cell-free DNA analyses in testing for tumors	256
12.18	Cell-free studies including transcriptome analyses	257
12.19	Somatic mutations in cancer	258
12.20	Breast cancer risk genes	258
12.21	Whole genome sequencing of metastatic solid tumors	259
12.22	Therapy-related genetic and genomic information: molecular profiling and cancer therapies	260
12.23	Synthetic lethality	262
12.24	Cancer immunotherapy	262
12.25	CAR-T cells	264
	12.25.1 CAR-T cell therapy and cytokine release syndrome (cytokine storm)	265
	References	265

CHAPTER 13 Benefits of the incorporation of genomic medicine in clinical practice

13.1	Genetic and genomic studies in congenital anomalies and/or neurodevelopmental anomalies	271
	13.1.1 Microarray analyses	271
	13.1.2 Exome sequencing	272
13.2	Rare disease medicine	273

13.3	Carrier screening	273
13.4	Diagnoses and management in disorders with phenotypic abnormalities	274
	13.4.1 Human Phenotype Ontology database	274
	13.4.2 Other resources	275
13.5	Databases of importance in searching for gene, genotype phenotype correlations	275
13.6	Genomic studies to guide diagnosis and therapy in epilepsy	276
	13.6.1 Metabolic pathways and epilepsy	276
13.7	Diagnostic testing for inborn errors of metabolism leading to seizures	277
13.8	Epilepsies, genetics, mechanisms, and therapy	278
13.9	Epilepsy genetics, genomics, and relevance to therapy	279
	13.9.1 Genes involved in developmental epileptic encephalopathies	279
	13.9.2 Chromatin modeling genes	280
	13.9.3 Dravet syndrome	280
13.10	Common epilepsies	280
	13.10.1 Genomic copy number variants and epilepsies	281
13.11	Cerebral palsy and genetic factors	281
13.12	Typical cerebral palsy	281
	13.12.1 Genes with mutations leading to autosomal dominant cerebral palsy	283
	13.12.2 Autosomal recessive (homozygous)	284
	13.12.3 Autosomal recessive compound heterozygous	284
	13.12.4 X-linked dominant	285
	13.12.5 X-linked recessive	285
13.13	Monoamine neurotransmitter disorders	285
13.14	Spastic paraplegias and ataxias	288
	13.14.1 General clinical features	289
13.15	Friedreich ataxia (FRDA)	290
13.16	Polyglutamine cerebellar ataxias	291
13.17	Autosomal recessive ataxias	291
13.18	Spinocerebellar ataxias	293
	13.18.1 Nonrepeat cerebellar ataxias	294
13.19	Spinocerebellar ataxias	294
13.20	Genomic medicine in common diseases in adults	295
	13.20.1 Hypertension	295
	13.20.2 Glucocorticoid remediable aldosteronism	295
	13.20.3 Apparent mineralocorticoid excess	296
13.21	Polygenic factors leading to hypertension	297

13.22	Coronary heart disease	299
	13.22.1 Heritability of coronary heart disease	300
	13.22.2 Nonmonogenic risk	300
13.23	Genetic-guided therapies	300
	13.23.1 Genetics-based therapeutic targets	300
13.24	Approaches to determining polygenic risk scores	301
13.25	Familial hypercholesterolemia	301
13.26	Undiagnosed diseases and application of genetic and genomic studies	303
	References	304
	Further reading	308

CHAPTER 14 Using insights from genomics to increase possibilities for treatment of genetic diseases

14.1	Introduction	309
14.2	Therapy lysosomal diseases	310
	14.2.1 Enzyme replacement therapy in clinical use for treatment of specific lysosomal storage diseases	311
14.3	Neuroimmune disorders, autophagy lysosomes, and treatment	312
	14.3.1 Chaperone therapies in the treatment of lysosomal storage diseases	312
14.4	Mucopolysaccharidosis II (Hunter syndrome)	313
	14.4.1 Precision medicine in lysosomal storage disease disorders	313
	14.4.2 Substrate reduction therapy	314
14.5	Strategies designed to increase the half-life of enzymes used in enzyme replacement strategies being considered in therapy	316
	14.5.1 Gene therapy in lysosomal storage diseases (Fabry disease)	316
14.6	Gene-directed treatments	316
14.7	Hemoglobinopathy treatment including gene therapy	318
14.8	Gene-directed therapies in clinical trials in hemoglobinopathies	320
14.9	Hemoglobinopathy treatment through genetic silencing of BCL11A expression using antisense strategy	320
14.10	Splice mutations and diseases	321
14.11	Pluripotent stem cells in investigations of disease therapies	322
14.12	Relevance to protein folding and secondary modifications	322
	14.12.1 Proteins that play roles in collagen folding and cross-linking	323

14.13	Defects in ossification and mineralization	324
14.14	Osteogenesis imperfecta treatment	325
14.15	Ongoing clinical trials related to cell and gene therapy	326
	14.15.1 Hematopoietic stem cells and progenitors	326
	14.15.2 Collection of hematopoietic stem cells for therapy	327
14.16	Coagulation disorders	328
	14.16.1 Hemophilia	328
14.17	Von Willebrand factor and disease	330
	14.17.1 Laboratory tests	330
14.18	Platelet receptors for Von Willebrand factor	331
14.19	Understanding mechanisms of rare diseases that may lead to therapy	331
14.20	Trinucleotide repeat disorders: progress toward therapy	332
14.21	Toward Huntington disease therapy	335
	14.21.1 RNA targeting and RNA interference approaches in gene therapy in Huntington disease	336
14.22	Protein clearance	336
14.23	Polyglutamine cerebellar ataxias	337
14.24	Duchenne muscular dystrophy	340
	14.24.1 Laboratory diagnosis of DMD	342
14.25	DMD therapy molecular approaches	342
14.26	Utrophin	344
14.27	Spinal muscular atrophy (autosomal recessive proximal muscular atrophy)	345
14.28	Antisense oligonucleotides in neurodegenerative diseases	347
14.29	Dynamic mutability of microsatellite repeats	348
	14.29.1 Genomic tandem repeats in autism	349
14.30	Ocular gene therapy	350
	References	352
	Further reading	357

Index 359

Preface

In the early chapters of this book I briefly present aspects of the early documentation of the family history of certain disorders, the history of discovery of genetic mechanisms, and their translation into clinical medicine. The growth of techniques to analyze the functions of gene products and application of such techniques to clinical diagnosis and disease management are then presented. In the latter chapters aspects of gene and genome analyses in clinical medicine are presented. Advances in aspects of therapy of genetic disorders including recent progress in gene therapy and possibilities for gene editing are also considered.

Acknowledgments

I am deeply grateful for the opportunities and learning experiences made available to me over the years through participating in patient care, in student teaching, and in attending lectures and conferences. I thank those who contribute to knowledge. I am particularly grateful for access to the University of California Library system and Internet resources and for resources made available through the National Library of Medicine and the National Institutes of Health, United States.

I wish to thank Peter Linsley and Kristi Andersen, Editors at Elsevier Academic Press, Peter Linsley for guidance as I planned this book, and Kristi Andersen for help throughout the process of submission of chapters. My thanks are also due to Kiruthika Govindaraju Sr. and Maharaj Rajendran for the work in the production phase of this book.

Epigraph

"To carefully observe the phenomena of life in all its phases, normal and perverted, to make perfect that most difficult of the arts the arts of observation, to call to aid the science of experimentation, to cultivate the reasoning faculty so as to be able to know the true from the false -these are our methods."

William Osler, Valedictory address: University of Pennsylvania, May 1, 1889.
William Osler, p. 267, in "Aequanimitas" Addresses to Medical Students Nurses and Practitioners of Medicine, published by McGraw Hill 1906.

PART I

History and Growth of Genetic Medicine

CHAPTER 1

Documentation of units of inheritance and their contribution to phenotype

1.1 Rediscovery of the laws of Mendel

We learned that at the turn of the century in 1900 three scientists in different countries rediscovered the laws of Gregor Mendel, they were Hugo De Vries in Holland, Carl Correns in Germany, and Carl von Tschermak in Austria. Mendel's work was published in 1866 in the Proceedings of the Natural History Society of Brünn.

In England, early in the 1900s, William Bateson drew attention to the rediscovery of the laws of Mendel. In subsequent studies he worked to integrate laws of heredity and his earlier work on biological variation, and published experimental studies in the physiology of heredity (Bateson and Punnett, 1905).

From his studies on artificial fertilization of plants (primarily peas) including crossing species with different measurable characteristics, color, shape, and measuring how specific characteristics were passed on to offspring, he recognized that specific characteristics were passed on through the parental gametes (germ cells) to their offspring and he generated Laws of Heredity.

Mendel's laws were followed by key discoveries by Theodor Boveri in Germany on the properties of nuclei. Boveri in 1887 reported that nuclei in gametes (sperm or egg) had half as many chromosomes as nuclei in somatic cells. In 1903 Sutton presented evidence that "hereditary particles" were borne on chromosomes. Together the work of Sutton (1903) and Boveri (1904a, 1904b) linked heredity to the nucleus and to chromosomes (the Sutton Boveri hypothesis).

Studies by Boveri were followed later by others including Janssens (1909). Thus processes of meiosis through which nuclei were modified to gametes became known. The processes of formation of gametes developed included elaborate nuclei changes, pairing of homologous chromosomes, cross-over of segments between aligned chromosomes and subsequently reduction division so that the gamete (germ cell) contained only one member of each chromosome pair.

Between 1905 and 1908, Bateson and Punnett generated information indicating that the different inherited "particles" and phenotype they determined segregated together to the next generation. Two phenotype determining particles that were inherited together were said to be in coupling, two that segregated independently were said to be in repulsion.

Studies by Morgan and colleagues between 1910 and 1915 led to generation of the chromosome theory of inheritance.

Morgan postulated that particles that determined phenotypes that were inherited together were likely on the homologous chromosomes. Those that were inherited separately were likely on different chromosomes or if on the same chromosome they were likely located at some distance from each other.

The ideas resulted in the concept that it might be possible to develop linkage maps of chromosomes.

Thomas H. Morgan was awarded a Nobel Prize in 1934. In his lecture he emphasized the constancy of the position of genes with respect to the other genes in a linear order on chromosomes and that that was deducible from genetic evidence and from cytological evidence. He noted the coming together of chromosomes: conjugating chromosome are like chromosomes, that is, chain of the same gene, it is the genes that come to lie side by side.

In considering the relationship of Mendelian inheritance to medicine Morgan wrote:

"I want to make it clear that the complexity of man makes it somewhat hazardous to apply only the simple roles of Mendelian inheritance for the development of many inherited characteristics."

1.2 Genes and genetics

First use of the word gene is ascribed to the Danish botanist Wilhelm Johannsen, in Danish and German the word used was gen.

There is some evidence that William Bateson used the word genetics in 1905 https://www.genome.gov/25520244/online-education-kit-1909-the-word-gene-coined.

In 1922 H.J. Muller published a paper entitled "Variation due to changes in the individual gene."In this paper Muller drew attention to the presence within the cell of "thousands of distinct substance the genes." He noted that genes existed as ultra-microscopic particles that played fundamental roles in determining cell substances and cell structures and that ultimately genes could affect the whole organism.

Muller emphasized that the chemical formulae of genes were then unknown and he addressed gene mutability and questioned what sort of structure genes possess that permitted mutability.

Now on to 1944 ...

In a recollection published in *Nature* in 2003, Maclyn Mc Carty wrote about the paper he published in 1944 along with his coworkers Avery and MacLeod. Mc Carty wrote "Experiments showed that the heritable property of virulence from one infectious strain of *Pneumococcus* could be transferred to a non-virulent infectious strain of *Pneumococcus* by pure DNA."

This conclusion was further supported by showing that the transforming activity could be destroyed by the DNA digesting enzyme DNAse.

Mc Carty wrote further (2003), "Our findings continued to receive little acceptance, scientists believed that DNA was too limited to carry the genetic information."

1.3 Nucleic acids

It is interesting to note that Albrecht Knossile had been awarded a Nobel Prize in 1910 for determining the chemical structure of the nucleus, adenine, thymine, guanine, cytosine. Erwin Chargaff and coworkers (1949) established that in DNA the levels of adenine equaled the levels of thymine, and the levels of cytosine equaled the levels of guanine.

1.4 The structure of DNA

A landmark paper was published by Watson and Crick (1953). In this paper they put forward "a structure of the salt of deoxyribonucleic acid." Their structure consisted of two helical chain coiled around a central axis. They emphasized that a novel feature of their structure was that the chains were linked to each other by the purine and pyrimidine bases. Hydrogen bonds joined the bases and importantly one member of a linked pair should be a purine and the other a pyrimidine. Thus, adenine would be linked to thymine and guanine would be linked to cytosine. They noted that it would not be possible to build such a structure with ribose, the double helix was dependent on the presence of deoxyribose.

The authors concluded this publication with the sentence:

"It has not escaped our notice that the specific pairing we have postulated suggests a copying mechanism for the genetic material."

1.5 DNA and chromatin

It is important to note that DNA exists in the cell within a complex that is rich in proteins. This complex is referred to as chromatin. Aaron Klug in 1982 received a Nobel Prize for elucidation of nucleic acid and protein complexes.

Information on the composition of chromatin and its functions have grown steadily over the years. It has become clear that the protein composition of chromatin changes in different stages of the cell cycle and as phases of gene expression changes.

1.5.1 Consequences of determination of DNA structure

The discovery of the structure of DNA inspired tremendous activity in the fields of molecular biology and molecular genetics. Finding out how DNA information was converted to a messenger, the nature of transcription and generation of MRNA sequences, processes of transcription termination, and establishing how MRNA transcripts generated proteins through activity of transfer RNA and aminoacyl-tRNAs and nucleosomes.

In due course it became possible to reliably sequence DNA, and to generate from mRNA cDNA sequences that could be sequenced and provide information on mRNA sequence. By 1978 it had become clear that the eukaryotic genes included protein coding segments and segments that did not encode protein (Leder, 1978). It had also become clear that for a particular gene not all exons were present in all transcripts and that in some cells and tissues, specific exons were spliced out, from primary transcripts. In 1993 Roberts and Sharp received a Nobel Prize for their discoveries related to gene splicing.

Haberle and Stark (2018) reviewed evidence for transcription that could initiate at core promoters or at alternate promoters.

Tian and Manley (2017) reviewed evidence that transcription termination could end at different 3′ sites and that the extent of polyadenylation at 3′sites could vary.

Initiation of gene transcription was shown to require a series of different transcription factors. Vaquerizas et al. (2009) manually curated more than 1000 different transcription factors.

1.5.2 Modifications of DNA sequences

Specific modification of DNA sequences were shown to significantly impact gene expression. An important modification involved methylation, particularly methylation at cytosine guanine nucleotides (CpG). Chromatin modification including histone modification, chromatin remodeling, and the binding of specific protein complexes to protein were shown to play important roles in regulation of gene expression.

Each of these processes, chromosome structure, gene organization on chromosomes, gene transcription, translation, aspects of chromatin composition, and remodeling of the genome through chromatin looping, were shown to be implicated in specific disorders.

1.6 Applications of studies of chromosomes, genomes, genes, and gene expression to clinical medicine

It is important to consider the chromosome as a unit of heredity. However, initiation of the study of human chromosomes seems relatively recent as it dates from the 1950s and publications of T.C Hsu (1952) and Tjio and Levan (1956). They treated proliferating human cells with colchicine that caused cell division to pause at metaphase. Harvested cells were then treated with hypotonic solution to swell nuclei, they were then fixed with acetic acid and alcohol and dropped onto glass slides, and stained with suitable dye. Microscopic analyses revealed that humans contained 46 chromosomes.

Studies were initially done on cultures of human fibroblasts. Subsequently, methods were developed for short-term culture, for example, 72 hours of human blood leukocytes followed by colchicine treatment and treatment as described above.

Careful studies were then done to arrange the chromosomes by size and also by position of the centromere in each member of the 23 pairs of chromosomes and differences between male and female cells were revealed. Subsequently, an international committee was established to develop a standardized nomenclature according to size and centromere position for 22 autosomes and the pair of sex chromosomes, XX in females and XY in males.

Initial clinical studies revealed that in some disorders a specific human chromosome was missing, leading to monosomy for that chromosome while in other cases an extra member of a specific chromosome pair was present leading to trisomy.

Earlier reports of chromosome abnormalities revealed abnormalities of sex chromosomes XXY in Klinefelter syndrome, XO in Turner syndrome, trisomy 21 in Down syndrome.

Harper (2004) noted a specific discovery that could be considered "as the starting point in clinical cytogenetics." It was the first report of the presence of 47 chromosomes in Down syndrome, with the supernumerary chromosomes being a small telomeric one, by Lejeune et al. Their paper was published in 1959.

In 1958 Polani et al. reported finding an extra X chromosome in Klinefelter syndrome; in 1959 Ford reported a case of Turner syndrome in a female with XO karyotype.

In 1965 Carr reported finding chromosome abnormalities in cases of spontaneous abortion.

In 1960 Nowell and Hungerford discovered an unusual chromosome in chronic granulocytic leukemia. This chromosome was subsequently shown to be a translocation between chromosome 9 and 22.

It is interesting to note that correlation of blood group and chromosome studies led to one of the earliest assignments of a human gene to a human autosome in 1968. This assignment was made by following inheritance of the Duffy blood group in a family and discovery of absence of an expected Duffy allele in an individual with variant of human chromosome 1 and was reported by Donahue et al. (1968).

In 2004 Ferguson-Smith reviewed history of human cytogenetics, noting early studies and subsequent development of chromosome banding techniques, either by trypsin-Giemsa or fluorescent techniques, standardization of nomenclature of chromosome segments visualized on banding that facilitated diagnosis of segmental chromosome defects. In 1971 Seabright introduced trypsin-Giemsa banding of metaphase chromosome and Caspersson et al. (1972) introduced fluorescent banding (Q Banding) of metaphase chromosomes.

Subsequent developments involved labeling of DNA probes corresponding to a specific gene or genomi segment and hybdrizing these to spreads of human chromosome.

Riegel (2014) reviewed development of molecular cytogenetics that included use of fragments of specific gene, labeled with fluorescent dye and the hybridized to human chromosomes. A specific gene probe ideally hybridized to a chromosome at the position where that gene was located and thus facilitated mapping of gene to their chromosomal locations. In situ hybridization could also facilitate detection of microdeletions or microduplications at specific positions on chromosomes. In situ hybridization could also potentially detect structural chromosome rearrangement including translocations or inversions. Fluorescence in situ hybridization was used to map the alpha globin gene cluster to human chromosome 16 (Deisseroth et al., 1977) (Fig. 1.1).

1.6.1 Chromosome microarray analyses

The first method developed was comparative genomic hybridization (CGH) (Fiegler et al., 2003) and subsequently mapped single nucleotide polymorphic markers were used (Haeri et al., 2015).

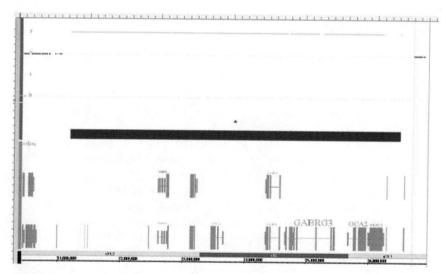

FIGURE 1.1

Image of microarray showing duplication on chromosome 15q11.2–15q11.3 with duplication of specific genes including one that encodes GABRG3, a Gamma aminobutyric acid subunit.

CGH involved the use of two sources of DNA, normal control DNA and DNA from the individual to be investigated. Each DNA sample was independently labeled with fluorescent probes, one with red fluorescence and the second with green fluorescence. DNA was denatured and then mixed in a 1:1 ratio. Fluorescent signals along the length of each chromosome were measured through fluorescent microscopy and computer analysis. Segments deleted in the test sample or duplicated segments in the test sample would display differences in the dye intensities. CGH was therefore primarily used to detect copy number variants in chromosomes.

Array-based CGH had reference genome chromosomal DNA segments hybridized to a solid matrix and the labeled test sample was then hybridized to the solid matrix with bound control DNA. DNA segments derived from across the whole genome could be used to generate the solid matrix platform or DNA from targeted genomic could be used (Pinkel and Albertson, 2005).

SNP (single nucleotide polymorphism) microarrays involve the use of short DNA segment corresponding to specific loci that are known to map to specific sites across the genome. These segments are then fixed to a solid matrix. Fluorescent labeled test DNA is then hybridized to the solid matrix and hybridization is measured through fluorescent microscopy and computer analyses to search for altered regions of hybridization (Miller et al., 2010). Clinical microarray analysis was reported to yield a higher diagnostic yield than G-banded karyotype analyses for individuals with developmental disabilities and congenital anomalies (Fig. 1.2).

1.7 Long-read sequencing for detection of genomic variants including structural chromosome abnormalities

Mantere et al. (2019) noted that standard next generation sequencing (NGS) protocols generate short-read sequences approximately 150–300 base pairs in length.

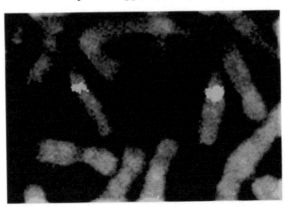

FIGURE 1.2

In situ hybridization indicated duplication in chromosome 15q11.2–15q11.3.

During alignment and analysis of sequence reads, significant problems emerge in that several regions have highly repetitive sequences. Problems in sequence generation and alignments are also presented in regions in which structural variation occurs on one member of the chromosome pair. Other sequence alignment problems occur in regions with high content of GC nucleotides.

Mantere et al. noted that long-read sequencing (LRS) can overcome some of the difficulties described above. LRS generates on average reads of 10 kb in length from a single-stranded DNA molecule. An advantage of LRS is that prior PCR amplification is not required. DNA is not modified by handling and methylation changes in DNA can be detected.

LRS methods currently in use include single molecule real-times sequencing using Pacific biosystems (PacBio) technologies and nanopore sequencing using Oxford nanopore technology systems. LRS is sometimes referred to as third generation sequencing.

PacBio sequencing is reported to capture sequence information during the DNA replication process. The sequencing process utilizes immobilized DNA polymerase and addition of four fluorescent labeled nucleotides (Rhoads and Au, 2015).

Oxford Nanopore sequencing is based on detection of electric charge differences of nucleotides as they pass through the nanopore (Lu et al., 2016).

As sequence data have been generated over the course of recent years, it has become clear that structural genome variants with lengths more than 50 bp are common in the human genome (Chaisson et al., 2019). A number of studies has provided evidence that second generation sequencing methods can readily detect single nucleotide variants and short insertion-deletion variants. However, LRS methods are superior in detecting longer structural variants.

Another important advantage of LRS is more accurate determination of haplotypes and whether or not specific haplotype variants occur on a single chromosome and are from one parent, while another set of variants are form another parent. This is referred to as haplotype phasing and can be particularly useful in cases of disorders due to compound heterozygous mutations. Haplotype phasing was also useful in trio sequencing to determine the parental origin of de novo mutation.

Another advantage of LRS was noted to be the distinction of sequence in functional genes from sequence in pseudogenes since pseudogenes are usually located on different segments of the genome than that of the corresponding functional gene.

LRS is also being applied in studies of transcription.

1.8 Determination of the significance of structural variants in the genome

Audano et al. (2019) carried out LRS on 15 human genomes and they then genotyped 440 additional genomes to confirm structural variants. They reported a ninefold structural variant bias within the last five megabases of human chromosomes.

Audano et al. noted that their data provided a framework for constructing a human reference database for structural variants.

1.8.1 Clinical significance of structural genomic variants

It is important to note that structural genomic variants can have significant phenotypic impact (Weischenfeldt et al., 2013). Nevertheless reference maps of nonpathogenic structural variants in human populations are required.

Collins et al. (2020) constructed maps of sequence-based structural variants in 14.891 genomes from different human populations. They determined that there is evidence for selection against structural variants that disrupt coding sequence. There was also evidence for modest selection against structural variant in cis regulatory genomic elements.

Data generated in the Collins et al. studies were contributed to the gnomAD database and was noted to provide clinical utility regarding interpretation of the clinical significance of structural genomic changes.

1.9 Mosaicism

In a 2015 review, Campbell et al. noted that as cells divide they can accumulate genomic changes, including single nucleotide variants, insertion-deletions (indels), and chromosome copy number variants so that each human is in fact a mosaic.

They noted that many of these genomic changes may not necessarily have functional effects. In addition, cells in which functional compromising changes occur may sometimes be removed from the organism. However, mutations occurring early in development may have significant effects.

Different forms of mosaicism have been defined. Postzygotic mosaicism refers to genomic changes that occur after fertilization of ovum. Somatic mosaicism referred to variation in cells in the body of the organism that are not present in the organs that produce gametes. Gonadal mosaicism refers to genetic and genomic changes that are present in cells that form the gametes and are therefore transmitted as germline mutations.

Placental mosaicism was reported to be present in 1%−2% of placentas tested. Specific genetic and genomic changes that are present in the placenta may not necessarily be present in the fetus.

Somatic mosaicism is defined as genetic/genomic changes in body cells and may be restricted to certain tissue or to certain cells, The cells and tissues affected are related to the developmental stage when the changes occurred. For example, if genetic or genomic changes occur in the postzygotic phase, identical twins may differ in some specific phenotypic features. Campbell et al. noted reports that revealed that mutations occurring after the period of left and right separation in the early embryo could lead to differences between left and right body tissues.

1.9.1 Chromosomal mosaicism

This can include aneuploidies, differences in numbers of a specific chromosome so that autosomes are not present in pairs. In monosomies only one member of the pair is present. In trisomies three members of a specific chromosome are present. Chromosomal aneuploidies can also be present in mosaic form, that is, not present in all cells or all tissues in the organism. Structural chromosomal abnormalities can also be present in mosaic form if they arise postzygotically.

Nucleotide variants, including single base changes and small insertion-deletions, are noted to sometimes arise postzygotically as a result of certain harmful environmental exposures.

Nucleotide repeat expansions, for example, trinucleotide or tetranucleotide repeats, can sometimes undergo postzygotic repeat expansion or contraction. Campbell et al. present an example where FMR1 repeat length was different in monozygotic twins.

Specific repetitive elements in the genome, especially LINE1 elements, were noted to undergo replication and to undergo replication. Questions remain regarding the mobility of these elements.

1.9.2 Mosaicism detection

Somatic mosaicism has been detected in cytogenetic studies and through DNA sequencing when DNA for testing was isolated from different tissues or from different cell types. Mangin et al. (2021) reported robust detection of trinucleotide repeat length and somatic mosaicism in myotonic dystrophy. Clearly, single cell sequencing may offer the most reliable method of mosaicism detection.

1.9.3 Mosaicism and genetic diseases

Campbell et al. (2015) and Buser et al. (2020) drew attention to Proteus syndrome that occurs only in the mosaic form. It occurs in individuals with a specific pathogenic mutation in the gene that encodes AKT1, c.49G > A. p.(Glu17Lys). The mutation occurs particularly in connective tissue and leads to overgrowth and abnormalities in bone. AKT1 is a serine threonine kinase.

Brain region overgrowth has been reported in individuals with specific activating mutation in PIK3CA, AKT3, and MTOR.

The AKT serine threonine kinases (e.g., AKT1, AKT3) are phosphorylated by phosphoinositide 3-kinase (PI3K). AKT/PI3K forms a key component of many signaling pathways. PIK3CA phosphatidylinositol-4,5-bisphosphate 3-kinase catalytic subunit alpha, MTOR, belongs to a family of phosphatidylinositol kinase-related kinases.

1.10 Germline mutations

It is important to note that postzygotic mutations early in development can lead to mosaic mutations in the cells that forms the gonads and in the germline.

Generation of new germline mutations has been shown to occur particularly in male gonads, likely due to the continuous mitosis that spermatogonia undergo. In females, primary oocytes in the developed ovary are known to remain arrested in prophase of meiosis 1. Rahbari et al. (2016) noted that, in males 610 genome replications were reported to have taken place by 40 years of age.

1.11 Genetic mosaicism in inborn errors of immunity

Inborn errors of immunity were reviewed by Aluri and Cooper (2021). They noted evidence that somatic mosaicism frequently occurred in these disorders, and investigations require analysis of multiple tissue and cell types and target sequencing of specific genes rather than NGS.

Aluri and Cooper emphasized that somatic mosaicism was particularly important to consider in cases of immunodeficiency where there was no family history of immunodeficiency.

Somatic mosaicism was noted to have been reported in adenosine deaminase deficiency, in X-linked combined immunodeficiency, and in auto-immune lymphoproliferative syndrome.

Aluri and Cooper reported genetic findings in auto-inflammatory disorders characterized by fever, skin findings, arthritis, gastro-intestinal manifestations, and lung disease. Defects in the gene that encodes NLRP3 can impact the regulation of inflammation, the immune response, and apoptosis. In some patients, variants were found to be restricted to myeloid cells.

References

Aluri J, Cooper MA: Genetic mosaicism as a cause of inborn errors of immunity, *J Clin Immunol* 41(4):718–728. Available from: https://doi.org/10.1007/s10875-021-01037-z, 2021.

Audano PA, Sulovari A, Graves-Lindsay TA, et al: Characterizing the major structural variant alleles of the human genome, *Cell* 176(3):663–675, 2019. Available from: https://doi.org/10.1016/j.cell.2018.12.019. e19.

Bateson W, Punnett RC. Experimental studies on the physiology of heredity report to the Evolutionary Committte of the Royal Society Reports 2, 3 and 5, pp 1905–1908

Boveri T., 1904a. Results concerning the chromosome substance of the cell nucleus, p 65, Fischer, Jena, Germany (in German) [Google Scholar]

Boveri, T.H. (1904b). Ergebnisse über die Konstitution der chromatischen Substanz des Zelkerns. Fisher, Jena.

Buser A, Lindhurst MJ, Kondolf HC, et al: Allelic heterogeneity of Proteus syndrome, *Cold Spring Harb Mol Case Stud* 6(3):a005181, 2020. Available from: https://doi.org/10.1101/mcs.a005181. Print 2020 Jun. PMID: 32327430.

Campbell IM, Shaw CA, Stankiewicz P, Lupski JR: Somatic mosaicism: implications for disease and transmission genetics, *Trends Genet* 31(7):382–392. Available from: https://doi.org/10.1016/j.tig.2015.03.013, 2015.

Caspersson T, Lommaka G, Zech L: The 24 fluorescence patterns of the human metaphase chromosomes - distinguishing characters and variability, *Hereditas* 67(1):89–102. Available from: https://doi.org/10.1111/j.1601-5223.1971.tb02363.x, 1972.

Chaisson MJP, Sanders AD, Zhao X, et al: Multi-platform discovery of haplotype-resolved structural variation in human genomes, *Nat Commun* 10(1):1784, 2019. Available from: https://doi.org/10.1038/s41467-018-08148-z. PMID: 30992455.

Chargaff E, Magasanik B, et al: The nucleotide composition of ribonucleic acids, *J Am Chem Soc* 71(4):1513, 1949. Available from: https://doi.org/10.1021/ja01172a525. PMID: 1812837.

Collins RL, Brand H, Karczewski KJ, et al: A structural variation reference for medical and population genetics, *Nature* 581(7809):444–451. Available from: https://doi.org/10.1038/s41586-020-2287-8, 2020.

Deisseroth A, Nienhuis A, Turner P, et al: Localization of the human alpha-globin structural gene to chromosome 16 in somatic cell hybrids by molecular hybridization assay, *Cell* 12(1):205–218, 1977. Available from: https://doi.org/10.1016/0092-8674(77)90198-2. PMID: 561664.

Donahue RP, Bias WB, Renwick JH, McKusick VA: Probable assignment of the Duffy blood group locus to chromosome 1 in man, *Proc Natl Acad Sci U S A* 61(3):949–955, 1968. Available from: https://doi.org/10.1073/pnas.61.3.949. PMID: 5246559.

Fiegler H, Carr P, Douglas EJ, et al: DNA microarrays for comparative genomic hybridization based on DOP-PCR amplification of BAC and PAC clones, *Genes Chromosomes Cancer* 36(4):361–374, 2003. Available from: https://doi.org/10.1002/gcc.10155. PMID: 12619160.

Haberle V, Stark A: Eukaryotic core promoters and the functional basis of transcription initiation, *Nat Rev Mol Cell Biol* 19(10):621–637, 2018. Available from: https://doi.org/10.1038/s41580-018-0028-8. PMID: 29946135.

Haeri M, Gelowani V, Beaudet AL: Chromosomal microarray analysis, or comparative genomic hybridization: a high throughput approach, *MethodsX* 3:8–18. Available from: https://doi.org/10.1016/j.mex.2015.11.005, 2015.

Harper P: *Landmarks in medical genetics*, 2004, Oxford University Press.

Hsu TC: Mammalian chromosomes in vitro—the karyotype of man, *J Hered* 43:167–172, 1952.

Janssens FA: The chiasmatype theory. A new interpretation of the maturation divisions, *Cellule* 25:389–411, 1909. Translated from the French; reprinted in Genetics 191: 319–346.

Leder P: Discontinuous genes, *N Engl J Med* 298(19):1079–1081, 1978. Available from: https://doi.org/10.1056/NEJM197805112981910. PMID: 643015.

Lu H, Giordano F, Ning Z: Oxford nanopore MinION sequencing and genome assembly, *Genomics Proteom Bioinforma* 14(5):265–279. Available from: https://doi.org/10.1016/j.gpb.2016.05.004, 2016.

Mangin A, de Pontual L, Tsai YC, et al: Robust detection of somatic mosaicism and repeat interruptions by long-read targeted sequencing in myotonic dystrophy Type 1, *Int JMol Sci* 22(5):2616, 2021. Available from: https://doi.org/10.3390/ijms22052616. PMID: 33807660.

Mantere T, Kersten S, Hoischen A: Long-read sequencing emerging in medical genetics, *Front Genet* 10:426, 2019. Available from: https://doi.org/10.3389/fgene.2019.00426. eCollection 2019.PMID: 31134132.

McCarty M: Discovering genes are made of DNA, *Nature*. 421(6921):406, 2003. Available from: https://doi.org/10.1038/nature01398. PMID: 12540908.

Miller DT, Adam MP, Aradhya S, et al: Consensus statement: chromosomal microarray is a first-tier clinical diagnostic test for individuals with developmental disabilities or congenital anomalies, *Am J Hum Genet* 86(5):749–764, 2010. Available from: https://doi.org/10.1016/j.ajhg.2010.04.006. PMID: 20466091.

Pinkel D, Albertson DG: Comparative genomic hybridization, *Annu Rev Genomics Hum Genet* 6:331–354, 2005. Available from: https://doi.org/10.1146/annurev.genom.6.080604.162140. PMID: 16124865.

Rahbari R, Wuster A, Lindsay SJ, et al: Timing, rates and spectra of human germline mutation, *Nat Genet* 48(2):126–133, 2016. Available from: https://doi.org/10.1038/ng.3469. PMID: 26656846.

Rhoads A, Au KF: PacBio sequencing and its applications, *Genomics Proteom Bioinforma* 13(5):278–289, 2015. Available from: https://doi.org/10.1016/j.gpb.2015.08.002. PMID: 26542840.

Riegel M: Human molecular cytogenetics: from cells to nucleotides, *Genet Mol Biol* 37(1):194–209, 2014. Available from: https://doi.org/10.1590/s1415-47572014000200006. PMID: 24764754.

Sutton WS: The chromosomes in heredity, *Bio Bull* 4:231–251, 1903. [Google Scholar].

Tian B, Manley JL: Alternative polyadenylation of mRNA precursors, *Nat Rev Mol Cell Biol* 18(1):18–30, 2017. Available from: https://doi.org/10.1038/nrm.2016.116. Epub 2016 Sep 28.PMID: 27677860.

Tjio JH, Levan A: The chromosome number in man, *Hereditas* 42:1–6, 1956. 1–2.

Vaquerizas JM, Kummerfeld SK, Teichmann SA, Luscombe NM: A census of human transcription factors: function, expression and evolution, *Nat Rev Genet* 10(4):252–263, 2009. Available from: https://doi.org/10.1038/nrg2538. PMID: 19274049.

Watson JD, Crick FH: Molecular structure of nucleic acids; a structure for deoxyribose nucleic acid, *Nature*. 171(4356):737–738, 1953. Available from: https://doi.org/10.1038/171737a0. PMID: 13054692.

Weischenfeldt J, Symmons O, Spitz F, Korbel JO: Phenotypic impact of genomic structural variation: insights from and for human disease, *Nat Rev Genet* 14(2):125–138, 2013. Available from: https://doi.org/10.1038/nrg3373. PMID: 23329113.

Further reading

Carr DH: Chromosome studies in spontaneous abortions, *Obstet Gynecol* 26:308–326, 1965. Sep.

Ferguson-Smith MA: *Chromosome genetics and evolution*, Chapter 8 in *A century of Mendelism in human genetics*, London, New York, 2004, CRC Press.

Ford CE, Jones K, Polani PE, De Almeida JC, Briggs JH: A sex-chromosome anomaly in a case of gonadal dysgenesis (Turner's syndrome), *Lancet* 273:711–713, 1959.

Klug A. From macromolecules to biological assemblies. Nobel Lecture, 8 December 1982. Biosci Rep. 1983 May;3(5):395–430. Available from: https://doi.org/10.1007/BF01121953. PMID: 6349708.

Kossel A. The chemical composition of the cell nucleus. The Nobel Prize in Physiology or Medicine 1910. <https://www.nobelprize.org/prizes/medicine/1910/kossel/lecture/>.

Lejeune J, Turpin R, Gautier M: Chromosomic diagnosis of Mongolism, *Arch Fr Pediatr* 16:962–963, 1959. PMID: 14415503.

Morgan TH: Sex limited inheritance in drosophila, *Science* :120–122, 1910.

Morgan TH The Nobel Prize in Physiology or Medicine 1933 <https://www.nobelprize.org/prizes/medicine/1933/morgan/facts/>.

Morgan TH, Sturtevant AH, Muller HJ, Bridges CB: *The mechanism of Mendelian heredity*, New York, 1915, Henry Holt and Co.

Morgan TH: Random segregation vs coupling in Mendelian inheritance, *Science* 384:1911.

Muller HJ: Variation due to changes in the individual gene, *Am Naturalist* 56:32–50, 1922.

Nowell PC, Hungerford DA: Chromosome studies on normal and leukemic human leukocytes, *J. Natl Cancer Inst* 25:85–109, 1960. Jul.

Polani PE, Bishop PM, Lennox B, et al: Colour vision studies and the x-chromosome constitution of patients with Klinefelter's syndrome, *Nature*. 182(4642):1092–1093, 1958. Available from: https://doi.org/10.1038/1821092a0. PMID: 13590234.

Roberts RJ, Shape PA. 1993 The Nobel Prize in Physiology or Medicine 1993. Accessed NobelPrize.org. Nobel Prize Outreach AB 2021. Tue. 27 July 2021. <https://www.nobelprize.org/prizes/medicine/1993/summary/.

Seabright M: A rapid banding technique for human chromosomes, *Lancet* 2(7731):971–972, 1971. Available from: https://doi.org/10.1016/s0140-6736(71)90287-x. PMID: 4107917.

Early documentation of inherited disorders through family studies

2.1 The Treasury of Human Inheritance

Publications in the series, entitled The Treasury of Human Inheritance, were initiated by Karl Pearson and the Francis Galton Laboratory at the University of London in 1909 and publication continued until 1958. During the years of the First and Second World Wars there were no publications in the series. Pearson wrote in the first edition in 1909, "The publication of family histories whether they concern physical abnormality, ability or achievement, whether they be new or old is the purpose of the treasury."

Specific entries were contributed by physicians who had studied families with specific disorders and articles also included extensive references to other publications pertaining to families with the specific disorder under discussion. In the first editions symbols to be used in pedigrees were provided. In the different editions of the Treasury dealing with different conditions, extensive pedigrees were included, often the pedigrees extended over five generations. Also included were descriptions of the disease manifestations.

I feel particularly fortunate to have editions of The Treasury of Human Inheritance published over 16 years.

2.2 Ectrodactyly

The 1909 edition generated by Bullock, Lewis, Rivers, and Horne included descriptions and pedigrees of families in which malformations of hands and feet were presented, with particular attention focused on a condition referred to as split hands and feet, a condition now known as ectrodactyly. In 2019 this condition was reviewed by Kantaputra and Carlson and was reported to arise due to mutations in any one of seven genes, TP63, DLX5, DLX6, FGF8, FGFR1, WNT10B, and BHLHA9 that were reported to lead to dysregulation of fibroblast growth factor FGF8 in the central portion of the apical ectodermal ridge (AER) and to disruption of Wnt-Bmp-Fgf signaling pathways in AER.

2.3 Deafness

Also included in Volume 1 of The Treasury of Human Inheritance in 1909 were pedigrees and discussions of impairment of hearing. Hearing impairment was noted to be congenital and acquired with the latter often being attributable to infectious causes. Striking pedigrees of hereditary deafness were presented. In addition, evidence was presented that deafness could also be present in individuals with other likely genetic conditions.

In 2002 Bitner-Glindzicz reviewed hereditary deafness and noted the extreme genetic heterogeneity of the condition with more than 40 genes identified for nonsyndromic dominant hereditary deafness and 30 for nonsyndromic autosomal recessive deafness.

2.4 Hemophilia

In the 1911 edition of The Treasury of Human Inheritance information on hemophilia was presented with extensive references, more than 800. Hemophilia was described as a "condition characterized by excessive and chronic liability to immoderate hemorrhage." Bullokh and Fildes noted that it was a condition confined to males. Affected individuals were referred to as bleeders. More than 400 different pedigrees were presented.

Two forms of hemophilia are now known. Hemophilia A due to factor VIII deficiency and Hemophilia B due to factor IX deficiency, both forms are due to mutations on the X chromosome (Bolton-Maggs et al., 2003). Other genetic factors predisposing to bleeding are now known and include Von Willebrand disease, known to be a relatively common bleeding disorder. It can arise due to mutations in specific autosomal genes (Goodeve, 2016).

2.5 Achondroplasia

In the 1912 edition of The Treasury of Human Inheritance information on dwarfism was presented by Prischbieth and Barrington with clear evidence that there were different forms of dwarfism. Achondroplasia (ACH) was noted to be a distinct form of dwarfism. Early studies on bone pathology in some forms of dwarfism were presented. ACH was noted to be a distinct form of dwarfism with short arms and legs relative to trunk length. Histologically ACH was noted to include sclerosis of the cartilage and "conjugation of the epiphyses." In some of the pedigrees presented there was evidence of inheritance of ACH while in others there was no clear evidence of inheritance of the condition from parents. In 1994 Shiang et al. reported that although ACH showed autosomal dominant inheritance in some pedigrees, the majority of cases were sporadic. Using linkage analyses they mapped a gene for ACH to chromosome 4p16.3. Importantly the ACH candidate region was shown to include the gene encoding fibroblast growth factor receptor 3 (FGFR3).

In 2007 Horton et al. reported that more than 95% of patients with ACH had the same point mutation in the gene for FGFR3 and more than 80% of these were found to be new mutations.

2.6 Color blindness

In 1926 an edition of The Treasury of Human Inheritance generated by Bell was devoted to color blindness. Information and references included extended to the year 1777 during which time written description of the condition first appeared. In 1794 the chemist John Dalton described aspects of his altered color vision. Dalton wrote, "I see only two at the most three distinctions, I call yellow, blue and purple."

Color blindness was noted to have been reported in many different countries and was noted to occur most frequently in males, females with the condition were very rarely encountered. Extended pedigrees included in the Treasury report did indicate that color blindness was transmitted from a female who herself had a color blind father.

The form of color blindness described by Dalton was most likely what is now referred to as deuteranomaly. In 1986 Nathans reported that red-green color blindness was due to defects in genes that encode specific opsins. These genes were mapped to chromosome Xq28. A specific gene, medium wave opsin 1 OPN1MW, was reported to encode green color pigment. A green and red sensitive opsin, OPSIN1LW, was also mapped in Xq28 and a key regulator of these two genes was identified (Deeb, 2005).

2.7 Blue sclerotics and fragility of bone

These topics were the subjects of the 1928 edition of The Treasury of Human Inheritance. Julia Bell raised the question, "may these anomalies be the result of general inability of the fibrous tissue of the body to develop normally?" Among the cases presented was that of a 14-year-old boy who had had seven major fractures of limbs between the ages of 4 years and 13 years. Bell noted that the apparent blueness of the sclera was likely due to increased translucency due to decreased thinness of the sclera and increased translucency of the choroidal pigment.

The condition was noted to be transmitted from males to males and in some cases to be transmitted by females to males and females.

Van Dyck and Silence (2014) reviewed osteogenesis imperfecta and emphasized the genetic heterogeneity of the condition and described five syndromic groups of the disorder. Online Mendelian Inheritance in Man currently lists 21 different forms of osteogenesis imperfecta, each with bone fragility and one or more additional manifestations, https://www.ncbi.nlm.nih.gov/omim/?term = Osteogenesis + imperfecta.

Currently more than 13 different chromosomes have been found to harbor genes that, when mutated, can lead to increased bone fragility.

Van Dijk and Silence noted that the majority of cases of osteogenesis imperfecta are due to defects in COl1A1 or COl1A2 mutations. They noted that bone fragility results from the primary gene defect and was accentuated by osteoporosis. Primary bone fragility was apparently due in many cases to increased activity of osteoclasts. Treatment with bisphosphonates were noted to be valuable in treatment of osteogenesis imperfecta.

2.8 Hereditary optic atrophy (Leber's disease)

One edition of The Treasury of Human Inheritance was focused on Hereditary Optic Atrophy, a condition referred to as Leber's disease (Bell, 1931). The disease was noted to first manifest as night blindness. This was followed by progressive reduction of the peripheral fields of vision. Central vision was maintained until the later stages of the disease. Eye examination eventually revealed atrophy of the optic disk. The disease was noted to occur more frequently in males than in females. This condition is now known as optic atrophy type 1 or autosomal dominant optic atrophy, caused by mutations in the mitochondrial cristae biogenesis and fusion protein optic atrophy 1 (Opa1), the gene maps to chromosome 3q29 (Xu et al., 2021).

2.9 Huntington's chorea

Part 1 of the 1934 edition of The Treasury of Human Inheritance includes details on Huntington's chorea. Bell noted that the first detailed descriptions of the disorder had originated from North America, with articles by Waters, Lyon, and Huntington. Waters in 1841 was noted to have described a condition referred to as the "marrams" involving spasmodic action of voluntary muscles with irregular motions of voluntary muscles of extremities, face, and trunk. The abnormal movements were noted to cease during sleep. Lyon in 1863 wrote of a similar condition referred to as "migroms" in his community. In 1872, George Huntington was noted to have written a detailed account of hereditary chorea. Huntington observed that the condition once present in a family, never skipped a generation to reappear in a subsequent generation. He provided details of the abnormal movements and noted that they were aimless and could be forceful. The movements were not apparently painful. A most serious complication of the disorders was reported to be mental deterioration that followed some years after initiation of the abnormal movements.

Pedigrees presented indicated that either a mother or a father of the patients could be affected. Multiple pedigrees presented in the Treasury indicated inheritance through five generations. There was clear evidence of dominant transmission of the disorder with both males and females affected.

In 1984 Gusella et al. reported that a polymorphic marker on human chromosome 4p segregated with Huntington's disease and emphasized that this was the first step in the process of utilizing recombinant DNA technology to identify the specific gene implicated in the disease. Through a strategy of positional cloning the Huntington disease gene, designated HTT, was identified by the Huntington's Disease Collaborative Research Group (1993). The disease was found to be associated with expansion of a trinucleotide repeat sequence in the gene. In unaffected individuals the repeat length varied between 11 and 34 while in Huntington' disease affected patients repeat lengths expanded beyond 42 −66 repeats. Zuccato (2001) reported that HTT was involved in the expression of brain-derived neurotrophic factor and that pathological repeat expansion in HTT impaired this regulatory function.

More recently modifying factors have been identified in Huntington disease and a number of these factors play roles in response to DNA damage. Modifying factors identified included FANCD2 and FANCI-associated nuclease 1 (FAN1), a nuclease involved in DNA interstrand cross link repair (Goold et al., 2019).

In 1939 The Treasury of Human Inheritance focused on different forms of hereditary ataxia and spastic paraplegia. One of the disorders reviewed was Friedreich's ataxia that was determined in most families to follow a recessive pattern of inheritance. This disorder is characterized by progressive muscle weakness that impacts the limbs and also loss of reflexes and diminished proprioception. Cardiac abnormalities may develop. This disorder was mapped to human chromosome 9 by Pandolfo et al. (1990). The defective gene in this disorder designated Frataxin FXN was identified by Campuzano et al. (1996). A number of cases were found to be compound heterozygotes for a mutation in FXN on one chromosome 9 and an abnormal expanded GAA nucleotide repeat sequence on the other member of the chromosome 9 pair.

In considering spastic hereditary ataxias and paraplegias Bell (1939) noted that in some families the disorders appear to follow a dominant inheritance pattern while in other families there was evidence for a recessive pattern of inheritance indicating that different genetic defects were likely involved in these disorders.

Ruano et al. (2014) reviewed global epidemiology of hereditary cerebellar ataxia and hereditary spastic paraplegia. They concluded that most families in population-based series remain without identified genetic mutation after extensive testing.

Volume IV (1947), (1948) issues of The Treasury of Human Inheritance focused on muscular disorders including peroneal muscular atrophy. Neurologists Charcot, Marie, and Tooth were noted to have published on this disorder and peroneal muscular atrophy is sometimes referred to as Charcot − Marie − Tooth disease. Tooth emphasized wasting of the small muscles. Prominent manifestations were reported to include wasting of muscle of the lower limbs and absent reflexes. Wasting that involved the shoulder-girdle muscles was also reported. Inheritance patterns of the disorders across different families were noted to not always be clear.

Different forms of peroneal muscular atrophy are now known to occur and genotypic and phenotypic heterogeneity has been observed (Pareyson & Marchesi, 2009). Genes on eight different autosomes and a gene on the X chromosome have been found to be mutated in different cases of this disease. Abnormalities of nerve conduction velocities are important features of these disorders.

Muscular dystrophies were reviewed by Bell in articles in The Treasury of Human Inheritance in 1947 and 1948. Several unique features of the disorder dystrophic myotonica (myotonic dystrophy) were presented. Myotonia features, referred to stiffness of the muscles of the hands and feet, included the cooccurrence of muscular dystrophy and cataracts in some cases and also evidence for anticipation, that is in specific pedigrees traced down the generations there was evidence for decreasing ages at which symptoms first appeared. Extensive pedigrees were included in the reports, in some the pattern of inheritance could not be judged.

Two forms of myotonic dystrophy are now distinguished. Thornton (2014) reviewed these disorders. Form 1 of the disorders is known to be due to CTG trinucleotide repeat expansion in the 3′ noncoding region of the gene on 19q13.32 that encodes a protein kinase referred to as DMPK. A second form of the disorder was more recently described and is attributed to a CCTG repeat expansion in the gene that encodes Zinc finger 9. Thornton noted that in both cases the repeat expansion is reported to lead to RNA toxicity. Both types of myotonic dystrophy were reported to lead to impaired muscle function, cardiac conduction defects, and cataracts and additional manifestations have been reported. The phenomenon reported earlier as anticipation is now considered to be due to intergenerational repeat expansion, and there is evidence that this occurs more frequently in case of maternal transmission.

2.10 Duchenne muscular dystrophy

In an article in 2004 Harper notes that descriptions of a muscular dystrophy appeared in the literature from the second half of the 19th century. Key contributions were from Duchenne who gave details on the impact of the disorder on muscle and provided evidence of altered histology of muscle in a specific form of muscular dystrophy. Harper also referred to detailed clinical descriptions of a muscular dystrophy disorder in males published in 1852 by Meryon, who described a muscular disorder in males with "loss of power" in lower limbs with later involvement of upper limbs and pectoral muscles. Meryon also drew attention to granular and fatty degeneration of muscles.

Mapping of the Duchenne muscular dystrophy (DMD) locus to a specific position of the X chromosome was facilitated through discovery of a boy with DMD who also had a deletion in the Xp21 region (Kunkel et al., 1985). The complete sequence of the very large DMD gene was reported by Koenig et al. (1988).

Intense studies have been carried out since then on the molecular characteristic of changes in the DMD gene that lead to muscular dystrophy.

2.11 Determination of genetic causes of specific diseases and family studies

Family studies in the past and in the present play important roles in diagnosis and gene discovery. Detailed documentation of clinical manifestations and their history of onset and progression are critically important. Increasingly specific imaging studies and laboratory studies are valuable in initial diagnosis and in following progression. Over the years incorporation of analyses of genes and genomes have proved to be of great value.

Confounding issues also continually emerge as we seek to correlate family history, clinical findings, and laboratory analyses. These confounders include variable penetrance of mutations, variable expressivity, and aspects of phenotypic expansion where "new" previously undescribed phenotypic manifestations arise and a number of genetically determined disorders turn out to have a wider range of phenotypic manifestations than were originally described for that disorders.

In the current age resources are being developed and expanded to provide data banks with details of phenotypic features in specific disorders and to use standardized terms for manifestations, phenotype ontology (Köhler et al., 2021).

Investigations are also being carried out to explore the possibilities of using clinical information documented in electronic medical records to generate phenotype risk scores to broaden possibilities for diagnosis of genetic disorders (Bastarache et al., 2019).

References

Bastarache et al., 2019Bastarache L, Hughey JJ, Goldstein JA, et al: Improving the phenotype risk score as a scalable approach to identifying patients with Mendelian disease, *J*

Am Med Inf Assoc 26(12):1437–1447, 2019. Available from: https://doi.org/10.1093/jamia/ocz179. PMID: 31609419.

Bell, 1939Bell J: *On hereditary ataxia and spastic paraplegia*, The treasury of human inheritance, 1939, Cambridge University Press.

Bell, 1931Bell J: *Hereditary optic atrophy, Leber's optic atrophy*, The treasury of human inheritance, 1931, Cambridge University Press.

Bolton-Maggs and Pasi, 2003Bolton-Maggs PH, Pasi KJ: Haemophilias A and B, *Lancet* 361(9371):1801–1809, 2003. Available from: https://doi.org/10.1016/S0140-6736(03)13405-8. PMID: 12781551.

Campuzano et al., 1996Campuzano V, Montermini L, Moltò MD, et al: Friedreich's ataxia: autosomal recessive disease caused by an intronic GAA triplet repeat expansion, *Science* 271(5254):1423–1427, 1996. Available from: https://doi.org/10.1126/science.271.5254.1423. PMID: 8596916.

Deeb, 2005Deeb SS: The molecular basis of variation in human color vision, *Clin Genet* 67(5):369–377, 2005. Available from: https://doi.org/10.1111/j.1399-0004.2004.00343.x. PMID: 15811001.

Goodeve, 2016Goodeve A: Diagnosing von Willebrand disease: genetic analysis, *Hematol Am Soc Hematol Educ Program* 2016(1):678–682. Available from: https://doi.org/10.1182/asheducation-2016.1.678, 2016.

Goold et al., 2019Goold R, Flower M, Moss DH, et al: FAN1 modifies Huntington's disease progression by stabilizing the expanded HTT CAG repeat, *Hum Mol Genet* 28(4):650–661, 2019. Available from: https://doi.org/10.1093/hmg/ddy375. PMID: 30358836.

Koenig et al., 1988Koenig M, Monaco AP, Kunkel LM: The complete sequence of dystrophin predicts a rod-shaped cytoskeletal protein, *Cell* 53(2):219–228, 1988. Available from: https://doi.org/10.1016/0092-8674(88)90383-2. PMID: 3282674.

Köhler et al., 2021Köhler S, Gargano M, Matentzoglu N, et al: The human phenotype ontology in 2021, *Nucleic Acids Res* 49(D1):D1207–D1217, 2021. Available from: https://doi.org/10.1093/nar/gkaa1043. PMID: 33264411.

Kunkel et al., 1985Kunkel LM, Monaco AP, Middlesworth W, et al: Specific cloning of DNA fragments absent from the DNA of a male patient with an X chromosome deletion, *Proc Natl Acad Sci U S A* 82(14):4778–4782, 1985. Available from: https://doi.org/10.1073/pnas.82.14.4778. PMID: 2991893.

Pandolfo et al., 1990Pandolfo M, Sirugo G, Antonelli A, et al: Friedreich ataxia in Italian families: genetic homogeneity and linkage disequilibrium with the marker loci D9S5 and D9S15, *Am J Hum Genet* 47(2):228–235, 1990. PMID: 2378348.

Pareyson and Marchesi, 2009Pareyson D, Marchesi C: Diagnosis, natural history, and management of Charcot–Marie–Tooth disease, *Lancet Neurol* 8(7):654–667, 2009. Available from: https://doi.org/10.1016/S1474-4422(09)70110-3. PMID: 19539237.

Ruano et al., 2014Ruano L, Melo C, Silva MC, Coutinho P: The global epidemiology of hereditary ataxia and spastic paraplegia: a systematic review of prevalence studies, *Neuroepidemiology* 42(3):174–183. Available from: https://doi.org/10.1159/000358801, 2014.

The Huntington's Disease Collaborative Research Group, 1993The Huntington's Disease Collaborative Research Group: A novel gene containing a trinucleotide repeat that is expanded and unstable on Huntington's disease chromosomes, *Cell* 72(6):971–983, 1993. Available from: https://doi.org/10.1016/0092-8674(93)90585-e. PMID: 8458085.

Thornton, 2014Thornton CA: Myotonic dystrophy, *Neurol Clin* 32(3):705–719, 2014. Available from: https://doi.org/10.1016/j.ncl.2014.04.011. viii.

Van Dijk and Sillence, 2014Van Dijk FS, Sillence DO: Osteogenesis imperfecta: clinical diagnosis, nomenclature and severity assessment, *Am J Med Genet A* 164A(6):1470–1481, 2014. Available from: https://doi.org/10.1002/ajmg.a.36545. Epub 2014 Apr 8.PMID: 24715559.

Xu et al., 2021Xu X, Wang P, Jia X, et al: Pathogenicity evaluation and the genotype-phenotype analysis of OPA1 variants, *Mol Genet Genomics* 296(4):845–862, 2021. Available from: https://doi.org/10.1007/s00438-021-01783-0. Epub 2021 Apr 21.PMID: 33884488.

Zuccato et al., 2001Zuccato C, Ciammola A, Rigamonti D, et al: Loss of huntingtin-mediated BDNF gene transcription in Huntington's disease, *Science* 293(5529):493–498. Available from: https://doi.org/10.1126/science.1059581, 2001.

Further reading

Bates et al., 1909Bates J, Raper W, Gilbey FW: *Deaf mutism, The treasury of human inheritance*, 1909, Cambridge University Press.

Bell, 1926Bell J: *Color blindness, The treasury of human inheritance*, 1926, Cambridge University Press.

Bell, 1935Bell J: *Peroneal type of progressive muscular atrophy, The treasury of human inheritance*, 1935, Cambridge University Press.

Bell, 1928Bell J: *Blue sclerotics and fragility of bone, The treasury of human inheritance*, 1928, Cambridge University Press.

Bell, 1947Bell J: *Dystrophica myotonica, The treasury of human inheritance*, 1947, Cambridge University Press.

Bell, 1934Bell J: *Huntington's chorea, The treasury of human inheritance*, 1934, Cambridge University Press.

Bitner-Glindzicz, 2002Bitner-Glindzicz M: Hereditary deafness and phenotyping in humans, *Br Med Bull* 63:73–94, 2002. Available from: https://doi.org/10.1093/bmb/63.1.73. PMID: 12324385.

Bullock and Fildes, 1911Bullock W, Fildes P: *Hemophilia, The treasury of human inheritance*, 1911, Cambridge University Press.

Bullock et al., 1909Bullock W, Lewis T, River WV, Horne J: *Split foot, deaf mutism, The treasury of human inheritance*, 1909, Cambridge University Press.

Gusella et al., 1983Gusella JF, Wexler NS, Conneally PM, et al: A polymorphic DNA marker genetically linked to Huntington's disease, *Nature* 306(5940):234–238, 1983. Available from: https://doi.org/10.1038/306234a0. PMID: 6316146.

Harper, 2004Harper P: *Landmarks in medical genetics classic papers with commentaries*, 2004, Oxford University Press.

Horton et al., 2007Horton WA, Hall JG, Hecht JT: Achondroplasia, *Lancet* 370(9582):162–172, 2007. Available from: https://doi.org/10.1016/S0140-6736(07)61090-3. PMID: 17630040.

Kantaputra and Carlson, 2019Kantaputra PN, Carlson BM: Genetic regulatory pathways of split-hand/foot malformation, *Clin Genet* 95(1):132–139, 2019. Available from: https://doi.org/10.1111/cge.13434. Epub 2018 Sep 10.PMID: 30101460.

Nathans et al., 1986Nathans J, Piantanida TP, Eddy RL, et al: Molecular genetics of inherited variation in human color vision, *Science* 232(4747):203–210, 1986. Available from: https://doi.org/10.1126/science. PMID: 3485310.

Prischbieth and Barrington, 1912Prischbieth H, Barrington A: *Dwarfism including achondroplasia, The treasury of human inheritance*, 1912, Cambridge University Press.

Shiang et al., 1994Shiang R, Thompson LM, Zhu YZ, et al: Mutations in the transmembrane domain of FGFR3 cause the most common genetic form of dwarfism, achondroplasia, *Cell* 78(2):335–342, 1994. Available from: https://doi.org/10.1016/0092-8674(94)90302-6. PMID: 7913883.

Bell, 2022Bell J: Color Blindness, *Treasury of Human Inheritance* II(Part 11):17–267, 2022.

CHAPTER 3

Discoveries in physiology, biochemistry, protein, and enzyme studies between 1920 and 1970

Clearly, progress between 1920 and 1970 was built on earlier efforts and achievements in the 19th century and in the early years of the 20th century. These included the work in the new sciences of organic chemistry and animal chemistry [Berzelius (1837) and Liebig (1839) and the work of Pasteur (1995)] Garrod (1908).

In his book, "The chemistry of life" Malcolm Dixon (1970) drew attention to early examples of catalytic processes. Berzelius (1837) wrote of "bodies that possess the property of exerting an influence on other complex bodies, thereby causing a rearrangement of the constituents of the body." Dixon noted that Pasteur adopted the same positions as Berzelius when he described catalysis.

Key findings in early biochemistry were reviewed by Teich and Needham (1992). Emil Fischer (1902) and Hofmeister (1902) were noted to have given talks about proteins and amino acids and they reported on amide linkages. The structures of most well-known amino acids were reported to have been discovered by 1922, and this was partly due to early methods of analysis of ammonia release.

Studies built on insights into key discoveries in biochemistry by Krebs, including the urea cycle (1932) and citric acid cycle (1938), and studies by Warburg on cellular metabolism, glycolysis and oxidation, and mitochondrial functions (1931).

Studies on proteins and earlier methods of separations of different proteins were achieved by different methods including centrifugation of proteins suspended in different solutions. Svedberg received the Nobel Prize in Chemistry in 1930 for his pioneering separation methods using ultra-centrifugation.

In the early decades of the 20th century, key studies were carried out on the importance of vitamins in nutrition (Hopkins, 1907). Key studies on vitamin D were carried out by Mellanby (1919) and the role of sunlight in counteracting rickets was established particularly after World War I.

Key work on vitamins essential for diet continued into the 1920s and through to the 1930s. In 1929 Castel and Townsend (1929) established that the intrinsic factor, a glycoprotein, was essential for absorption of vitamin B12. The structure of vitamin C was reported to have been defined by Haworth (1937a,b).

Vitamin A structure and associated aldehydes and the role in the visual cycle were established by Wald (1935a,b).

In a chapter on the history of enzymology, Malcolm Dixon noted that purifications of enzymes began after the 1920s and perhaps the earliest purification was that of xanthine oxidase.

It is, however, important to note that activities of enzymes and aspects of their functions can be determined without them being purified. Studies on enzyme kinetic benefited greatly from the work of Michaelis (1913).

Key to progress in biochemistry and enzymology were developments in spectrophotometry. Spectrophotometry is defined as the quantitative measurement of reflection or transmission properties of a substance at a specific wavelength (nist.gov.programs-projects/spectrophotometry).

By the 1940s these included transition from use of the mercury vapor lamp that had limited wavelength of light emission to inclusion of ultraviolet light sources with wave lengths between 250 and 400 nanometers. This was particularly useful for analysis of reactions that led to conversion of pyridine nucleotides NAD to NADH and conversion of NADP to NADPH.

Another key instrument that facilitated in developments in biochemistry and enzymology was the fluorometer, an instrument designed to measure the intensity of light emitted by fluorophores or fluorochromes (Udenfriend, 1969). Methods were developed to attach fluorochromes to certain chemicals to visualize reactions, one example was the fluorochrome methylumbelliferone.

A key development that contributed to progress in separation and isolation of specific proteins and enzymes from mixtures was chromatography. It could be used to separate components in a mixture by their size and flow speed (Martin and Synge, 1952). Column chromatography also included ion exchange chromatography useful in separated components with different charge.

In 1948 Tiselius received a Nobel Prize for pioneering electrophoretic and chromatographic techniques.

In 1951 Pauling and Corey analyzed aspects of protein folding. Folding was reported to place specific domains of the protein in positions for specific functions and for interactions with other molecules.

Fred Sanger made several important contributions to biology, biochemistry, and medical sciences. In 1959 he reported establishing the linear amino acid sequence of insulin. This was achieved by carrying out sequential partial hydrolysis of the purified protein and labeling terminal amino acids with a specific chemical 1-fluoro2-4nitrobenzene.

Throughout the 1950s and 1960s, studies were also carried out on the structure of specific purified biochemical molecules though use of X-ray crystallography. In 1956, Dorothy Hodgkin and coworkers published the structure of vitamin B12 using X-ray crystallography.

Methods evolved for the analyses of the properties of enzymes. These included:

1. Definition of the substrates with which an enzymes interacted and cofactors needed for enzyme activity and pH optimum for enzyme activity.
2. Products that resulted from activity of a specific enzyme.
3. Inhibitors of enzyme and relative stability of enzyme.
4. Physical properties of the enzyme or proteins size and charge.
5. Levels of the enzyme or protein in different cells and tissues.
6. Quantitative or qualitative differences in enzyme or protein between different individuals.

Analyses of these parameters could provide evidence as to whether in different individuals mutations possibly occurred that altered specific properties of the enzyme.

Information could also be evaluated to determine whether an enzyme or protein was likely encoded by different loci in the genome. Such evidence might emerge from studies in different tissues, studies at different stages of development, and differences in properties such as electrophoretic mobility, substrate specificity, and pH optima.

Fig. 3.1 illustrates use of electrophoresis to demonstrate developmental changes in expression of the enzyme alcohol dehydrogenase genes and encoded isozymes in liver samples obtained at autopsy from fetuses and infants of different gestational age and postnatal age. ADH1A (alpha alpha) is present in early

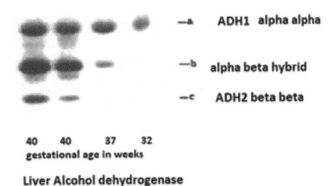

FIGURE 3.1

Illustrates use of electrophoresis to demonstrate developmental changes in expression of the enzyme alcohol dehydrogenase genes and encoded isozymes in liver samples obtained at autopsy from fetuses and infants of different gestational age and postnatal age. ADH1A (alpha alpha) is present in early fetus with subsequent expression [also of ADH1B also known as ADH2 (beta isozyme)] present initially in alpha/beta heterozygous band and later a beta band is also present.

fetus with subsequent expression [also of ADH1B also known as ADH2 (beta isozyme)] present initially in alpha/beta heterozygous band and later a beta band is also present.

Fig. 3.2 demonstrates allelic variation in ADH1C (also known as ADH3) gamma polypeptides in stomach samples with isozymes gamma1 gamma1 homozygote, gamma1 gamma 2 heterozygote, and gamma2 gamma2 homozygote.

Harry Harris (1963) emphasized the importance of using techniques of enzymology to demonstrate alterations that could lead to metabolic changes.

Harris and Hopkinson described ingenious methods for examining specific enzymes following electrophoresis of suspension of cell or tissue extracts, particularly in starch gels. Chromogenic staining methods following electrophoresis, involved exposing of the gel surface to a specific enzyme substrate and cofactors and accessory chemicals that led to a change in color on contact with the enzyme.

Fluorescent staining methods included use of a substrate coupled to a fluorescent dye that changes on exposure to enzymes. Other staining methods included use of electron transfer dyes, for example, tetrazolium salts, methyl thiazolyl tetrazolium.

Ingenious staining methods were developed that involved inclusion of secondary enzymes as exogenous agents in staining reaction (Hopkinson and Harris, 1965; Harris and Hopkinson, 1976).

Analysis of levels of amino acids and organic acids, for example, in urine evolved initially through chromatography and later through mass spectrometry.

A number of important inborn errors of metabolism were diagnosed through discovery of abnormal urine levels of amino acids or organic acids (Stanbury et al., 1966).

Specific methods were developed to detect levels of urinary metabolites derived from the breakdown of complex glycosaminoglycans and sialic containing complexes, thus facilitating diagnosis of lysosomal storage disease (DeDuve, 1964; Dean and Barrett, 1976).

The first edition of "The Metabolic Basis of Inherited Diseases" was published in 1960 and a second edition was generated in 1966 by Stanbury, Wyngaarden, and Frederickson. Attention was focused on metabolic and genetic factors involved in carbohydrate amino acids and lipid metabolism. In addition, new

FIGURE 3.2

Demonstrates allelic variation in ADH1C (also known as ADH3) gamma polypeptides in stomach samples with isozymes gamma1 gamma1 homozygote, gamma1 gamma 2 heterozygote, and gamma2 gamma2 homozygote.

insights into biochemical and genetic factors in red cell and connective tissue disorders and porphyrias were presented.

In the years between 1920 and 1970 discoveries of metabolic disorders in humans needed to take into account biochemical findings and possible Mendelian genetic mechanisms were given some consideration.

This was to change dramatically following discoveries of the genes and gene regulation.

References

Berzelius JJ, editor: *Lehrbuch d.Chem. Arnoldischen Buchhandlung* 6:19−25, 1837.

Castel WB, Townsend WC: The effect of administration to pateitns with pernicious anemia of beef muscle after incubation with normal human gastric juice, *Am J Med Sc* 178:764, 1929.

Dean RT, Barrett AJ: Lysosomes, *Essays Biochem* 12:1−40, 1976.

DeDuve C: From cytases to lysosomes, *Fed Proc* 23:1045−1049, 1964.

Dixon M: The history of enzymes and biological oxidations. Chap. 2 In Needham J, editor: *The chemistry of life*, 1970, Cambridge University Press.

Fischer E: Ueber die Hydrolyse derProteinstoffe, *Chem. Atg* 26:939, 1902.

Garrod AE: *Inborn errors of Metabolism Croonian Lectures, Croonian Lectures*, Oxfors, 1997, Oxford University Press.

Harris H, Hopkinson DA: *Handbook of enzyme electrophoresis in human genetics*, Amsterdam, London New York, 1976, North Holland Publishing Company.

Harris H: *Supplement to Garrod's inborn errors of metabolism, Oxford monographs on medical genetics*, London, New York, Toronto, 1963, Oxford University Press.

Haworth N: *The Nobel Prize in Chemistry 1937 was divided equally between Walter Norman Haworth "for his investigations on carbohydrates and vitamin C" and Paul Karrer "for his investigations on carotenoids, flavins and vitamins A and B2"*, 1937a. <https://www.nobelprize.org/prizes/chemistry/1937/summary/>.

Haworth, W.N. Investigations on carbohydrate and vitamin C. 1937, 37b. <https://www.nobelprize.org/prizes/chemistry/1937/haworth/biographical/>.

Hofmeister F: Uber bau und gruppierung der eiweisskorper, *Erg. Physiol.* 1:759, 1902.

Hopkins FG: Four lectures on the significance of variations in the constituents of urine Guy's Hosp, *Gazette* :. 21: 327, 382, 403, 423.

Hopkinson DA, Harris H: Evidence for a second structural locus determining human phosphoglucomutase, *Nature* 208:410, 1965.

Liebig J: *Liebig's Ann.* 30:250, 1839.

Martin, A.J.P. , Synge, R.L.M. The Nobel Prize in Chemistry 1952 was awarded jointly to Archer John Porter Martin and Richard Laurence Millington Synge "for their invention of partition chromatography." 1952. <https://www.nobelprize.org/prizes/chemistry/1952/summary/>.

Mellanby E: An experimental investigation on rickets, *Nutr Rev* 34(11):338−340. Available from: https://doi.org/10.1111/j.1753-4887.1976.tb05815.x, 1919.

Michaelis ML: Menten Kinetik der Invertinwirkung Biochem, *Zeitung* 49:333−369, 1913.

Pasteur L: Mémoire sur la fermentation appelée lactique (Extrait par l'auteur), *Mol Med* 1 (6):599–601, 1995. 1857; Article Reprinted. PMCID: PMC2229983.

Pauling L, Corey RB: The structure of synthetic polypeptides, *Proc Natl Acad Sci U S A* 37(5):241–250. Available from: https://doi.org/10.1073/pnas.37.5.241, 1951.

Stanbury JB, Wyngaarden JB, Fredrickson DS: *The metabolic basis of inherited disease (1960)*, New York, Toronto, Sydney, London, 1966, Blakiston Division McGraw Hill Book Company.

Svedberg T. (1930) Measurement of ultra-centrifugal dispersity in protein solutions. Koll Z 57:10, 1908.

Teich M, Needham SM: *A documentary history of biochemistry 1770-1940*, UK, 1992, Leicester University press.

Tiselius, A. Nobel prize for pioneering electrophoretic techniques, 1948. <https://www.nobelprize.org/prizes/chemistry/1948/tiselius/biographical/>.

Udenfriend S: . (1962) *Fluorescence assay in biology and medicine*, I, New York, 1969, Academic Press.

Wald G: Carotenoids and the visual cycle, *J Gen Physiol* 19(2):351–371, 1935a. Available from: https://doi.org/10.1085/jgp.19.2.351. PMID: 19872932.

Wald G: Vitamin A in eye tissues, *J Gen Physiol* 18(6):905–915, 1935b. Available from: https://doi.org/10.1085/jgp.18.6.905. PMID: 19872899.

Further reading

Garrod AE: *The Croonian Lectures. 1929 Publ. l Henry Frowde. Hodder and Stroughton London*, 1908.

Harris H: Enzyme polymorphisms in man, *Proc. Roy. Soc. B* 164:298, 1966.

Harris H: *The principles of human biochemical genetics*, Amsterdam, London, 2014, North-Holland Publishing, . American Elsevier Publishing Company Inc, New York.

Hartman G: *Group antigens in human organs*, Copenhagen, 1941, Munksgard.

Hodgkin DA, Pickworth J, Robertson JH, et al: The crystal structure of the hexacarboxylic acid derived from B12 and the molecular structure of the vitamin, *Nature*. 176(4477):325–328, 1955. Available from: https://doi.org/10.1038/176325a0. PMID: 13253565.

Hopkins FG: The practical importance of vitamins, *Brit Med J* 1:507, 1919.

Hopkins FG, Wolf CG: *Purine metabolism and its relation to gout*, Oxford Medicine, Vol 4, 1921, Oxford University Press, p 97.

Krebs HA, Henseleit K: Untersuchungen über die Harnstoffbildung im Tierkörper. [Hoppe-Seyler's] Zeitschrift für physiologische Chemie, *Strasbourg* 210:33–46, 1932.

Krebs HA, Salvin E, Johnson WA: The formation of citric and alpha-ketoglutaric acids in the mammalian body, *Biochem J* 32(1):113–117, 1938. Available from: https://doi.org/10.1042/bj0320113. PMID: 16746585.

Sanger F: Chemistry of insulin: determination of the structure of insulin opens the way to greater understanding of life processes, *Science* 128(3359):1340–1344, 1959. 10.1126/science.129.3359.1340 PMID: 13658959.

Svedberg T: Measurement of ultra-centrifugal dispersity in protein solutions, *Koll Z* 57:10, 1930.

Warburg OH: *Nobel Prize in chemistry*, 1931. <https://www.nobelprize.org/prizes/medicine/1931/summary/>.

CHAPTER 4

Early translation of biochemical, metabolic, and genetic discoveries into clinical medicine

In reviewing early studies on specific proteins that have maintained clinical importance even until the present it seems particularly relevant to consider the ABO and Rh blood group systems.

4.1 ABO

In 1901 Karl Landsteiner reported that when serum from one human individual was added to blood and blood cells from another individual it could sometimes cause the red cells to clump and coalesce. His tests on a number of different individuals caused him to propose that there were two types of individuals and to conclude that there were two different blood groups that he designated A and B.

In 1902 Von Decastello and Sturli recognized that there was a third group of individuals that had blood group designated O because their serum did not aggregate red blood cells from other individuals.

Later Von Dugern and Hirschfeld in Zurich (1910) established Mendelian inheritance of the ABO system.

The first successful human blood transfusion is sometimes attributed to Reuben Ottenberg at Mount Sinai Hospital (1907).

Richard Lewisohn (1916) of Mount Sinai Hospital in New York is credited with discovery that addition of citrate to whole blood would prevent coagulation and thus storage of blood under conditions of refrigeration was enabled.

Following his move from Europe to the Rockefeller Institute in New York, Landsteiner was initially involved in discovery of minor blood group antigens in the MN and P systems (Landsteiner & Levine, 1927).

4.2 Further information on the ABO blood group system

An unusual feature of the ABO system is that without prior exposure to other blood types individuals with blood group A have anti-B activity in their serum and individuals with blood group B have anti-A antibodies in their serum. Individuals with blood group O can be regarded as universal blood donors since they have no A or B antigens on their red cells and can donate to individuals of blood groups A, B, or O. Individuals with blood group O should receive blood from blood group O individuals since they have anti-A and anti-B factors in their serum. Individuals with blood group AB have no A or B antibodies in their serum and then can receive blood from individuals with blood groups A, B, or O and are known as universal recipients.

The A and B antigens were subsequently found to be present not only on red cells but also on certain epithelial cells.

In 1956 Kabat reported that blood group substances could be extracted from membranes with alcohol solutions. In 1966 Szulman reported that AB blood group antigens were complex macromolecules that contained carbohydrate and were firmly bound to red cell membranes.

Specific carbohydrate chains were identified that contained some but not all of the polysaccharides present in ABO. These were designated as H chains and they specifically lacked the terminal carbohydrate sequences present in blood groups A and B.

Detailed studies by Watkins (1966) led to definition of the exact terminal carbohydrate containing sequences of A and B.

A: alpha GalNAC-beta gal-GNac terminal sequence in A is N-acetyl galactosamine.

B: alpha gal-beta gal-GNac terminal sequence in B is galactose.

H: beta-gal terminal sequence in H is beta gal.

These early studies suggested that ABO specificity was likely determined by specific glycosyltransferase. Soluble forms of A or B and H specificities were found to be present in saliva and other secretions form mucosal cells but were present as glycoproteins. Their presence was also determined by a separate genetic locus designated Secretor Se.

4.3 Secretor status

Glycoprotein water-soluble derivatives of H, A, and B blood group substances were found to be present in 80% of Europeans (Landsteiner & Harte, 1941). Their presence was shown to be dependent on an active form of a locus designated Secretor that is distant from the ABO locus. The secretor locus was later discovered to encode a fucosyl transferase FUT2 and it maps to chromosome 19q13.33 (Kelly et al., 1995).

In 1976 Badet et al. studied glycosyl transferase both in individuals with blood group A and blood group B and they determined that there was a single amino acid difference between the blood group A glycosyl transferase and the blood group B glycosyl transferase.

Their results suggested that mutation in a specific glycosyl transferase determined whether that enzyme would transfer galactose or N-acetyl galactosamine as the terminal sugar.

4.4 Mapping of the ABO locus to a chromosome

Genetic linkage studies in families revealed that the ABO blood group locus was linked to the locus that encoded adenylate kinase 1 (AK1). Chromosome studies by Ferguson-Smith et al. (1976) assigned ABO and AK1 to chromosome 9q34.

In 1990 Yamamoto et al. (1990), through isolation of CDNA clones, were subsequently able to generate sequence of the ABO gene locus. They determined that the product of the ABO locus in individuals with blood group O lacked glycosyl transferase activity. Their studies in 1990 revealed four key amino acid differences between A and B loci.

In a 2019 report, Westhoff noted that DNA typing, for example, with use of arrayed DNA sequences, was not routinely used for ABO or for Rh typing because serological typing was accurate. However, DNA typing was useful in situations where it was important to use samples other than blood to determine blood group status, for example, in organ transplantation cases and in individuals who had received multiple blood transfusions. Blood group typing could then be done on buccal scrapings or cell from tissues.

4.5 Rh blood group system

Landsteiner and Wiener in 1940 discovered the Rh blood group system. It was named Rh because this particular factor also occurred on the surface of red cells of the monkey species *Maccacca rhesus*. Although this species was later referred to as *Maccaca mulatta*, the name Rhesus remained in use for the blood group system.

A year prior to this Levine and Stetson (1939) reported an unusual case of a woman who gave birth to a stillborn fetus with hydrops and the woman hemorrhaged severely and was to receive transfusion from her husband's ABO matched blood. However, her blood agglutinated her husband's red cells. The authors postulated that she had been immunized during pregnancy with an antigen that the fetus inherited from the father.

In 1941 Levine et al. (1941) reported a follow-up paper on isoimmunization during pregnancy, its possible bearing on the etiology of erythroblastosis fetalis.

An important aspect of this system was the discovery by Levine that in a woman who was negative for Rh carried a fetus and gave birth to an infant who was Rh positive could develop antibodies to Rh and in subsequent pregnancies, if the fetus was Rh positive, the antibodies from the mother could lead to hemolysis of fetal red cells and give rise to a condition known as hemolytic disease of the newborn. This hemolysis was shown to lead to significant increases in the levels of unconjugated bilirubin that could be deposited in specific brain regions and give rise to a condition known as kern icterus that led to brain damage and spasticity.

Wiener and coworkers (Wiener et al., 1947) initially proposed a single locus for Rh. However, later studies revealed more than one locus. However, Fisher and Race proposed three pairs of Rh antigens Dd, Cc, Ee (1948), and the presence of D determined Rh positivity (Race, 1948).

Rh positivity was later determined to be positive for the RhD antigen.

In their review of the Rh system in 2000 Aventi and Reid noted that three different genes are involved in production of proteins that constitute Rh antigens. Genes were designated RHDd, RHCc, and RHEe and were noted to be encoded on chromosome 1p36.11. A related gene RHAG was mapped to chromosome 6p11.

RhCc, RhEe peptides were noted to be expressed on the surface of red cell only if Rhag peptides were present.

The Rh proteins were reported to have 12 transmembrane segments with N-terminal and C-terminal regions reaching the cytoplasm. In addition other proteins, referred to as Rh accessory proteins, were found to be associated with Rh proteins and together with Rh proteins they formed the Rh complex. Rh proteins were shown to be linked to the cell membrane via palmitic acid. Rh accessory proteins include LW (Landsteiner Wiener) glycoprotein and integrin-associated protein, Glycophorin B and Fy glycoproteins which constitute determinants for the blood group referred to as Duffy.

Rh genes were reported to each have 10 exons and to be highly homologous and to have approximately 40% homology with RHAG gene.

An important discovery was that specific individuals have deletion of the RHD gene and are RhD antigen negative. More comprehensive studies have revealed that some individuals are RhD antigen negative because of specific mutation in the Rh gene.

The RHC and RHE genes give rise to antigens Rh C and c and Rh E or e; difference between C and c and E and e were reported to be due to allelic variations.

RhD-related hemolytic disease of the newborn continues to receive much attention. Anti-RhD reagents were developed to immunize Rh negative (RhD negative) women immediately following delivery of a fetus or an infant. Fyfe et al. (2014) reviewed development and use of anti-D IgG.

4.6 RHD genotyping

Saramago et al. (2018) reviewed value of noninvasive prenatal testing of fetal cell DNA to guide antinatal prophylaxis with anti-RhD immunoglobulin.

4.7 Chemical analyses and metabolism incorporating information gathered across the decades

Very early chemical studies in human focused often on analysis of urine. Early studies of F. Gowland Hopkins dealt with measurement of uric acid in urine and on relationship of uric acid excretion to dietary content. In 1907 Gowland Hopkins published four lectures on constituents of urine.

He also focused considerable attention on dietary components and their relation to growth and he studied proteins and amino acids. In 1906 Wilcock and Hopkins published a paper on the importance of individual amino acids in metabolism. In 1919 Hopkins and Chick wrote about "accessory factors" in food; these later came to be known as vitamins.

Hopkins continued his studies on diet and gout and in 1921 Hopkins and Wolf published a paper on purine metabolism and its relation to gout.

In 1908 Archibald Garrod gave his highly influential Croonian lectures that focused on "derangements of metabolic processes in ill-health." He noted further that interpretations of these changes were impacted by our scant knowledge of metabolism.

He noted uncertainty as to whether the accumulation of uric acid in blood and deposition of sodium biurate in tissue in gout were caused by rearrangements in metabolism or by excretory defects.

Garrod drew attention to diabetes mellitus, to glycosuria, and proposed that these could initially be due to a diet rich in carbohydrates but eventually the glycosuria was noted to be continuous and was accompanied by destruction of fats leading to the production of acetone bodies.

Garrod emphasized that abnormalities that gain attention "are those that results in unusual appearance of surface tissues or excretions, detected by available laboratory tests." He emphasized though that there were disorders "that did not readily advertise their presence and escaped out notice."

In reviewing Garrod's Croonian lectures, Harry Harris drew attention to Garrod's fundamental ideas regarding inborn errors of metabolism as a block that occurred at a specific step in metabolism that resulted from defects in the function of a specific enzyme.

Garrod also drew attention to the unique features of specific inborn errors of metabolism and to the insights that these unique features provided into chemical processes.

Garrod further emphasized that even as we concentrated on "extreme examples of variations in chemical behavior" it was likely that there are also more subtle chemical differences between individuals.

Intense efforts to elucidate biochemical processes, identify biochemical abnormalities associated with specific diseases occurred throughout the following decades (Stanbury et al., 1966).

These then merged with increasing abilities to analyze genes, their function, and their sequence so that today biochemical studies, genetics genomics, DNA, and gene transcript analysis merge to provide a comprehensive approach to genetic disease and inborn errors of metabolism.

It is interesting to note the most current versions of the esteemed textbook originally designed by Stanbury et al. (1966) in versions since 2001 are entitled, "The Metabolic and Molecular Basis of Inherited Disease" (Scriver et al., 2001).

However, biochemical studies and enzyme analyses are still of great value in determining the precise functional effects of gene changes discovered through DNA analysis.

References

Ferguson-Smith MA, Aitken DA, Turleau C, de Grouchy J: Localisation of the human ABO: Np-1: AK-1 linkage group by regional assignment of AK-1 to 9q34, *Hum Genet* 34(1):35–43, 1976. Available from: https://doi.org/10.1007/BF00284432. Sep 10.

Fyfe TM, Ritchey MJ, Taruc C, et al: Appropriate provision of anti-D prophylaxis to RhD negative pregnant women: a scoping review, *BMC Preg. Childbirth* 14:411, 2014. Available from: https://doi.org/10.1186/s12884-014-0411-1. Dec 10.

Kelly RJ, Rouquier S, Giorgi D, et al: Sequence and expression of a candidate for the human Secretor blood group alpha(1,2)fucosyltransferase gene (FUT2). Homozygosity for an enzyme-inactivating nonsense mutation commonly correlates with the non-secretor phenotype, *J Biol Chem* 270(9):4640–4649, 1995. Available from: https://doi.org/10.1074/jbc.270.9.4640. Mar 3.

Landsteiner K, Harte R: A group-specific substances in human saliva, *J Biol Chem* 140:673–674, 1941.

Landsteiner K, Levine P: Further onservations on individual differences in human bllod, *Proc Soc Exp Biol Med* 24:941, 1927.

Levine P, Katzin EM, Burnham L: Isoimmunization in pregnancy: its possible bearing on the etiology of erythroblastosis fetalis, *JAMA* 116(9):825–827. Available from: https://doi.org/10.1001/jama.1941.02820090025006, 1941.

Levine P, Stetson RE: An unusual case of intra-group agglutination, *JAMA* 113:126–127, 1939.

Lewisohn R: The importance of the proper dosage of sodium citrate in blood transfusion, *Ann Surg* 64(5):618–623, 1916. Available from: https://doi.org/10.1097/00000658-191611000-00017. Nov.

Ottenberg R: Reuben Ottenberg of Mount Sinai Hospital in New York performed the first successful transfusions, 1907. Available from <https://www.rockefeller.edu/our-scientists/karl-landsteiner/2554-nobel-prize/>

Race RR: The Rh genotype and Fisher's theory, *Blood* 3:27–42, 1948.

Saramago P, Yang H, Llewellyn A, et al: High-throughput, non-invasive prenatal testing for fetal Rhesus D genotype to guide antenatal prophylaxis with anti-D immunoglobulin: a cost-effectiveness analysis, *BJOG* 125(11):1414–1422, 2018. Available from: https://doi.org/10.1111/1471-0528.15152. Oct.

Scriver CR, Beudet AL, Sly W, Valle D, Childs B, Kinzler KW, et al: *The metabolic and molecular basis of inherited disease*, ed 3, 2001, Mc Graw Hill.

Stanbury JB, Wyngaarden JB, Fredrickson DS: *The metabolic and molecular basis of inherited disease*, ed 2, New York, London, Toronto Sydney, 1966, McGraw Hill book Company.

von Dungern E, Hirschfeld L: Über Vererbung gruppenspezifischer Strukturen des Blutes. Zeitschrift für Immunitätsforschung und Experimentelle Therapie, 1910.

Watkins WM: Blood group subsances, *Science* 152:172, 1966.

Wiener AS, Sonn-Gordon EB, Handman EB: Heredity of the Rh blood types; additional family studies, with special reference to the theory of multiple allelic genes, *J. Immunol.* 57(3):203–210, 1947. Nov. Available from: 20268415.

Yamamoto F, Clausen H, White T, et al: Molecular genetic basis of the histo-blood group ABO system, *Nature* 345(6272):229–233, 1990. Available from: https://doi.org/10.1038/345229a0. May 17.

Further reading

Avent ND, Reid ME: The Rh blood group system: a review, *Blood* 95(2):375–387, 2000. January 15. Available from: 10627438.

Badet J: Serum glycosyltransferase activity associated with antigen biosynthesis in blood groups A and B. Study of normal B group and cis AB group subjects, *Rev. Fr. Transfus. Immunohematol.* 19(1):105–116, 1976. Available from: https://doi.org/10.1016/s0338-4535(76)80091-8. Mar.

Garrod AE: The Croonian Lectures. 1929 Publ. L Henry Frowde. Hodder and Stroughton London, 1908.

Harris H: Supplement to Garrod's Inborn Errors of Metabolism 1963 Oxcord Monographs on Medical Genetics osxford University Press London, New York, Toronto.

Hopkins FG: Four lectures on the significance of variations in the constituents of urine Guy's Hosp, *Gazette* 327(382):403–423, 1907.

Hopkins FG, Chick H: Accessory factors in food, *Lancet* 2:28, 1919.

Hopkins FG, Wolf CG: *Purine metabolism and its relation to Gout Oxford Medicine*, Vol 4, 1921, Oxford Univesity Press, p 97.

Kabat EA: *Blood group substances: their chemistry and immunochemistry*, New York, 1956, Academic Press.

Landsteiner K: Nobel Lectures Physiology or Medicine 1930 On individual differences in human blood. Available from <https://www.nobelprize.org/uploads/2018/06/landsteiner-lecture.pdf>

Landsteiner K: Ueber Agglutinationserscheinungen normalen menschlichen Blutes, *Wien klin Wochschr* 14:1132–1134, 1901.

Landsteiner K, Wiener AS: An agglutinable factor in human blood recognized by immune sera for Rhesus blood, *Proc Soc Exp Biol Med* 43:223–224, 1940.

Landsteiner K, Wiener AS: Studies on agglutinogen (Rh) in human blood reacting with anti-rhesus sera and with human isoantibodies, *J Exp Med* 74:309–320, 1941.

Martin JP, Synge RLM: 1952 were awarded a Nobel Prize for their work on Partition Chromatogrphy. Available from <https://www.nobelprize.org/prizes/chemistry/1952/summary/>

Mellanby E: An experimental investigation on rickets, *Nutr Rev* 34(11):338–340. Available from: https://doi.org/10.1111/j.1753-4887.1976.tb05815.x, 1919.

Needham J, Dixon M: Hopkins and Biochemistry Papers Publ 1949 W. Heffer and Sons, Cambridge.

Szulman AE: Chemistry, distribution, and function of blood group substances, *Ann Rev Med* 17:307, 1966.

Tiselius A: Nobel Prize for Electrophoresis and Chromatography, 1948. Available from <https://www.nobelprize.org/prizes/chemistry/1948/tiselius/biographical/>

Von Decastello A, Sturli A: Alfred von Decastello and Adriano Sturli discovered the fourth blood group, AB, further elucidating the differences in compatibility among blood types. Available from <https://www.rockefeller.edu/our-scientists/karl-landsteiner/2554-nobel-prize/>

Westhoff CM: Blood group genotyping, *Blood* 133(17):1814–1820, 2019. Available from: https://doi.org/10.1182/blood-2018-11-833954. April 25.

Willcock EG, Hopkins FG: The importance of individual amino acids in metabolism, *J Physiol* 35:88, 1906.

CHAPTER 5

Advances in methods of genome analyses, nucleotide analyses, and implications of variants

5.1 Introduction

Clinical human genetics began with analyses of the patterns of inheritance of specific diseases. Later methods emerged to analyze human chromosomes. With growing expertise in ability to study proteins and enzymes and their variations (polymorphisms) in individuals, linkage patterns were derived by analyzing polymorphic genetic markers and determining if within a particular family with a specific genetic disease one or more polymorphic markers co-segregated with the disease. Through growing expertise of protein purification, amino acid determination, and by combining knowledge of the genetic code for generation of specific amino acids, it became increasingly possible to determine the sequence of genes that encode specific proteins. Combination of techniques to isolate gene-specific DNA segments and to label these with marker dyes facilitated the mapping of specific genes to their position on chromosomes.

Advanced knowledge of the DNA sequence of specific genes also led to initiation of DNA analyses in specific genetic diseases.

5.2 DNA sequencing

Early DNA sequencing methods were developed by Sanger et al. (1977) and Maxam and Gilbert (1992). The Maxam − Gilbert sequencing method involved chemical degradation of DNA.

In sequencing human DNA, the DNA was cloned into bacteriophage or phage vector. Cultivation of the organisms enabled amplification of the cloned DNA segments.

The Sanger method utilized the polymerase chain reaction to synthesize new strands of DNA. To initiate the polymerase chain reaction, an oligonucleotide that corresponded to bacteriophage sequence adjacent to the cloning site was used as primer polymerase and deoxynucleotides were added including labeled nucleotides.

Sanger devised a method to add dideoxynucleotides, which terminate chain synthesis. In sequencing experiments four tubes were set up each with a different dideoxynucleotide: ddATP, ddTTP, ddCTP, or ddGTP. Chain synthesis is terminated when a dideoxynucleotide is incorporated.

The Sanger method therefore involved use of a DNA template, new synthesis of DNA in a polymerase chain reaction with inhibition of elongation and chain termination with specific dideoxynucleotides.

Sequencing reactions led to generation of a series of different sized fragments depending on the position where the dideoxynucleotide was incorporated. Different sized fragments were separated by electrophoresis on denaturing polyacrylamide gels. Visualization could be achieved through use of radiolabeled nucleotides or fluorescent labeled nucleotides. The different sized fragments led to staggered pattern on gel electrophoresis and the specific nucleotide at the position where termination occurred could be assessed.

Development of libraries with cloned DNA segments corresponding to coding regions of genes provided resources that could be utilized in the development of the gene map of human chromosomes.

Early stage of the Human Genome Projects involved identification of polymorphic DNA markers, for example, microsatellite markers and their assignment on individual chromosome. The HapMap consortium genotyped and mapped more than one million single nucleotide polymorphisms, https://www.ncbi.nlm.nih.gov/variation/news/NCBI_retiring_HapMap/.

If a series of polymorphic markers were mapped to a specific chromosomal region, specific co-segregation of markers and haplotype could be determined. A haplotype is defined as a series of linked genetic markers that segregate as a unit in meiosis.

Linkage maps were derived by analyzing polymorphic genetic markers and determining if within a particular family with a specific genetic disease one or more polymorphic markers co-segregated with the disease.

Physical maps were generated by mapping specific genes to human chromosomes through hybridization of labeled DNA corresponding to a specific gene, to human chromosome spreads and their visualization, for example, through fluorescent microscopy (Fig. 5.1).

If a specific disease was shown to be linked to a series of polymorphic markers on a specific human chromosome, that information could subsequently be used to carry out extensive cloning and sequencing of DNA in that chromosomal segment to isolate the disease gene (positional cloning) (Collins, 1995).

Cloned segments of DNA that were known to be mapped to specific positions on human chromosome were also used to generate microarrays, for example, platform with attached DNA probes known to map to specific chromosome positions. The specific DNA probes can be arranged on the platform in an order corresponding to their map position on human chromosomes. Labeled short segments of test DNA, for example, from specific patients can then be hybridized to the array. If deletions or duplication are present in the test DNA, for example, from a patient, these will be visible when the array is subsequently scanned (Fig. 5.2).

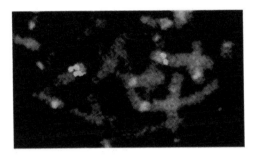

In situ hybridization illustrating duplication of a gene on chromsome 15q11.2

FIGURE 5.1

One member of the chromosome 15 pair shows fluorescent signal for probe in 15q11.2 region. On a second member of the chromosome 15 pair the signal is duplicate indicating a duplication of 15q11.2.

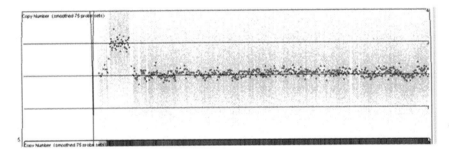

Microarray data showing extra copies of chromosome 15q12

FIGURE 5.2

Annotation microarray with distribution of signals along the length of chromosome 15. Note abnormally increased distribution of signal in the proximal regions of chromosome indicating duplication in 15q12 region.

Availability of segments of human DNA representing specific genes, or DNA mapped to a specific chromosomal position were cloned, for example, into bacteriophage, DNA could be isolated and labeled and used in in situ hybridization on metaphase spreads. This then facilitated analysis of human structural rearrangements, for example, inversions or translocation that did not alter dose and that were therefore not detectable in microarray experiments.

Following mapping and isolation of a gene that when mutated gives rise to a specific disease, it was necessary to determine the function of that gene and its role in physiology. Elucidation of the physiological roles of a definitive disease gene can potentially facilitate development of specific therapies (Collins, 1999).

van Dijk et al. (2018) reviewed revolutions in sequencing technologies from Sanger dideoxynucleotide sequencing to next generation sequencing (NGS) to third generation sequencing.

Second generation sequencing, sometimes referred to as massively parallel sequencing, involved breaking DNA into fragments, preparing single stranded templates, ligating these with specific primers at fragment ends. Singled stranded templates were then immobilized on beads or platforms. Sequencing reactions were then carried out on immobilized DNA fragments.

Disadvantages of massively parallel sequencing included generation of short reads, need to align reads, possibilities for incorrect assembly of sequence contigs; sequence alignment of repeat sequences was often problematic. In second generation sequencing haplotype phasing was problematic.

van Dijk et al. also noted that early methods of NGS required use of polymerase chain reaction that could lead to artifacts.

Third generation sequencing was noted to involve single molecule sequencing and it also allowed generation of long sequence reads. Specific protocols for long-read sequencing were developed by PACBIO and Oxford Nanopore Technology. Length of sequence reads achieved with these methods was reported initially as in the 10–15 kb range. However, a higher nucleotide error rate was reported in long-read sequencing.

It is interesting to note that new sequencing techniques can detect base modifications, for example, methylation without using additional sequencing reactions. Earlier techniques required an additional separate pre-sequence bisulfite sequence reaction.

5.3 Applications of long-read sequencing

Logsdon et al. (2020) reviewed applications of long-read sequencing noting its capacity to achieve assembly of diploid genomes and to detect the full spectrum of human genetic variation. They noted that long-read sequencing methods could achieve genetic sequence reads of length 10 kb − 1 MB. Furthermore, long-read sequencing methods were reported to read through repetitive sequence elements.

They emphasized that long-read sequencing data analysis led to clearer understanding of more complex forms of genetic variation including large insertions and deletions. They noted work of Huddleston et al. (2017) in identifying structural variation in a haploid genome.

Miga et al. (2020) achieved telomere to telomere sequencing of the human X chromosome.

Chaisson et al. (2015) documented enhanced discovery of structural variants, including inversion, complex insertion, and long tracts of tandem repeats, through single molecule sequencing.

Logsdon et al. also noted that long-read sequencing can be applied to full-length sequencing of RNA transcripts and is particularly useful in providing information on different isoforms derived in transcription from a single gene. Long-range sequencing was noted to capture information on full-length isoforms that were often missed when cDNA was generated from RNA to achieve sequencing. Accurate long-read sequencing was noted to lead to generate new models of specific genes.

Xie et al. (2020) reported use of long-read sequencing technologies to facilitate genetic diagnoses in dystrophinopathies. They drew attention to patients with Duchenne muscular dystrophies where genetic diagnosis was not achieved through regular sequencing due to structural rearrangements or deep intronic variants. The DMD gene was reported to span 2.5 MB and more than 99% of the sequence was noted to be in introns. They presented an example of a patient in whom DMD resulted from an intragenic inversion.

Ascari et al. (2021) reported the utility of long-read sequencing in discovery of complex structural variants in the gene CEP78 centrosomal protein. Defects in this gene lead to defects in primary cilia that result in sensorineural hearing loss retinal degeneration associated with cone-rod degeneration disease (one form of Usher syndrome).

5.4 Sequence variant interpretation

Strande et al. (2018) addressed aspects of nucleic acid sequence variant interpretation and noted the importance of distinguishing between variant level interpretation and clinical care interpretation.

They noted that currently a binary distinction is made in genetic disease classification, with genetic disorders being considered Mendelian or polygenic. They noted that additional factors need to be taken into account in clinical genetics, including differences in penetrance levels and variable expressivity. In 2015 a five-tier classification of variants was recommended by the American College of Medical Genetics (ACMG). Basic levels of classification of variants included classification as pathogenic, likely pathogenic, uncertain significance, likely benign, benign. National Institutes of Health (NIH) sponsored resources that incorporated these guidelines include the Clinical Genome Resource (ClinGen) and ClinVar.

For interpretation of the clinical significance of sequence variants it is also important to consider the population frequency of certain allelic variants. This information is available in databases including gnomAD. There are, however, concerns about limited population diversity in databases.

Strande et al. noted that the relative risk for a specific Mendelian disease can also be determined based on the known incidence of that specific disease in the specific population to which the patient belongs.

They also noted that segregation of a specific allele in family members with the same disorder can also be taken into account in assessing variant significance.

Computational data, in silico predictors to be taken into account include the likely impact of a variant on protein function, and/or protein structure and evidence pertaining to the evolutionary conservation of the variant across species.

Functional changes that may be significant include loss of function, gain of function, impaired interactions with cofactors or other proteins within a complex.

Variant types considered to be likely pathogenic include nonsense variants, frame shift variants, premature termination variants, or alterations in canonical splice sites (Fig. 5.3).

Other data to be taken into account include the position of the variant, for example, is it located within an important region of the protein, and is it located in a mutation hotspot.

Information available on laboratory assessment of protein function may also be of value.

Strande et al. noted that Variant Curation Expert Panels have been established with data on assessment of variants reported in a specific disease, and their impact on phenotype.

Beyond these considerations Strande et al. noted importance of considering clinical history in a specific patient and family history, bearing in mind aspects of variable expressivity and possibilities of phenotype expansion.

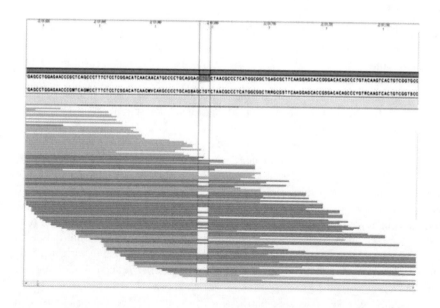

Deletion of 4 base pairs CTGT leading to frameshift and transcription termination in exon 31.
Note TSC2 has 40 exons.

FIGURE 5.3

Deletion in sequence.

Strande et al. also noted importance of being aware of new gene − disease associations and new gene variant − disease associations.

Lappalainen et al. (2019) reviewed genomic analyses in the age of genome sequencing noting that data analysis remained challenging. They presented protocols to follow genome sequencing, variant detection, genotype calling analysis, and annotation of the impact of variants, distinguishing rare and common variants, reviewing associations and correlations with phenotype.

Lappalainen et al. stressed aspects of coverage noting that as depth of coverage increases, the capacity to decrease errors in interpretation increases. However, depth of coverage utilized does impact the cost of sequencing.

In presenting a landscape of genomic variation, variant type, and size they noted that these included single nucleotide variants 1 bp, small deletions or insertions 1−49 bp, structural variations greater than 50 bp, insertions and balanced rearrangements.

In considering repeats they noted occurrence of short tandem repeats, microsatellites and minisatellites 7−49 bp. Critical improvements in methods of long-read sequencing were noted.

Lappalainen et al. reviewed information on genetic variation in different human populations and variant frequencies. Rare variants were noted to have a population frequency less than 1%; low frequency variants had frequencies between 1% and 5%; common variants were noted to have frequencies greater than 5%.

Variants that have a greater effect on function were those that induced quantitative or qualitative effects on protein. Variants in noncoding region were noted to include quantitative trait loci that impact levels of gene expression.

5.4.1 Sequencing in clinical diagnosis: reinterpretation of data and secondary findings

There are cases where a patient with defined phenotypic features was referred for sequencing analysis to determine the genetic cause of a specific disorder but where results of sequence analysis reveals the presence of mutations in genes unrelated to the patient's phenotypic features. These findings are referred to as secondary findings and questions arise as to the responsibilities of diagnostic laboratories and clinical personnel to report these findings to caregivers and patients. Deignan et al. (2019) noted that the ACMG periodically evaluates genetic variants that may have clinical relevance. As data accumulate potential reclassification of the clinical relevance of variants may be required. Updated information on the clinical significance of variants is made available by the ACMG and utilization of this information in clinical reporting is a responsibility of laboratories.

Communication to patients is also important when information on reinterpretation of the clinical significance of variants becomes available through additional studies or new information reported in the literature or databases. It is particularly important that information be made available regarding new discoveries of variants that alter therapeutics and patient management (Miller et al., 2021).

5.4.2 Long-range sequencing relevance to diagnosis of rare disorders

Posey (2019) reviewed genome sequencing and implication for diagnosis of rare disorders, noting that increased sensitivity of long-range sequencing and techniques such as transcriptomics and metabolomics need to be more fully explored. She noted, however, that additional factors and mechanisms need to be more fully explored. These included allelic heterogeneity, multilocus determination of diseases, and the impact of rare and common variation at a particular locus.

Posey cited evidence that 8000 Mendelian disorders were known by 2020 and that as of 2019 only 20% of protein coding genes had been linked to a specific disorder.

Variant types associated with genetic diseases include nucleotide variants, insertions, and deletions, complex genomic variants and repeat expansions. Posey noted that additional mechanisms to be considered included digenic inheritance where two separate loci are involved and this could include two rare variants in different loci, or one common and one rare variant. Other possibilities to consider include mutational burden that involves two highly penetrant alleles or mutational burden that includes one highly penetrant allele and one less penetrant allele.

Evidence for digenic inheritance of a form of retinitis pigmentosa was reported by Kajiwara et al. (1994). All affected individuals in a study had variants in PRPH2, peripherin 2, a cell surface glycoprotein found in the outer segment of both rod and cone photoreceptor cells. Some individuals in a family had the same variant but were unaffected. All affected members were also found to have pathogenic variants at a second locus ROM1 retinal outer segment membrane protein 1, reported to be essential for disk morphogenesis, and may also function as an adhesion molecule involved in the stabilization and compaction of outer segment disks or in the maintenance of the curvature of the rim.

Digenic inheritance was also demonstrated in facioscapulohumeral dystrophy with defects in DUX1 in chromosome 4 and SMCHD1 on chromosome 18. Balog et al. (2018) reported that deletion on chromosome 18p that leads to deletion of the gene that encodes SMCHD1 (structural maintenance of chromosomes flexible hinge domain containing) and a moderately sized D4Z4 repeat on chromosome 4 can lead to facioscapulohumeral dystrophy.

Posey also drew attention to specific genes associated with more than one genetic disease. A specific example presented included the gene that encodes lamin protein LMNA. Monoallelic variants at any one of three different positions in this gene were reported to lead to dominant monoallelic conditions including dilated cardiomyopathy type 1a, Hutchison-Gilbert Progeria, Emery Dreifuss muscular dystrophy.

Biallelic recessive mutation in LMNA was reported to lead to Charcot−Marie−Tooth disease, axonal type, and Emery Dreifuss muscular dystrophy.

Regions of homozygosity identified on single nucleotide polymorphisms (SNP) microarrays or on exome sequencing must be distinguished from regions with deletions of one chromosome. Occurrence of regions of homozygosity can lead to identification of disease-associated gene defects.

Uniparental disomy (UPD) can also lead to phenotypic defects. UPD is defined as the occurrence of two copies of a chromosome, or two copies of a chromosome region derived from one parent and no copy from the other parent. It can best be detected if DNA from both parents is available for testing.

Posey et al. emphasized the utility of genome sequencing in demonstrating copy number neutral structural genomic variants.

Further applications of long-range sequencing included analysis of the regulatory genome. Abnormalities potentially leading to abnormal gene expression have been identified through alteration in the intronic regions of genes. It is important to note that variants in an intronic polypyridine tract that impacts splicing of the CFTR (cystic fibrosis) transcript was shown to impact clinical manifestations in individuals with CFTR variants p. Arg117His (R117H). Individuals with R117H on an IVS8−7T background were reported to not manifest cystic fibrosis. Individuals with R117 H on an IVS8−5T background manifested clinical signs of cystic fibrosis (Massie et al., 2001).

5.5 Additional evidence for digenic or complex inheritance

Duerinckx et al. (2020) 31606992 reported that genes with defects responsible for primary microcephaly have been identified on the basis of analyses in humans with this condition and on the basis of studies in zebra fish models. These studies have led to the identification of 75 genes implicated in primary microcephaly.

They noted that disorders that manifest Mendelian inheritance can be oligogenic with digenic inheritance being the most common.

Microcephaly in humans was noted to be syndromic in many cases. However, 18 different genes were reported to cause nonsyndromic primary microcephaly. In several cases, products of these genes were reported to be involved in centrosome function. Other genes were shown to be involved in mechanisms related to cell proliferation, for example, CASC5 also known as KNL1 kinetochore scaffold 1 protein.

Studies in mice have provided some evidence for digenic/oligogenic inheritance of primary microcephaly.

Duerinckx et al. carried out studies in a first cohort with 47 patients with primary microcephaly and 140 controls. In a replication cohort, studies were carried out on an additional 47 primary microcephaly patients. In a second cohort, studies were carried out on 64 patients and 63 controls. Based on initial studies 75 genes were selected for detailed analysis.

In analysis of association, Duerinckx et al. measured the burden of variants beyond simple Mendelian inheritance. Genes that manifested an excess of variants in primary microcephaly patients beyond that documented in controls were identified.

Variants were filtered out if they had a general population frequency of 5% or higher.

Data analyses revealed that the burden of variants was higher in six centrosomal genes in patients than in controls. These results suggested a digenic inheritance mode in primary microcephaly that involved centrosomal function-related genes.

5.6 Variants in nonprotein coding genomic regions

Zhang and Lupski (2015) reviewed likely mechanisms through which variants in nonprotein coding genomic regions influence gene functions. They also considered technologies for investigating variants in noncoding regions.

Zhang and Lupski considered potential functional effects of variants in noncoding regions. Enhancers and suppressors of gene expression and elements in DNAse 1 hypersensitivity regions received special attention. An enhancer that received special attention impacted expression of nitric oxide synthase 1 adaptor protein. In cardiac tissue this protein functions as an adaptor protein that links to nitric oxide synthase and defects in the enhancer were reported to be linked to specific forms of cardiac arrhythmia.

Another intergenic enhancer element was shown to impact expression of MYB, a transcriptional activator involved in critical regulation of erythroid development.

Variants in 3′ untranslated regions of gene have also been reported to influence expression of specific gene products in part through impacting micro-RNA binding sites.

Zhang and Lupski documented additional specific noncoding region variants implicated in human disease. They included POU3F4 in deafness. In a patient with deafness Aristidou et al. (2018) identified a chromosomal breakpoint that mapped 87 kb upstream of an X-linked gene that encodes POU3F4 a neural transmitter, thus suggesting an underlying long-range position effect mechanism. Zhang and Lupski noted publications that reported Charcot−Marie−Tooth disorder. CMT1 neuropathy was shown to be associated with a copy number variant upstream of the gene that encodes peripheral myelin protein 22. A genomic disruption upstream of the SHOX homeobox gene that likely interrupted an enhancer, was shown to be associated with short stature.

Zhang and Lupski also presented evidence for compound inheritance that included inheritance of a genomic variant on one chromosome along with a structural variant, for example, deletion in the corresponding region on the homologous chromosome.

Various computational resources were reported to have been established that include information on noncoding and potential regulatory elements. These include HGMD, the Human Gene Mutation Database, and GWAS, Genome Wide Annotation of Variants.

5.7 Haplotype phasing

If heterozygous pathogenic variants at different positions within a particular gene are detected in an individual, it is important to know if they are present on one chromosome or on two homologous chromosomes. This question can potentially be resolved if both parents are available and if each variant can be shown to have been contributed by a specific parent. However, both parents are not always available. Also in some cases variants are de novo in origin.

5.7.1 Noninvasive prenatal screening and haplotype phasing

Che et al. (2020) reported on the use of haplotype analyses in cell-free plasma DNA and its use in prenatal diagnosis. They noted the increased need for prenatal noninvasive screening methods for monogenic disorders. They presented applications of haplotyping of cell-free fetal DNA for presence of monogenic disorders.

Che et al. noted that while detection of paternally inherited alleles in cell-free fetal DNA (cff DNA) was relatively straightforward, assessment of maternally inherited alleles in cff DNA was more complicated because of the excess of maternal DNA in cell-free plasma.

They noted that haplotype dosage levels were used. Genome sequencing may enable mutational profile to be detected. Che et al. noted that target-based haplotyping with definition of the variants along a specific locus can be valuable.

They carried out a study to demonstrate the use of haplotyping of cff DNA in nine families with known genetic disorders. The study involved samples collected during pregnancy and subsequent analyses of neonatal samples.

5.8 Haplotype analysis

Haplotype analysis involves analysis of a series of genetic variants. Extensive haplotype data were derived through analysis of microsatellite repeats at defined loci throughout the genome. Haplotypes can also be generated through analysis of single nucleotide variants.

Earlier use of haplotyping was reported by Lemna et al. (1990) who reported that a study revealed that a specific 3 bp deletion was a common cause of cystic fibrosis and that this deletion occurred on a single haplotype in 96% of 877 cases they studied.

Vermeulen et al. (2017) reported difficulties experienced in monogenic heritable disease diagnosis using cff DNA in maternal plasma because of high levels of maternal cell-free DNA. They promoted target-based capture methods for specific loci and incorporation of parental haplotyping. They reported that in haplotype phasing the heterozygous SNPs around the maternal disease allele can be readily assessed.

Chiu et al. (2018) reported on the use of haplotype analysis in noninvasive prenatal diagnosis. They noted that optimally haplotypes of both parents should be determined prior to haplotyping of cell-free DNA recovered from maternal blood. Haplotype association with the disease locus can also be facilitated if there are other children with the disorder and if family studies are available. In couples known to be at risk for bearing a child with a monogenic disorder haplotype analysis within the genomic region surrounding the disease allele could facilitate molecular diagnosis during noninvasive prenatal diagnosis.

Haplotype disease associations are more readily determined using long-read DNA sequencing. Zhang et al. (2019) reported use of long-read sequencing and haplotype analysis to determine the presence of a pathogenic structural chromosome abnormality in preimplantation genetic testing.

Gigante et al. (2019) reported that long-read nanopore sequencing enabled both haplotype analysis and analysis of methylation of cytosine at CpG sites.

Chen et al. (2021) reported on use of parental haplotyping in noninvasive prenatal testing (NIPT) for alpha and beta thalassemia. They proposed use of haplotyping in NIPT analyses of these thalassemia. They noted that in certain populations development of a reference panel of genomic haplotypes associated with thalassemia mutations would be useful.

Scotchman et al. (2020) reviewed NIPT in monogenic disorders. They noted that cff DNA is released into the maternal circulation at approximately 4 weeks of gestation. However, it constitutes only between 5% and 20% of the cell-free DNA present in maternal plasma. The fraction of cff DNA from male fetuses is determined through assay of sequences on Y chromosome sequences. The female fraction of cff DNA in the maternal circulation has been assessed by HLA typing, and determination of the fraction not corresponding to maternal alleles.

cff DNA was reported to be short length segments, and this was noted to complicate detection of large genomic deletion, or duplication and to complicated detection of long triplet repeat disorders.

Assessment in a particular fetus may be complicated by the existence of DNA from a vanishing twin.

Scotchman et al. noted that in the UK panel-based screening has been developed for screening of certain dominant disorders including FGFR2 and FGFR3 skeletal dysplasia and craniosynostosis syndrome.

Scotchman et al. noted that NIPT autosomal recessive and X-linked disorders presented more of a challenge, because of complication of detecting the maternally derived mutations. However, if maternal and paternal mutations are different and if the variants in the maternal wild type gene copy can be determined, diagnosis can be facilitated. Haplotype analyses can also be applied to achieve diagnosis. However, the recombination rate between the disease allele and the haplotypes assessed must be known.

Other approaches being applied include target-based amplification of a specific region. Use of long-read sequencing that can determine sequences of lengths 20 kb or greater can also be applied.

Other possibilities for noninvasive prenatal screening include analysis of circulating fetal cells in maternal blood and analysis of trophoblastic cells isolated from the external cervix. Scotchman et al. noted that in these methods distinguishing cells of maternal origin remained a possibility.

5.9 Long-range sequencing and identification of structural genomic variants leading to disease

Merker et al. (2018) reported use of long-range sequencing to attempt to identify the underlying defect in a patient with a clinical diagnosis of Carney complex. The patient was reported to have a cardiac myxoma at 7 years of age, a testicular mass found to be a Sertoli-Leydig cell tumor at 10 years of age, a pituitary tumor at 13 years, and an adrenal microadenoma at 16 years.

The diagnosis of Carney complex was suspected in this patient but the PRKAR1A gene was not found to have abnormalities on sequencing. PRKAR1A protein kinase cAMP-dependent type I regulatory subunit alpha has been associated with this autosomal dominant syndrome. Short-read sequencing did not reveal any other relevant defects.

Merker et al. reported undertaking long-read sequencing with a ninefold depth of coverage. This revealed the deletions within PRKAR1A. Other defects found on long-range sequencing were heterozygous and located in genes associated with autosomal recessive conditions.

Subsequent studies on the patient revealed low levels of expression of PRKAR1A.

Thibodeau et al. (2020) reported use of long-read sequencing using the Oxford nanopore system to investigate possibility of rearrangements in hereditary cancer syndromes. They identified an 8.5 kb inversion on chromosome 16p13.3 that was flanked by a deletion leading to partial loss of NTHL1 and TSC2 in a patient with tuberous sclerosis. NTHL1 is adjacent to the centromeric end of TSC2.

5.10 Investigations of chromatin structure and genomic function

Shashikant and Ettensohn (2019) emphasized that changes in chromatin impact regulatory programs in cells. Chromatin changes include alterations in nucleosome positioning and density and changes in the three-dimensional topology of chromatin domains, alterations in the organization of chromosomes in the nucleus, and epigenetic modifications.

They noted that local accessibility of chromatin was particularly important in gene expression. Methods to assess patterns of chromatin accessibility are based on methods that include accessibility of chromatin to specific enzymes, for example, DNase I, or enhanced susceptibility of chromatin to mechanical shearing.

Open chromatin regions are noted to be often associated with active enhancers, promoters, and transcription factors. The period in which chromatin in a specific region is open is hard to measure. ATAC (assay for transposase accessible chromatin) is a frequently used method to assess chromatin accessibility. ATAC utilizes a modified version of the enzyme Tn5 transposase that can cleave DNA in regions of open chromatin and tag it with specific sequencing primers.

Shashikant and Ettensohn emphasized the roles of cis acting DNA elements brought together as a result of chromatin looping in regulation of gene expression. These interactions included enhancer − promoter interactions. Additional factors included DNA − protein interactions, including transcription factor − DNA interactions.

Functional genome analysis was reviewed by Guigo and de Hoon (2018). They focused on sequence-based methods of assessment genome function, and also noted techniques that involved genome perturbation to analyze specific impacts on function.

In considering function they reviewed methods to identify regions of active chromatin and methods to define methylation status. The latter includes bisulfite sequencing used to convert unmethylated cytosine to uracil that is then recognized on sequencing as thymine (Fig. 5.4).

Methylation analysis of specific regions can also be carried out with use of methylation-specific antibodies. Antibodies can also be used to assay histone

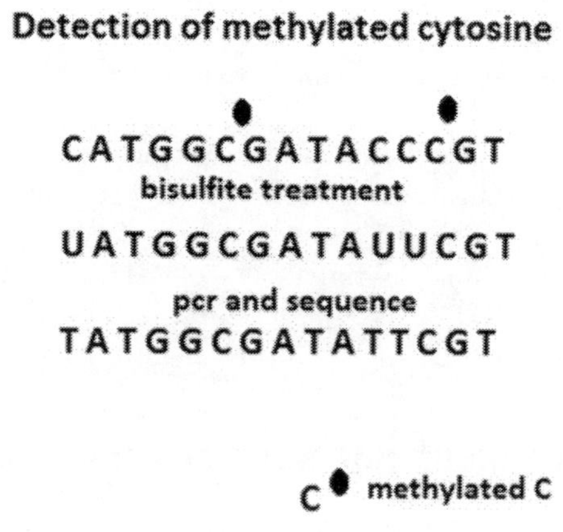

Detection of methylated cytosine

CATGGCGATACCCGT
bisulfite treatment

UATGGCGATAUUCGT
pcr and sequence
TATGGCGATATTCGT

C● methylated C

FIGURE 5.4

Bisulfite sequencing for detection of methylated cyrosine.

modifications and antibodies and immunoprecipitation can be used to identify transcription factor – DNA interactions. Methods to carry out sequencing after immunoprecipitation of modified chromatin are referred to as CHIP seq.

Special methods to sequence the 5' end of sequences using primers corresponding to the transcription start site are referred to as CAP analyses. Sequences at the 3' end of the gene can utilize oligonucleotide primers corresponding to sequence adjacent to the polyadenylation site.

5.10.1 Chromatin conformation capture

Special methods are designed to investigate chromatin interactions, looping, and folding, which bring together chromatin from different regions of the genome. These interactions are important because they impact regulation of gene expression. Promoters may be brought into contact with enhancers of expression or promoters may be brought into contact with repressors. Different techniques have been applied.

Belton et al. (2012) reported application of formaldehyde that leads to crosslinking of interacting genome regions. Chromatin and DNA are then digested, and fragments are cloned into vectors to generate a library. The clones in the library can then be amplified using primers specific to regions of interest. Other analysis methods include application of NGS to analyze ligation products. The goal of chromatin conformation capture is to identify genomic regions that interact.

5.11 Methylation analyses

Rauluseviciute et al. (2019) reviewed available tools for DNA methylation analysis, noting that high throughput methylation analyses are available through many different laboratories involved in clinical diagnoses.

Methods for methylation analysis include restriction endonuclease-based studies, affinity enrichment studies, and bisulfite conversion analyses, and analyses can be carried out in selected genomic regions or genome wide.

The 5'methyl cytosine conversion was noted to be the most common epigenetic mark. Different analyses methods reported included methylated DNA precipitation and DNA hydromethylation sequencing. Sequencing data analyses were noted to include analyses of differential methylated region. Comparisons can also be made to determine if differences occur between different tissues or between cancer tissue and normal tissue.

In vertebrates, DNA methylation occurs primarily in the context of CpG dinucleotides. Promoters in approximately 70% of annotated genes were noted to be unmethylated, thus facilitating expression. Intergenic regions were noted to more frequently exhibit methylation indicating nonexpression.

Array-based methylation protocols were noted to be in use particularly in clinical settings. These include Illumina Infinium arrays. Sequencing methods were noted to be more useful for single site methylation analyses.

Use of restriction enzyme digestion to examine DNA methylation was noted to be primarily applied to analyses of specific genomic regions. Methylation-sensitive restriction enzymes include Msp1, HpaII, Not1, Sma1, Bstu1. These enzymes only cleave methylated sequences.

Rauluseviciute et al. noted that specific proteins are sometimes used to bind to methylcytosine and to immunoprecipitated methylated genomic segments.

Bisulfite sequencing can be applied to segments of the genome or to the whole genome. In utilizing this method it is important that sequence be generated from both bisulfite treated and nonbisulfite treated DNA, and that these then be compared.

Rauluseviciute et al. noted that specific tools are available to facilitate interpretation of bisulfite sequencing data, for example, BS Seeker, BS Map.

Cerrato et al. (2020) reviewed methylation abnormalities associated with specific monogenic or oligogenic disorders.

In reviewing methylation, they noted that in primordial germ cells and in the preimplantation stage, waves of methylation erasure occur. Subsequently during embryogenesis methylation occurs through activity of methyltransferases DNMT3A, DNMT3B, DNMT3L. They noted that methylation that impacts promoters and enhancers is particularly important, since this impacts gene expression. Methylation can inhibit transcription factor binding and can inhibit chromatin modification.

Some regions were known to evade methylation. These regions were referred to as differentially methylated regions (DMR) and specific DMR evade methylation on a gamete of origin basis.

DNA methylation variations also occur in postnatal life and are known to occasionally be altered in specific monogenic diseases. There is also evidence for age-related alterations in methylation.

5.12 Imprinted genomic regions

DNA methylation differences define specific regions in the genome imprinted regions. In imprinted regions expression occurs either from the maternally inherited allele or from the paternally inherited allele. Imprinting is imposed early in embryogenesis and is dependent in part on methylation that occurs at specific loci. In general, imprinting is an epigenetic process dependent largely on methylation. Structural genomic abnormalities and specific mutations can lead to abnormal imprinting.

In 1997 Bartolomei and Tilghman (Bartolomei and Tilghman, 1997a,b) wrote: "a handful of autosomal genes in the mammalian genome are inherited in a silent state from one of the two parents and in a fully active state from the other, rendering the organism functionally hemizygous for the imprinted genes."

5.13 Genetic disorders where analysis of methylation is important

Cerrato et al. (2020) documented five genetic disorders where analysis of methylation is particularly important.

In one disorder, Fragile X syndrome abnormal methylation was associated with abnormal CGG repeat expansion and abnormal methylation in the 5' regions of the FMR1 gene.

In four of the five disorders DNA methylation defects were widespread despite the fact that the specific gene defect was in one particular gene. Loss of methylation in multiple gene regions occurred in the Claes Jensen syndrome due to KDM5C mutations. This syndrome is associated with cognitive and neurological abnormalities. Widespread abnormal patterns of methylation also occur in Sotos syndrome due to pathogenic NSD1 mutation; in Kabuki syndrome due to pathogenic mutations in KMT2D or KDM6A. Abnormal patterns of methylation were also observed in Charge syndrome due to CHD7 variants.

KDM5C lysine demethylase 5C.
NSD1 nuclear receptor binding SET domain protein 1.
KMT2D lysine methyltransferase 2D.
KDM6A lysine demethylase 6A.
CHD7 chromodomain helicase DNA binding protein 7.

Abnormalities of methylation can also be observed in other syndromes including Beckwith−Wiedemann syndrome, transient neonatal diabetes, pseudohypoparathyroidism, multiple locus imprinting syndrome, and in facioscapulohumeral muscular dystrophy.

Methylation analyses have been used to diagnose imprinting disorders, for example, Prader−Willi syndrome and Beckwith−Wiedemann syndrome.

Facilitation of methylation analyses through array studies has led to the discovery that a number of genetic disorders have methylation episignatures. Aref-Eshghi et al. (2020) reported identification of 34 robust disease-specific episignatures detected through analysis of blood samples.

5.13.1 Methylation and cancer

It is interesting to note that DNA methylation studies particularly through methylation microarrays are increasingly being applied to detection and surveillance in cancer. Chen et al. (2020) reported utilization of DNA methylation analysis of cells and cell-free DNA in urine for early detection and monitoring of bladder cancer. They reported discovery of 26 significant markers of bladder cancer based on utilization of methylation microarray studies.

Altered methylation patterns have also been documented in colon cancer to define the impact of methylation in facilitating expression of cancer driver gene expression and to analyze decreased expression cancer suppressor genes (Liang et al., 2019).

In 2019 de Almeida et al. reported analyses of DNA methylation and gene expression in breast cancer. They identified differential expression in normal and tumor tissue at 368 different CpG sites. CpG methylation at sites in three genes was shown to be of diagnostic and prognostic importance. These three genes included TDRD10 Tudor domain containing 10, PRAC2 small nuclear protein, and TMEM132C transmembrane protein 132C.

References

Aref-Eshghi E, Kerkhof J, Pedro VP, et al: Evaluation of DNA methylation episignatures for diagnosis and phenotype correlations in 42 Mendelian neurodevelopmental disorders, *Am. J. Hum. Genet.* 106(3):356–370, 2020. Available from: https://doi.org/10.1016/j.ajhg.2020.01.019. Epub 2020 Feb 27. PMID: 32109418.

Aristidou C, Theodosiou A, Bak M, et al: Position effect, cryptic complexity, and direct gene disruption as disease mechanisms in de novo apparently balanced translocation cases, *PLoS One* 13(10):e0205298, 2018. Available from: https://doi.org/10.1371/journal.pone.0205298. eCollection 2018. PMID: 30289920.

Ascari G, Rendtorff ND, De Bruyne M, et al: Long-read sequencing to unravel complex structural variants of CEP78 leading to cone-rod dystrophy and hearing loss, *Front. Cell Dev. Biol* 9:664317, 2021. Available from: https://doi.org/10.3389/fcell.2021.664317. eCollection 2021. PMID: 33968938.

Balog J, Goossens R, Lemmers RJLF, et al: Monosomy 18p is a risk factor for facioscapulohumeral dystrophy, *J. Med. Genet* 55(7):469–478, 2018. Available from: https://doi.org/10.1136/jmedgenet-2017-105153.

Bartolomei M, Tilghman S: Genomic imprinting in mammals., *Annu Rev Genet.* 31:493–525, 1997a. Available from: https://doi.org/10.1146/annurev.genet.31.1.493.

Bartolomei M, Tilghman S: Genomic Imprinting in mammals, *Annual Rev. Genet.* 268–276, 1997b. Available from: https://doi.org/10.1016/j.ymeth.2012.05.001.

Belton JM, McCord RP, Gibcus JH, et al: Hi-C: a comprehensive technique to capture the conformation of genomes, *Methods.* 58(3):268–276, 2012. Available from: https://doi.org/10.1016/j.ymeth.2012.05.001.

Cerrato F, Sparago A, Ariani F, et al: DNA methylation in the diagnosis of monogenic diseases, *Genes (Basel)* 11(4):355, 2020. Available from: https://doi.org/10.3390/genes11040355. PMID: 32224912.

Chaisson MJ, Huddleston J, Dennis MY, et al: Resolving the complexity of the human genome using single-molecule sequencing, *Nature.* 517(7536):608–611, 2015. Available from: https://doi.org/10.1038/nature13907.

Che H, Villela D, Dimitriadou E, et al: Noninvasive prenatal diagnosis by genome-wide haplotyping of cell-free plasma DNA, *Genet. Med* 22(5):962–973, 2020. Available from: https://doi.org/10.1038/s41436-019-0748-y. PMID: 32024963.

Chen C, Li R, Sun J, et al: Noninvasive prenatal testing of α-thalassemia and β-thalassemia through population-based parental haplotyping, *Genome Med* 13(1):18, 2021. Available from: https://doi.org/10.1186/s13073-021-00836-8. PMID: 33546747.

Chen X, Zhang J, Ruan W, et al: Urine DNA methylation assay enables early detection and recurrence monitoring for bladder cancer, *J. Clin. Invest* 130(12):6278–6289, 2020. Available from: https://doi.org/10.1172/JCI139597. PMID: 32817589.

Chiu EKL, Hui WWI, Chiu RWK: cfDNA screening and diagnosis of monogenic disorders—where are we heading? *Prenat. Diagn* 38(1):52–58, 2018. Available from: https://doi.org/10.1002/pd.5207.

Collins FS: Positional cloning moves from perditional to traditional, *Nat. Genet* 9(4):347–350, 1995. Available from: https://doi.org/10.1038/ng0495-347. PMID: 7795639.

Collins FS: Shattuck lecture—medical and societal consequences of the Human Genome Project, *N. Engl. J. Med* 341(1):28–37, 1999. Available from: https://doi.org/10.1056/NEJM199907013410106. PMID: 10387940.

de Almeida BP, Apolónio JD, Binnie A, Castelo-Branco P: Roadmap of DNA methylation in breast cancer identifies novel prognostic biomarkers, *BMC Cancer* 19(1):219, 2019. Available from: https://doi.org/10.1186/s12885-019-5403-0.

Deignan JL, Chung WK, Kearney HM: Points to consider in the reevaluation and reanalysis of genomic test results: a statement of the American College of Medical Genetics and Genomics (ACMG), *Genet. Med* 21(6):1267–1270, 2019. Available from: https://doi.org/10.1038/s41436-019-0478-1.

Duerinckx S, Jacquemin V, Drunat S, et al: Digenic inheritance of human primary microcephaly delineates centrosomal and non-centrosomal pathways, *Hum. Mutat* 41(2):512–524, 2020. Available from: https://doi.org/10.1002/humu.23948. PMID: 31696992.

Gigante S, Gouil Q, Lucattini A, et al: Using long-read sequencing to detect imprinted DNA methylation, *Nucleic Acids Res.* 47(8):e46, 2019. Available from: https://doi.org/10.1093/nar/gkz107. PMID: 30793194.

Guigo R, de Hoon M: Recent advances in functional genome analysis, *F1000Res* 21(7). F1000 Faculty Rev-1968. Available from: https://doi.org/10.12688/f1000research.15274.17. eCollection 2018. PMID: 30613379.

Huddleston J, Chaisson MJP, Steinberg KM, et al: Discovery and genotyping of structural variation from long-read haploid genome sequence data, *Genome Res* 27(5):677–685, 2017. Available from: https://doi.org/10.1101/gr.214007.116.

Kajiwara K, Berson EL, Dryja TP: Digenic retinitis pigmentosa due to mutations at the unlinked peripherin/RDS and ROM1 loci, *Science* 264(5165):1604–1608, 1994. Available from: https://doi.org/10.1126/science.8202715. PMID: 8202715.

Lappalainen T, Scott AJ, Brandt M, Hall IM: Genomic analysis in the age of human genome sequencing, *Cell.* 177(1):70–84, 2019. Available from: https://doi.org/10.1016/j.cell.2019.02.032. PMID: 3090155.

Lemna WK, Feldman GL, Kerem B, et al: Mutation analysis for heterozygote detection and the prenatal diagnosis of cystic fibrosis, *N. Engl. J. Med* 322(5):291–296, 1990. Available from: https://doi.org/10.1056/NEJM199002013220503. PMID: 2296270.

Liang Y, Zhang C, Dai DQ: Identification of differentially expressed genes regulated by methylation in colon cancer based on bioinformatics analysis, *World J. Gastroenterol* 25(26):3392–3407, 2019. Available from: https://doi.org/10.3748/wjg.v25.i26.3392. PMID: 31341364.

Logsdon GA, Vollger MR, Eichler EE: Long-read human genome sequencing and its applications, *Nat. Rev. Genet* 21(10):597–614, 2020. Available from: https://doi.org/10.1038/s41576-020-0236-x.

Massie RJ, Poplawski N, Wilcken B, et al: Intron-8 polythymidine sequence in Australasian individuals with CF mutations R117H and R117C, *Eur. Respir. J.* 17(6):1195–1200, 2001. Available from: https://doi.org/10.1183/09031936.01.00057001. PMID: 11491164.

Maxam AM, Gilbert W: A new method for sequencing DNA. 1977, *Biotechnology* 24:99–103, 1992. PMID: 1422074.

Merker JD, Wenger AM, Sneddon T, et al: Long-read genome sequencing identifies causal structural variation in a Mendelian disease, *Genet. Med* 20(1):159–163, 2018. Available from: https://doi.org/10.1038/gim.2017.86. PMID: 28640241.

Miga KH, Koren S, Rhie A, et al: Telomere-to-telomere assembly of a complete human X chromosome, *Nature.* 585(7823):79–84, 2020. Available from: https://doi.org/10.1038/s41586-020-2547-7.

Miller DT, Lee K, Gordon AS, et al: Recommendations for reporting of secondary findings in clinical exome and genome sequencing, 2021 update: a policy statement of the American College of Medical Genetics and Genomics (ACMG), *Genet. Med.* Available from: https://doi.org/10.1038/s41436-021-01171-4. PMID: 34012069.

Posey JE: Genome sequencing and implications for rare disorders, *Orphanet J. Rare Dis* 14(1):153, 2019. Available from: https://doi.org/10.1186/s13023-019-1127-0. PMID: 31234920.

Rauluseviciute I, Drabløs F, Rye MB: DNA methylation data by sequencing: experimental approaches and recommendations for tools and pipelines for data analysis, *Clin. Epigenetics* 11(1):193, 2019. Available from: https://doi.org/10.1186/s13148-019-0795-x. PMID: 31831061.

Sanger F, Nicklen S, Coulson AR: DNA sequencing with chain-terminating inhibitors, *Proc. Natl. Acad. Sci. U S A* 74(12):5463–5467, 1977. Available from: https://doi.org/10.1073/pnas.74.12.5463. PMID: 271968.

Scotchman E, Shaw J, Paternoster B, et al: Non-invasive prenatal diagnosis and screening for monogenic disorders, *Eur. J. Obstet. Gynecol. Reprod. Biol* 253:320–327, 2020. Available from: https://doi.org/10.1016/j.ejogrb.2020.08.001. PMID: 32907778.

Shashikant T, Ettensohn CA: Genome-wide analysis of chromatin accessibility using ATAC-seq, *Methods Cell Biol* 151:219–235, 2019. Available from: https://doi.org/10.1016/bs.mcb.2018.11.002. PMID: 30948010.

Strande NT, Brnich SE, Roman TS, Berg JS: Navigating the nuances of clinical sequence variant interpretation in Mendelian disease, *Genet. Med* 20(9):918–926, 2018. Available from: https://doi.org/10.1038/s41436-018-0100-y. PMID: 29988079.

Thibodeau ML, O'Neill K, Dixon K, et al: Improved structural variant interpretation for hereditary cancer susceptibility using long-read sequencing, *Genet. Med* 22(11):1892–1897, 2020. Available from: https://doi.org/10.1038/s41436-020-0880-8. PMID: 32624572.

van Dijk EL, Jaszczyszyn Y, Naquin D, Thermes C: The third revolution in sequencing technology, *Trends Genet* 34(9):666–681, 2018. Available from: https://doi.org/10.1016/j.tig.2018.05.008.

Vermeulen C, Geeven G, de Wit E, et al: Sensitive monogenic noninvasive prenatal diagnosis by targeted haplotyping, *Am. J. Hum. Genet* 101(3):326–339, 2017. Available from: https://doi.org/10.1016/j.ajhg.2017.07.012.

Xie Z, Sun C, Zhang S, et al: Long-read whole-genome sequencing for the genetic diagnosis of dystrophinopathies, *Ann. Clin. Transl. Neurol* 7(10):2041−2046, 2020. Available from: https://doi.org/10.1002/acn3.51201. PMID: 32951359.

Zhang F, Lupski JR: Noncoding genetic variants in human disease, *Hum. Mol. Genet* 24(R1):R102−R110, 2015. Available from: https://doi.org/10.1093/hmg/ddv259.

Zhang S, Liang F, Lei C, et al: Long-read sequencing and haplotype linkage analysis enabled preimplantation genetic testing for patients carrying pathogenic inversions, *J. Med. Genet* 56(11):741−749, 2019. Available from: https://doi.org/10.1136/jmedgenet-2018-105976.

PART II

Clinical Applications of Genomic Medicine

CHAPTER 6

Expansion of use of genome analyses and sequencing in diagnosis of genetic diseases

6.1 Measurement toolkit for assessing the clinical utility of whole genome sequencing

Hayeems et al. (2020) presented a measurement toolkit for assessing the clinical utility of whole genome sequencing. Parameters included in this toolkit were based on findings of the Medical Genome initiative working group. They included diagnostic efficacy, therapeutic efficacy, patient outcome efficacy, and societal efficacy. This study focused primarily on diagnosis of rare germline diseases. But the working group noted that genome sequencing had broader applications. Provision of evidence of utility of whole genome sequencing was important.

Parameters for measurement of clinical utility formulated by the Centers for Disease Control were noted to include the following components: analytical validity, clinical validity, clinical utility, and ethical implications.

The American College of Medical Genetics and Genomics was reported to define clinical utility as the effect of a genetic test on the following: diagnostic and therapeutic management and prognosis; health and psychological impact on patients and families; economic impacts on healthcare systems.

Hayeems et al. summarized key factors in the clinical utility chain of evidence as technical efficacy; diagnostic accuracy efficacy; diagnostic thinking efficacy; therapeutic efficacy; patient outcome efficacy; societal efficacy.

Diagnostic thinking efficacy was noted to refer to how testing would impact clinicians' thinking regarding differential diagnosis and decision making. Particularly important was how discovery of a particular genetic variant was relevant to diagnosis. Important in this context, was also deep knowledge of the patient's phenotype.

Genetic variants were noted to potentially lead to specific functional testing. Variant discovery could lead to additional in-depth clinical evaluation including subspecialty referrals.

Therapeutic efficacy was defined as ways in which relevant genomic variant could lead to application of specific forms of management and to targeted therapies.

In some cases family testing and reproductive counseling may be indicated.

Hayeems et al. noted that Health Outcomes Measurement and Quality of Life Measurements are important to take into account and require continued data collection and assessment.

6.1.1 Next generation sequencing in clinical neurology

Rexach et al. (2019) reviewed the clinical application of next generation sequencing (NGS) in neurology. They noted that genetic factors have long been known to be important in neurological diseases. They also noted that more than 20,000 different genes have been reported to be expressed in brain. Furthermore, they noted evidence that of the approximately 5000 genetic disorders reported in Online Mendelian disease in Man (OMIM) 40% were reported to involve brain or nervous system.

Rexach et al. undertook to review successes and challenges encountered as NGS technologies were applied to diagnosis of neurological diseases. They documented specific diseases and percentage of cases where diagnosis was achieved using exome sequencing.

Disorder	Patients evaluated	Molecular diagnosis achieved	%
Dystonia	16	6	37.5
Leukodystrophy	71	25	35.2
Peripheral neuropathy	42	13	30.8
Ataxia	294	84	28.5
Developmental delay	407	127	31.2
Birth defect, MCA	898	328	36.5
Autism	699	149	21.3
Epilepsy	1224	387	31.6
Intellectual disability	1966	601	30.5
Neurol. dev. disorder	1951	508	38.4

MCA; multiple congenital abnormalities.

Rexach et al. stressed the need for computationally sophisticated bioinformatic analyses of genomic sequence data. The human genome was reported to have approximately 30,000 variants on exome sequencing and 3−4 million variants on whole genome sequencing. Key then is comparison of variants in patients with those found in the general population. Specifically for a specific nucleotide variant data derived from patients are compared with the minor allele frequency at that nucleotide in the general population.

One present barrier results from the low number of different world population for whom DNA sequence data are available. Minor allele frequency data can be assessed in databases, for example, ExAC (now transferred to Gnom AD database). Disorder-specific databases have also been developed. Examples include muscular dystrophy databases.

Rexach et al. emphasized that on exome sequencing pathological sequence variants may go undetected if they occurred in specific genomic regions, for example, nonprotein coding regions, regions with nucleotide repeat expansion. Also on exome sequencing, structural chromosome changes or copy number variants (CNVs), that are potentially disease causing, could go undetected. Another occurrence that impacts molecular diagnosis relates to somatic mosaicism and the particular tissue utilized for DNA extraction for exome sequencing.

Regarding use of molecular testing through use of gene panels developed for specific diseases, Rexach et al. noted that there is a lack of consensus as to which genes should be included in a panel designed to test for a specific genetic disease.

They noted that the value of whole genome sequencing in neurological disease diagnosis was being investigated in a specific study.

Rexach presented a case that demonstrated the benefits of NGS. A patient who presented with headaches and apparent ischemic attacks; similar manifestation occurred in her mother, sister, and brother.

The patient was found to have multiple lacunae infarcts on brain MRI. Clinical diagnosis suggested a diagnosis of CADASIL (Cerebral Autosomal Dominant Arteriopathy with Subcortical Infarcts and Leukoencephalopathy). However, on genetic testing no mutation was found in the gene associated with CADASIL (NOTCH 3). Exome sequencing revealed a pathogenic missense mutation in the gene that encodes HTRA1, a member of the trypsin family of serine proteases. It is a secreted enzyme that is proposed to regulate the availability of insulin-like growth factors (IGFs) by cleaving IGF-binding proteins. Mutations in this gene have been found in an autosomal dominant condition associated ischemic attacks and subcortical infarcts on brain MRI and designated CADASIL 2. The finding on sequencing enables genetic counseling in the family.

Rexach et al. stressed the importance of a multidisciplinary team approach to diagnosis in neurological conditions including neurologists, pathologists, laboratory scientists, bioinformaticians, geneticists, and genetic counselors. They stressed the importance of validation of potentially disease-causing variants found on sequence analysis.

In general, molecular diagnosis was reported to be achieved in less than 40% of cases studied. Rexach et al. stressed that on exome sequencing pathological defects may go undetected particularly if they occur in specific genomic regions including nonprotein coding region, and regions with nucleotide repeat expansions. Also complex structural chromosome changes or CNVs that are potentially disease causing may go undetected. Another problem relates to chromosome mosaicism that restricts disease detection.

Other important issues requiring consideration include ethical issues and concerns regarding secondary findings not relevant to the specific indication for sequencing.

6.2 One phenotype many genes

Gannamani et al. (2021) discussed challenges in clinical and genetic correlations in movement disorders and considering the challenge of "one phenotype many genes." They noted for example that dystonia, cerebellar ataxia, and myoclonus syndromes have each been reported to be due to mutation in approximately 100 different genes.

Gannamani et al. discussed deep phenotyping in neurological diseases, noting that this includes history, physical examination, electrophysiology, neuro-imaging, metabolic analysis of blood, and cerebrospinal fluid. More recently data from wearable devices have been included in patient assessment. They noted that movement disorders can be inherited or acquired and therefore genetic studies are often indicated.

Gannamani et al. noted importance of close correlation between clinicians and genetic laboratories. In addition, ongoing curation of genetic test results and correlation with phenotypic features were important as well as consideration of reanalysis of sequencing findings as more information becomes available through expanded studies.

In the research setting they proposed attention be focused also on correlation of phenotypes and on gene networks. In this regard they presented data on pathway and network analyses relative to dystonia phenotype. Pathways to be considered also included metabolism and mitochondrial function, metabolism of vitamins and cofactors, signal transduction, and DNA damage repair.

Magrinelli et al. (2021) reviewed clinicogenetic correlations in movement disorders and neurodegenerative diseases. They also considered underlying mechanism of phenotypic heterogeneity. With respect to phenotypic heterogeneity they considered incomplete penetrance, variable expressivity, and pleiotropy.

Phenotypic heterogeneity could arise from different mutation types in a particular gene, somatic mosaicism, modifiers, interacting genes, epigenetic factors, and environmental factors. In the case of mitochondrial disorders mitochondrial heteroplasmy impacts phenotype.

Magrinelli et al. defined phenotype as "the observable and quantifiable characteristics in an individual." In considering phenotypic heterogeneity in monogenic disorders incomplete penetrance was important to consider. Incomplete penetrance can be defined as the situation where a specific individual with a defined mutation in a specific gene does not manifest the specific manifestations known to be associated with that specific mutation in that specific gene.

Pleiotropy referred to the different manifestations that could arise due to a specific gene mutation. Variable expressivity was noted to be related to penetrance and pleiotropy.

6.2.1 Incomplete penetrance

Incomplete penetrance was noted to be most frequently encountered in autosomal dominant disorders. Specific examples of condition that manifested incomplete penetrance referred to by Magrinelli included the autosomal dominant dystonia

DYT1 (OMIM 128100) due to defects in the TOR1A encoding gene and SGCE myoclonus syndrome also known as DYT11 (OMIM 159900).

TOR1A is a member of the AAA family of adenosine triphosphatases (ATPases). SGCE is sarcoglycan epsilon, sarcoglycans are transmembrane proteins that are components of the dystrophin-glycoprotein complex.

Incomplete penetrance was also noted to be a factor in many disorders due to nucleotide repeat expansions.

Incomplete penetrance was also noted to be a feature of many of the nucleotide repeat expansion disorders. X-linked disorders (e.g., Fragile X mental retardations), some autosomal recessive disorders (e.g., Wilson disease), and mitochondrial disorders were also known to manifest incomplete penetrance.

Magrinelli also noted age-related penetrance changed in Huntington disease and in the Fragile X tremor associated syndrome.

Variable expressivity was noted to refer to quantifiable differences in phenotypic effects of a specific mutation. In repeat expansion disorders this could be due to changes in repeat length with ongoing cell division. The molecular underpinnings of variable expressivity were noted to be poorly defined. A specific condition noted to manifest variable expressivity was the disorder due to mutations in ATP1A3. ATPase Na+/K+ transporting subunit alpha 3 belongs to the family of P-type cation transport ATPases, and to the subfamily of Na+/K+-ATPases. Na+/K+-ATPases are defined as an integral membrane protein responsible for establishing and maintaining the electrochemical gradients of Na and K ions across the plasma membrane. These gradients are essential for osmoregulation, for sodium-coupled transport of a variety of organic and inorganic molecules. Pathogenic variants in this gene can lead to ataxia, alternating hemiplegia, dystonia, optic atrophy, or sensorineural hearing loss.

6.3 Genome sequencing in pediatric developmental defects

Neu et al. (2019) reviewed the clinical utility of genome sequencing in evaluation of neonatal and pediatric developmental disorders. They tabulated diagnostic yield of specific genomic evaluation methodologies in prenatal and pediatric populations.

Method	Diagnostic yield (%)
Chromosome microarray	10
Single gene or gene panel analysis	20
Whole exome	25
Whole genome	>25

In studies that utilized genomic sequencing and patient phenotype information together, diagnoses were reported to be between 25% and 40%.

Neu et al. stressed that several rare disorders including Beckwith – Wiedemann syndrome (BWS), Silver – Russell disorder, and Angelman syndrome (AS) result from defects in imprinted chromosomal region. Pathogenic mechanisms in these disorders include uniparental disomy, CNVs, single nucleotide variants (SNVs), insertion/deletion, or epigenetic mutations. They emphasized that suspicion of defects due to defects in imprinted genomic region require methylation studies.

6.4 Optical DNA mapping in human genome studies

Müller et al. (2019) reported information on use of optical mapping to enable examination of long-range sequence along a single DNA molecule. They reported on the value of optical DNA mapping (ODM) in the detection of structural genomic variations. It was also reported to be useful in mapping DNA damage and epigenetic modifications.

ODM was reported to be based on use of specific DNA cleavage enzyme and particular DNA stain Yoyo1. Specific steps in ODM were described as follows:

Extract large fragments of DNA, initiate optical enzymatic pre-labeling, add Yoyo1 and Netrospin, stretch DNA in nanofluidic channels, carry our fluorescence detection.

It was noted that specific pre-labeling of DNA could be used to detect epigenetic changes or DNA damage.

The specific enzyme used was reported to nick DNA at 7 base pair intervals. These nicks were then repaired using polymerase and ligase and a fluorescent labeled nucleotide. Muller et al. reported that this generated labeled dots along the genome. The pattern of the dots on test DNA could be compared with patterns in control genomic DNA to determine if structural variants were present in the test DNA.

The two small molecules used in ODM included Yoyo1 and Netrospin. Netrospin was reported to block AT regions from binding to the fluorescent dye Yoyo1.

The ODM system developed by Bionano was reported. It was also reported to be able to detect chromosomal insertions and deletions larger than 500 base pairs, balanced or unbalanced translocation larger than 50 kb, inversion duplications larger than 30 kb, and CNVs larger than 500 kb.

6.5 Transcriptome sequencing

Lee et al. (2020) reviewed the utility of transcriptome sequencing for diagnosis of rare Mendelian disorders. Data were generated from 113 cases selected for in-depth analysis by the Undiagnosed Disease Network Program in Los Angeles. They reported that the diagnostic yield of genome sequencing was 31%. RNA seq diagnosis was reported to achieve diagnosis in additional seven cases. For RNA sequencing whole blood, fibroblasts, muscle, or bone marrow tissue was used to derive DNA.

Sequence depth in their study ranged between 50X and 150X. In data analyses of SNVs, small insertion deletions were identified and compared with occurrence and frequency in public databases. Variants with a frequency greater than 1% in public data were not considered relevant to diagnosis.

Rare variants in sequence data from patients were classified using a specific resource, Variant Annotate Extra, that provided information on the likely effect of the variant on protein function, population allele frequency, prior evidence for disease causality. This resource utilized information from the Ensemble Variant Effect Predictor. Information on variant classification was also gathered from the ACMG and Association of Molecular pathology websites.

ACMG guidelines were built on a report published by Richards et al. (2015). They noted that variants with strong evidence of pathogenicity included null variants, nonsense, frameshift, splice site and initiation codon variants, and deletion of single or multiple exons. Also important was information on whether loss of function in a specific gene was known to be associated with a disease mechanism.

Lee et al. noted that variants with a frequency above 1% in population data were not defined as polymorphic.

Non-polymorphic variants were classified into 1 of 5 categories: pathogenic, likely pathogenetic, uncertain significance, likely benign, benign.

A specific database that includes population sequence data is GnomAD. In silico predictive algorithms are designed to estimate likely effects of a given sequence variant on amino acid and predictive algorithms may also take evolutionary stability. Information on impact of variants on splicing can also be gathered from databases.

Lee et al. (2020) reported RNA sequencing on 48 families. Through integration of DNA and RNA sequencing, data diagnoses were made on seven additional cases. In these cases splice region variants or deep intronic variants were identified that caused exon skipping, retention of intron, or generation of an intronic pseudogene.

They reported that specific gene changes and disorders identified on the basis of DNA and RNA studies included the following:

Gene product	Clinical diagnosis
SEPSECS Sec (selenocysteine) tRNA synthase	Pontocerebellar hypoplasia
LMNA lamin A/C component of proteins next to nuclear membrane	Progressive muscular disorder
SLC25A46 solute carrier family 25 member 46 (mitochondrial)	Seizures developmental delay

(Continued)

Continued

Gene product	Clinical diagnosis
Myopathy COL6A1 collagen type VI alpha 1 chain, extra cellular matrix protein	
DMD (dystrophin) bridges inner cytoskeleton and extracellular matrix	Muscular dystrophy
SRS2 seryl-tRNA synthetase 2, mitochondrial (Silver – Russell syndrome)	Postnatal growth delay
MPV17 mitochondrial seryl-tRNA synthase precursor	Mitochondrial depletion

Cummings et al. (2017) reported improving the diagnosis of Mendelian disorders through use of transcriptome sequencing. They specifically reported the value of muscle RNA sequencing in patients with rare presentations of muscle disease.

They emphasized the importance of comparing patient mRNA sequence data with a muscle RNA reference panel. They specifically undertook muscle RNA analyses on patients where DNA sequencing studies had failed to return a genetic diagnosis. In defining the pathogenic changes discovered on RNA sequencing, they noted that these included exon skipping, exon extension, exon splice site gain (novel splice site generated by variant), and introduction of a premature stop codon.

Gene product impacted	Clinical diagnosis
NEB Nebulin, giant protein in cytoskeletal matrix and sarcomere filaments	Nemaline myopathy
TTN Titin a large abundant protein of striated muscle	Fetal akinesia
DMD Dystrophin bridges inner cytoskeleton and extracellular matrix	Duchenne dystrophy (MD)
COL6A1 Collagen VI is a major structural component of microfibrils	Bethlem/Ulrich MD
COL6A2 binds extracellular matrix proteins	Bethlem/Ulrich MD
COL6A3 binds extracellular matrix proteins	Bethlem/Ulrich MD

6.6 Imprinting

In a review of imprinting disorders in humans Iglesias-Platas et al. (2014) noted that imprinting refers to a number of genes associated with monoallelic expression. They particularly emphasized importance of the roles that imprinted genes in humans play in placental development and in fetal growth. Multiple layers of epigenetic regulation were noted to impact expression of imprinted genes. These included DNA methylation and histone tail modifications.

Iglesias-Platas et al. reviewed studies on the paternally expressed gene PLAG1 (also known as ZAC1) that encodes a zinc finger transcription factor. It is located on chromosome 6q24.2. A separate gene HYMA1 was reported to be derived from a different promoter within the same gene. They carried out studies on PLAG1 and on occurrence of different splice forms in human placenta. Studies revealed that the HYMA expression levels were higher in placentas and that this effect was more pronounced in infants with intra-uterine growth retardation. Expression levels of PLAG1 did not correlate with intra-uterine growth retardation.

A specific promoter in PLAG1 was noted to be imprinted and to preferentially lead to paternal expression.

Arima et al. (2001) reported that some forms of transient neonatal diabetes were found to be associated with uniparental disomy of 6q24.

Mackay and Temple (2010) reported that transient neonatal diabetes often resolves after a few months of life but may recur later in life. They reported that expression of PLAG1 derived from the paternal allele and that overexpression can derive from uniparental duplication in the 6q24 region.

Monk et al. (2019) reviewed the epigenome and imprinting disorders. They noted that in humans approximately 100 imprinted genes had been identified. They emphasized that many of these genes had important functions during development.

The molecular changes in imprinting that lead to disorders include sequence variants, copy number changes, uniparental disomy, epigenetic changes, and epimutations. Primary epimutations were noted to occur without DNA changes in the imprinted region and were noted to include changes in the imprinted region. Secondary epimutations were referred to changes that occurred in transacting regions.

Monk et al. noted that advances in genome sequence analyses and in single cell analyses have increased insights into epigenetic and imprinting errors.

A key question is whether imprinting of a gene primarily affects a specific tissue or if it applies widely to different cells and tissues and cell types. There is evidence that UBE3A imprinting varies in different tissues. There is also some evidence that it may differ in different developmental stages (Fig. 6.1).

6.7 Imprinting disorders

In a review of imprinting disorders in humans, Butler (2020) stressed that an imprinted gene is expressed from only one member of a chromosome pair, either the paternal or the maternal-derived chromosome. Specific disorders arise in humans due to deviation from normal imprinting. The best known imprinting disorders in human are described below.

Chromosome and region	Disorder/s
15q11-q12	Prader – Willi syndrome, Angelman syndrome
11p15.5	Beckwith – Wiedemann syndrome
20q13.3	Albright hereditary osteodystrophy, pseudo-hypoparathyroidism

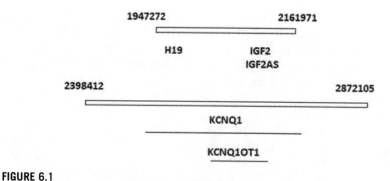

FIGURE 6.1

Imprinted genes on chromosome 11p15.5.

6.8 Prader–Willi syndrome and Angelman syndrome

Prader–Willi syndrome (PWS) was noted to result from hypermethylation of the 15q11.2-q13 critical region or from loss of paternally expressed genes in this region including SNRPN.

AS was reported to be due to loss of expression of UBE3A from the maternal chromosome.

Both PWS and AS could result from imprinting defects, from point mutations, microdeletions gene disrupting translation, or uniparental disomy. Aref-Eshghi et al. (2017) reported that AS resulted from mutations in the UBE3A gene in 10% of cases. They noted that methylation differences in the imprint control region on 15q12-q13 was primarily impacted in both PWS and AS and that the SNURF SNPRN genes were key.

Aref-Eshghi et al. emphasized the different mechanisms that lead to imprinted gene defects and that they include chromosome changes, deletions and duplication, coding region differences, epimutations or uniparental disomy implying that both homologs of a specific chromosome or of a specific chromosome region are inherited from one parent.

Single locus methylation analyses have been carried out to detect abnormalities typical of PWS and AS. Aref-Eshghi noted that analyses also include determination of methylation levels on chromosome 15q and comparison of methylation levels with those in a reference panel.

Aref-Eshghi et al. (2017) reported clinical validation of genome-wide methylation for molecular diagnosis of imprinting disorders. They carried out genome-wide DNA analyses and analyses of epigenetic profiles in cases of AS, Silver Russell syndrome, Beckwith Wiedemann syndrome and compared results with methylation analyses in 361 unaffected individuals.

They noted that imprinting disorders could arise in either of two regions in 11p15.5. These include ICR1 with H19 and IGF2 2 genes that undergo methylation on the paternal alleles and the ICR2 region with KCNQOT1 and CDKN1 that are methylated on the maternal allele.

If IGF2 is expressed from both parental alleles (e.g., methylation fails on the paternal allele) BWS results.

Biallelic expression of H19 only leads to Russell − Silver syndrome.

6.9 Silver − Russell syndrome

The Silver − Russell (Russell − Silver) syndrome has been found to be due to defects on any one of five different chromosomes. SRS1 11p15.5, SRS2 7p13-q32, SRS3 11p15.5, SRS4; 8q12.1, SRS5 12q14.3. Russell Silver Partington syndrome was assigned to the X chromosome.

In SR1 syndrome 20%−60% of cases were reported to involve epigenetic changes in the genes and genomic segments in H19/IGF2 imprint control region maternally expressed. H19 is a nonprotein coding transcript. In SR3 the chromosome 11p15, the paternally derived IGF2 insulin-like growth factor 2 is impacted. Normal functions of IGF2 include involvement in cell growth, proliferation, and migration.

Particularly important in generation of the 11p15 disorder BWS, is the KCNQ1OT1 transcript a long noncoding transcript that undergoes differential methylation. KCNQ1OT1 gives rise to a transcript LIT1 that is preferentially expressed from the paternal allele. Loss of this transcript was reported in BWS cases by Mitsuya et al. (1999). Expression of KCNQ1OT1 was reported to be silent in BWS by Lee et al. (1999).

The KCNQ1 and p57(KIP2) (CDKN1C) domain was reported to be impacted in BWS.

KCNQ1 potassium voltage-gated channel subfamily Q member 1, gene exhibits tissue-specific imprinting, with preferential expression from the maternal allele in some tissues, and biallelic expression in others. CDKN1C gene encodes p57(KIP2), a potent tight-binding inhibitor of several G1 cyclin/Cdk complexes and a negative regulator of cell proliferation and is reported to be paternally imprinted Fig. 6.2).

Clinical manifestations in SRS1 and SRS3 include severe intra-uterine growth retardation and postnatal growth impairment with some degree of facial dysmorphism.

Õunap (2016) noted that Silver − Russell syndrome and BWS are opposite phenotypes both due to abnormalities on chromosome 11p15 with Silver − Russell syndrome being associated with growth retardation and BWS being associated with overgrowth.

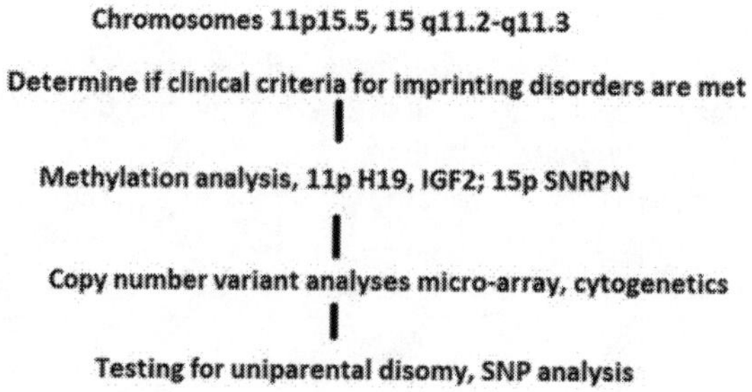

Chromosomes 11p15.5, 15 q11.2-q11.3

Determine if clinical criteria for imprinting disorders are met
|
Methylation analysis, 11p H19, IGF2; 15p SNRPN
|
Copy number variant analyses micro-array, cytogenetics
|
Testing for uniparental disomy, SNP analysis

FIGURE 6.2

Flow chart for diagnosis imprinting disorders.

Molecular diagnosis of the 11p15 imprinting disorders were reviewed by Russo et al. (2016). They reported utility of determination of methylation levels by methylation-specific multiplex ligation-dependent probe amplification.

Brioude et al. (2018) reported an international consensus statement on BWS. This syndrome was noted to be a genomic imprinting disorder that manifests phenotypic heterogeneity. Manifestations include overgrowth, macroglossia, hypoglycemia, predisposition to embryonic tumors, and abdominal wall defects; hemihyperplasia was sometimes observed. The embryonal tumors that occur in BWS include Wilms tumor and hepatoblastoma.

Three categories of BWS syndrome were observed, isolated lateralized growth syndrome can include BWS or other disorders; atypical BWS and typical BWS that could include patients with molecular diagnosis and patients with clinical diagnosis only. Hypoglycemia in BWS was reported to be associated with elevated insulin levels.

Brioude et al. developed a flow chart for molecular diagnosis of BWS. They noted that the molecular diagnosis of BWS is challenging; however, it has value for family counseling and for guidance regarding tumor surveillance.

In the differential diagnosis of BWS are other conditions associated with overgrowth. These were noted to include defects in PIK3CA phosphatidylinositol-4,5-bisphosphate 3-kinase catalytic subunit alpha and AKT1, AKT serine/threonine kinase 1.

Simpson − Golabi − Behmel syndrome 1 is an overgrowth syndrome due to defects in the *GPC3* gene on chromosome X26.3 that encodes glypican 3, a heparan sulfate proteoglycan. Simpson − Golabi − Behmel syndrome 2 is due to a defect in CXORF5 gene (OFD1) (open reading frame) on chromosome Xp22.

Brioude et al. defined the 11p15.5−11p15.4 regions as having two domains, one centromeric and one telomeric, with each domain having its own imprint control region and differentially methylated region. The telomeric domain was reported to include the insulin-like growth factor gene IGF2 and H19 that gives rise to a long noncoding RNA (lncRNA) IGF2 and H19 manifested differential methylation and constitute imprint control region 1 ICR1.

Cell cycle inhibitor CDKN1C and lncRNA KCNQ1 were reported to be located in the centromeric domain and constitute the imprint control region ICR2.

BWS was defined as due to loss of methylation in the maternal ICR2 in 50% of cases; gain of methylation in maternal ICR1 was reported to occur in 5%—10% of cases. Uniparental paternal isodisomy in 11p15.5 occurred in 20% of cases. Intragenic mutations in CDKN1C occurred in 5%—40% of cases.

The frequency of twinning was reported to be much higher in BWS than in the general population. Female twins were most frequently described as often one twin was affected and the other was unaffected.

6.10 GNAS locus

Genomic analyses have revealed that GNAS locus on chromosome 20q13.32 has complex regulation of expression. Transcripts are derived from four alternate promoters and there are thus alternative 5'exons. Transcripts vary since some are maternally expressed and others are paternally expressed. There is also an antisense transcript derived from the GNAS locus (Bastepe and Jüppner, 2005).

Specific disorders involving GNAS include Albright hereditary osteodystrophy, pseudo-hypoparathyroidism, McCune – Albright syndrome, progressive fibrous osteodysplasia, polyostotic fibrous dysplasia, and pituitary tumor.

This disorder is noted to be due to decreased production of products of the GNAS gene that has complicated imprinting.

Specific disorders in which GNAS changes are observed include Albright hereditary osteodystrophy, McCune – Albright syndrome, progressive osseous heteroplasia, polycystic fibrous dysplasia, and pituitary tumor.

McCune – Albright syndrome can manifest endocrine abnormalities like hyperthyroidism, hyperparathyroidism, Cushing syndrome, precocious puberty, skeletal abnormalities, and unusual skin pigmentation (Spencer et al., 2019). They reported that treatment for endocrinopathies can mitigate some skeletal morbidities.

Another imprinted region that is associated with abnormalities is the DLK1 locus that encodes delta like noncanonical Notch ligand 1. It is expressed from placenta and adrenal gland DLK1 maps to chromosome 14 q32.2.

Barbosa et al. (2018) compared DNA methylation profiles of 489 individuals with neurodevelopmental disorders and congenital anomalies against a reference panel of 1534 controls. These studies led them to conclude that epivariation occurs frequently in the human genomes and are enriched in certain patients. They also carried out RNA analyses and reported that epivariants impacted gene expression. Studies were carried out using the Illumina Infinium bead chip 450 array.

In some cases there was a 2.8-fold increase in epivariants. 5% of individuals had two or more epivariants. In follow-up studies they carried out PCR bisulfite sequencing to confirm epivariants and concordant analyses were identified in 55 of 58 assays.

Epivariants occurring in the promoter regions of MEG3 and FMR1 were defined as epimutations based on follow-up studies including clinical phenotype. MEG3, maternally expressed gene 3, maps to chromosome 14q32.2 and its expression is reported to be lost in certain cancerous tumors; there is evidence that the gene gives rise to a lncRNA that acts as a tumor suppressor.

Aref-Eshghi et al. (2019) reviewed utility of genome-wide methylation analyses in undiagnosed cases with suspected genetic disorders. Motivation for their study was the fact that DNA sequence analyses and copy variants analyses left a significant number of cases with neurodevelopmental defects and congenital anomalies undiagnosed. They noted that some undiagnosed cases have single locus methylation defects while others had multilocus syndrome-specific methylation defects referred to as episignatures.

They reported findings of genome-wide methylation analyses in 67 individuals with uncertain clinical diagnoses and also developed a computation analysis model. Subsequently they applied this approach to 965 cases with neurodevelopmental defects.

Aref-Eshghi et al. drew attention to genetic conditions that manifest unique combinations of DNA methylation changes that constitute episignatures that could lead to diagnoses. They documented 14 syndromes that were found to display episignatures; these are listed below.

ADCADN autosomal dominant cerebellar ataxia with deafness and narcolepsy. Gene locus DNMT1 map location 19p13. One previous study identified three different heterozygous mutations in exon 21 of the DNMT1 gene.

Coffin – Siris syndrome (CSS): There are 11 different chromosome regions implicated in this syndrome in prior studies. It is defined in OMIM as "a congenital malformation syndrome characterized by developmental delay, intellectual disability, coarse facial features, feeding difficulties, and hypoplastic or absent fifth fingernails and fifth distal phalanges."

It is interesting to note that the genes involved in nine of the different forms of CSS encode products in involved chromatin modification, chromatin remodeling; two genes encode products involved in transcription.

Locus	Chromosome	Gene product
Coffin – Siris syndrome 1	6q25.3	ARID1B
Coffin – Siris syndrome 2	1p26.3	ARID1A
Coffin – Siris syndrome 3	22q11.23	SMARCB1
Coffin – Siris syndrome 4	19p13.2	SMARCA4
Coffin – Siris syndrome 5	17q21.2	SMARCE1
Coffin – Siris syndrome 6	12q12	ARID2
Coffin – Siris syndrome 7	11q13.1	DPF2 (BAF noncatalytic)
Coffin – Siris syndrome 8	12q13.2	SMARCC2

(Continued)

Continued

Locus	Chromosome	Gene product
Coffin – Siris syndrome 9	2p25.2	SOX11 (transcription factor)
Coffin – Siris syndrome 10	6p22.3	SOX4 (transcription factor)
Coffin – Siris syndrome 11	12q13.12	SMARCD1

Other syndromes found to harbor episignatures are listed below. The products produced by involved genes include products involved in histone modification, and also transcription and transcription activation.

Nicolaides – Baraitser syndrome	9p24.3	SMARCA2
Charge syndrome	7q21.11, 8q12.2	CHD2, SEMA3E
Claes – Jenson syndrome	Xp11.2	KDM5C
Genitopatellar syndrome	10 p22.2	KAT6B
Floating Harbor syndrome	16p11.2	SRCAP
Kabuki syndrome 1	12q13.2	KMT2D
Kabuki syndrome 2	Xp11	KDM6A
Sotos syndrome 1	5q35	NSD1
Sotos syndrome 2	19p13.3	NFIX
Sotos syndrome 3	19p13.3	APC2

6.11 Epivariations

Garg et al. (2020) reported results of a study on 26116 human genomes and provided information on disease-relevant epivariations and CGG nucleotide expansions. Studies were carried out on Illumina 450 methylation arrays and in addition bisulfite DNA sequencing was carried out to determine if methylation microarray and bisulfite sequencing data concurred.

They reported identification of 4452 autosomal epivariations. Importantly, potentially inactive mutations were found in 384 genes known to be associated with specific human diseases. They also used expression analyses to establish impact of epimutation on gene expression.

Garg et al. also reported identification of rare hypermethylation in unstable CGG repeats. In addition, epimutations were identified in known folate-sensitive fragile sites in the genome, thus providing insight into molecular mechanisms that lead to generation of fragile sites.

Importantly, they also identified promoter methylation changes in BRCA1 with a population frequency of 1 in 3000 and promoter hypermethylation in the low density lipoprotein receptor (LDLR) gene with a population frequency of 1 in 6000. Promoter hypermethylation of BRCA1 and gene silencing has been reported in association with breast and ovarian cancer by Evans et al. (2018).

Ghose et al. (2019) reported altered LDLR gene promoter methylation in coronary heart disease patients.

Garg et al. also reported that epivariations were frequently discordant in monozygotic twins. Concordance rates were reported to be 63% and epigenetic concordance decreased with advances in age. Imprinted gene loci were reported to be more prone to epivariation.

Another important aspect of the Garg study included their analyses of CGG repeat methylation and differences in methylation frequencies in these regions in the general population.

Garg et al. concluded that epigenetic variation analyses provide additional insights into gene expression variations.

Godler and Amor (2019) reviewed DNA methylation diagnostic testing in neurodevelopmental disorders. They noted that diagnoses in leading neurodevelopmental disorders benefitted from DNA methylation analyses. They included five disorders. Initial studies for analysis of PWS and AS involved methylation PCR. It is, however, important to note that AS may result from nucleotide variants and in these cases methylation abnormalities may not be detected.

Disorder	Chromosome region	Key gene
Prader – Willi syndrome	15q11-q13	SNRPN promoter
Fragile X syndrome	Xq27.3	FMR1
Kagami syndrome	14q32.2	MEG
Temple syndrome	14q32	Between DLK1 and MEG3
Rett syndrome	Xq28	MECP2

Kagami – Ogata syndrome is also known as uniparental disomy syndrome since MEG3 is expressed from the maternal chromosome.

6.12 Multilocus imprinting disorders

Begemann et al. (2018) drew attention to patients with multilocus imprinting disorders. These patients can manifest abnormalities in well-known imprinted regions. For example, 30% of patients with BWS and KCNQ1OT1 differentially imprinted region defects also manifest abnormal imprinting at the H19 differentially methylated site.

Begemann noted that an important group of genes are maternally expressed and encode the proteins, these include the NLPR gene family.

NLPR5 19q13.43 NLR family pyrin domain containing 5. Expression restricted to the oocyte.

TLE6 19p13.3 TLE family member 6, subcortical maternal complex member.

OOEP 6q13 oocyte expressed protein also expressed in placenta.

KHDC3L 6q13 KH domain containing 3 like, subcortical maternal complex member.

PADI6 1p36.13 peptidyl arginine deiminase 6, converts arginine residues into citrullines in the presence of calcium ions.

These gene products were reported to form a complex expressed in oocytes and to be important in preimplantation development. A related complex with KHDC3L, NLRP5, OOEP, and TLE6 was reported to be present in fetal embryos at the precleavage stage.

Begemann et al. also cited evidence that maternal mutation in NLRP5 were reported to lead to reproductive wastage (Docherty et al., 2015).

NLRP7 mutation was reported to be found in biparental hydatidiform moles with absence of fetal development. NLRP7 mutations and NLRP2 mutations occurred in cases of multilocus imprinting disorders. KHDC3L mutation has also been found in cases of multilocus printing disorders.

Begemann et al. carried out studies in 38 families referred because of multilocus imprinting disorders. Exome sequencing was carried out to search for nucleotide variants. Pathogenicity of variants was analyzed with variant effect parameters. Epigenetic and epigenomic analyses were carried out using methylation-specific PCR or by methylation-specific multiplex ligation probe-dependent amplification assay for differential methylation determination at the TSS DMR methylation regions in the following loci:

DIRAS3 DIRAS family GTPase 3 imprinted gene with monoallelic expression of the paternal allele 1p13.3.

PLAGL1 zinc finger protein downstream promoter imprinted, expression from the paternal allele 6q24.

IGF2R Int 2 insulin like growth factor 2 receptor, imprinting of the human gene may be polymorphic,6q24.

GRB10 growth factor receptor bound protein 10, gene is imprinted in i tissue-specific manner,7p12.

MEST mesoderm specific hydrolase, preferential expression from the paternal allele in fetal tissue 7q32.

H19 only expressed from the maternally inherited chromosome 11p15.5.

KCNQ1OT1 KCNQ1 antisense transcript 1 methylated on the maternal chromosome11p15.

IGF1R Int2 insulin like growth factor 1 receptor, Int2 15q26.

MEG3 maternally expressed 3 lncRNA s (lncRNAs transcribed 14q32.2).

SNURF SNRPN upstream open reading frame Transcripts are paternally imprinted. 15q11.2.

PEG3 paternally expressed 3, zinc finger protein. Paternally expressed, 19q13.43.

GNAS-AS1 antisense RNA 1, paternally imprinted antisense RNA transcript regulates GNAS 20q13.32.

GNASA/B GNAS complex locus maternally, paternally, and biallelically expressed transcripts that are derived from four alternative promoters and 5' exons. 20q13.32.

Begemann et al. reported that clinical manifestations included history of early miscarriages. In these cases coding variants were identified in the following genes:

NLRP7 NLRP family 7 pyrin domain containing (5 mutations).
PADI6 peptidyl arginine deiminase 6, 4 mutations.
OOEP oocyte expressed protein 1 mutation.
UHRF1 ubiquitin like with PHD and ring finger domains 1 1 mutation.
ZAR1L zygote arrest 1, expressed predominantly in early embryos and oocytes.
Clinical manifestation included a history of early miscarriages.

Clinical manifestations also occurred in children with NLRP2 mutation. These included omphalocele, macroglossia, neonatal hypoglycemia, radial defects, and growth abnormalities.

6.13 Chromosomes genomes and sequence

Deakin et al. (2019) stressed the importance of bridging the gap between chromosomes and genomes. They stressed the need for interaction between individuals involved in cytogenetics, genomics, and bioinformatics. Early advances included molecular characterization of chromosome breakpoints through sequence analyses. More recent advances include sequence determination for an entire human chromosomes. Other strategies include analysis of chromosome territories within the nucleus of a specific cell type.

Grubert et al. (2020) reported advances in analyses of the landscape of cohesin mediated loops in the human genome. They emphasized that these interactions facilitated by loops vary in different cell types. Analyses of chromatin looping and interactions are expedited by the CHIA-PET technique that involves paired-end sequencing.

6.14 Structural genomic variants

Roca et al. (2019) reported specific variant detection tools available for analyses of structural genomic changes that lead to copy number differences between 1 Kb and 5 Mb. Some methods of CNV detection involve depth of sequence reads in specific genomic segments. These are described as depth of coverage detection methods. Duplications will manifest greater number of reads and deletions will manifest fewer reads. It is, however, also important to bear in mind that the number of reads can also be influenced by other factors, for example, the CG content of the region.

CNV information was noted to also be gathered by assembling overlapping reads and determining how these compared in the test sample versus the control sample.

6.14.1 Population analyses

Abel et al. (2020) reported use of whole genome sequencing for analysis of structural genomic variation in human genes through studies on 17,795 human genomes. Significant data sources used for this study included the NHGRI Centers for Common Disease Genomics and the Simons Genomic diversity panel.

Abel et al. reported that per genome 4442 high confidence structural variants were identified, of these 35% were deletions, 27% were mobile element insertions, and 11% were tandem duplications.

They noted that the burden of deleterious rare structural variants remained unclear.

Rare structural variants that altered gene dosage were primarily deletions 54%, and duplications 42%; other rare gene altering or gene rearranging variants included primarily inversions.

Analyses of loss of function structural variants revealed that 23.4 of these included multiple genes.

Overall structural variants were reported to account for between 4% and 11% of high impact gene alteration in a population sample.

6.14.2 Dosage sensitivity and haploinsufficiency

Johnson et al. (2019) reviewed the relationships between haploinsufficiency dosage sensitivity and genetic dominance. Haploinsufficiency was noted to refer to diploid organisms where loss of function of one allele at a specific locus led to impairment. They noted that the exact number of haploinsufficient genes in humans was not known.

For haplosufficient genes loss of function of one allele does not lead to impairment. Haploinsufficiency and dominant disorders are therefore related.

In considering mechanisms involved in haploinsufficiency, Johnson et al. considered protein insufficiency and subunit imbalance. These were considered relevant to transcription factors and in the light of evidence that half of the normal quantity of a specific transcription factor was not sufficient to trigger promoter activation. Transcription factor haploinsufficiency has been shown to be involved in certain developmental disorders. Johnson et al. noted reports of haploinsufficiency of PAX6 leading to generation of eye abnormalities.

Johnson et al. also referred to examples of macromolecular complexes and impact of haploinsufficiency of one member of the complex. Examples included collagen disorders, for example, a form Ehlers − Danlos syndrome (vascular type) where insufficiency of one member of a complex COL3A1 had deleterious consequences.

An important resource for information on haploinsufficient gene in the human genome is the ClinGen database, https://clinicalgenome.org. Categories included in this database are:

D dosage sensitivity; G gene disease validity; V variant pathogenicity; A actionability P; S somatic variants (cancer); B baseline annotation evidence from literature.

Bartha et al. (2018) noted that several computer models are available to assess haploinsufficiency. They noted that more than 3000 human genes manifest haploinsufficiency.

6.14.3 ACMG recommendations regarding genomic copy number variants analysis and reporting

These ACMG recommendations were reviewed by Riggs et al. (2020). They noted introduction of a quantitative scoring framework and a five-tier classification system. Specific categories used for CNV classification included genomic content, dosage sensitivity prediction, functional effect overlap with repeats, overlap with reports in the literature, and inheritance pattern. A committee evaluated 114 CNVs including 58 deletions and 56 duplications. A semiquantitative scoring system was developed, and the components are listed below.

Specific information regarding genomic variant:

Copy number loss: 1A contains functionally important elements or 1B. No functionally important elements. Overlap with benign gene, no critical gene.

Haploinsufficiency expression of half as much product as normal is not sufficient for product to function normally. Partial overlap with coding region, with untranslated region with last exon; partial overlap with 5″ gene region; partial overlap with 5′ untranslated region.

Evaluation of number of genes impacted by CNVs.

There is a reported phenotype specific to genome region included in copy number change or there are no reports of specific phenotype related to copy number change.

There is a specific phenotype related to the specific CNV; or there are no statistically significant differences documented in cases or controls with this specific variant. There is overlap of the CNV with common population variation.

CNV segregates with a specific phenotype in the family or variant is inherited from an unaffected parent.

Gurbich and Ilinsky (2020) developed a tool to determine significance of CNVs based on ACMG criteria. For genomic deletions, the key questions were: Do any genes, gene promoters, or enhancers overlap the CNV?; number of protein coding genes in CNV; overlap with haploinsufficient genes.

6.15 Assessment of copy number changes in different conditions and at different life stages

Coe et al. (2014) carried out a study to identify genomic and DNA sequence variants implicated in developmental delay. They carried out CNV analyses taking into account information on haploinsufficiency of genes. They developed CNV morbidity map based on studies of 29,085 children with developmental delay and 19,584 healthy controls. These studies led to identification of 70 significant CNVs.

A subset of the gene involved in CNVs in affected children were sequenced in other affected children. This led to identification of 10 genes that manifested sequence variants leading to loss of function.

Some of the CNVs involved gene loci that had been identified in specific named disorders. It is useful to list regions where copy number changes have distinct phenotypic effects that can include developmental delay. Regions with CNV associated with named disorders will be listed separately.

1q21.1 del; 1p36; 2p15–16 del; 2q11.2del; 2q37del; 3q29del; 6q16; 8q23.1del; 9q34del; 15q24 A to C del; 15q24 B to C; 15q26; 16p13.11; 17p13.3; 22q11.2

CNV associated with named syndromes in which developmental delay may be included:

Cri du Chat 5p15.33–15.32 del.
Williams syndrome 7q11.23 del.
Sotos syndrome 5q35.3 deletion.
Smith – Magenis syndrome 17p11.2 deletion.
PWS/AS 15q11.2 deletions.
Rubinstein – Taybi 16p13.3 deletion.
NF1 deletion syndrome17q11.2.
DiGeorge syndrome 22q11.2 deletion.

Other CNV associated with developmental delay reported in Coe et al. (2014) included: 1p32;1q21.1; 2p15–13; 2q27; 10q23; 12q14; 13p12; 15q11.3; 15q25.2;16p12.1;16p11.2/17q21.1;17 q23.1; 17 q23.2

Specific genes identified with abnormalities in the Coe study of developmental disabilities included:

ACACA acetyl-CoA carboxylase alpha biotin containing enzyme, catalyzes first step in fatty acid synthesis.

ADNP activity-dependent neuroprotector homeobox, likely functions as a transcription factor.

ARID1B component of the SWI/SNF chromatin remodeling complex may be involved in cell-cycle activity.

CHD1L chromodomain helicase DNA binding protein 1 like, helicase involved in DNA damage repair.

CYFIP1 cytoplasmic FMR1 interacting protein 1, protein translation regulation.

DIP2B disco interacting protein 2 homolog B, likely participates in DNA methylation.

DNMT3A DNA methyltransferase 3 alpha, important for embryonic development, and imprinting.

DYRK1A dual specificity tyrosine phosphorylation regulated kinase 1A, t role in a signaling pathway.

FOXP1 forkhead box P1 transcription family member.

GRIN2B glutamate ionotropic receptor NMDA type subunit 2B has agonist binding site for glutamate.

KANSL1 KAT8 regulatory NSL complex subunit 1, involved in histone acetylation.

MAPT microtubule-associated protein tau, involved in neuronal maturation.

MBD5 methyl-CpG binding domain protein 5.

NRG3 neuregulin 3 encodes ligands for the transmembrane tyrosine kinase receptors ERBB3, ERBB4.

NRXN1 neurexin 1 binds neuroligins at synapse to form $Ca(2+)$-dependent neurexin/neuroligin complexes.

PTEN phosphatase and tensin homolog negatively regulates levels of phosphatidylinositol-triphosphate.

SCN1A sodium voltage-gated channel alpha subunit 1, Voltage-dependent sodium channel.

SCN2A sodium voltage-gated channel alpha subunit 2 generation and propagation of action potentials.

SETBP1 SET binding protein 1, involved in DNA replication.

SHANK2 synaptic protein that may function as molecular scaffolds in the postsynaptic density.

SIN3A SIN3 transcription regulator family member A.

SOX5 SRY-box transcription factor 5, regulation of embryonic development and cell fate.

TTC21(C)tetratricopeptide repeat domain 21, ciliary function.

ZMYND11 zinc finger MYND-type containing 11, functions as a transcriptional repressor.

6.16 Prenatal exome sequence analysis

Lord et al. (2019) reported results of prenatal exome analysis in cases where fetal structural anomalies were detected on ultrasound. They noted that testing for genomic aneuploidy in such cases is often done using chromosomal microarrays.

In the reported study, data were gathered from 34 fetal medicine units in England and Scotland. The cases selected for further study had fetal structural abnormalities including altered nuchal translucency. In cases where fetuses were found to have chromosomal aneuploidy and CNVs matched parents' samples were also analyzed. Following exome sequencing, variants identified were compared with the lists of genes and variants likely associated with developmental, in the developmental disabilities database generated by Wright et al. (2015).

In this database six factors were included:

1. Frequency and prevalence of sequence variant in the general population.
2. Function and most severe functional consequences of disruption of altered gene.
3. Consequence of altered sequence: transcript ablation, splice donor variant, splice acceptor variant, stop gained, stop lost, initiator codon variant, in frame deletion, missense variant, coding sequence variant, transcription alteration.
4. Variant type, heterozygous, homozygous.
5. Loss or gain of function due to small CNVs.
6. Inheritance information though studies of parental sequence.
7. Patient phenotype compared with published phenotype for defects in that particular gene.

In the Lord et al. study of (2019) of 1133 prenatal cases, the diagnostic rate in sequencing was 27% with 311 diagnostic or strongly contributory variants.

6.17 Deciphering Developmental Disorders Study

The Deciphering Developmental Disorders Study group (2017) reported results of a study of 2293 families in which there were individuals with developmental delay. Subsequently they carried out exome sequencing on 3287 individuals with similar developmental disorders.

These studies led to identification of 94 genes with de novo damaging variants. The diagnostic yield was 42%. Analyses revealed that significant variants change gene function and some variants changed protein function.

Variants were classified into three classes that included the following:

Class 1: protein truncating variants, splicing variants, splice donor variants, splice acceptor variants, frame shift, initiator codon variants, and exon terminus variant.

Class 2: missense variants, stop lost variants, inframe deletions, inframe insertion, coding sequence, and protein altering variant.

Class 3. Silent variants, synonymous variants.

For variant, validation primers were designed and PCR analyses were carried out to confirm variant.

Detailed phenotype data were obtained and also appropriate consent photography was carried out.

Analyses revealed an excess of de novo sequence changes in affected probands. Studies led to identification of 150 haploinsufficient genes.

Phenotypic summaries were prepared of cases with definitive mutations. Phenotypic features were coded according to Human Phenotype Ontology parameters, https://hpo.jax.org/app/.

The key conclusions of this study were that 42% of the cohort had de novo damaging mutations that disrupted gene function.

The average prevalence rate of developmental disabilities due to damaging mutations was reported to be between 1 in 213 and 1in 448 births.

6.18 Investigations of causes of recurrent miscarriage

Dong et al. (2019) reported on use of genomic sequencing in exploration of genomic abnormalities in cases of recurrent miscarriage.

Recurrent miscarriage was defined as loss of two or more clinical pregnancies. They noted that chromosome abnormalities were reported to be responsible for approximately 60% of cases. Furthermore, chromosome abnormalities were reported to be present in 2%−4% of couples with history of recurrent miscarriage. Studies revealed that balanced translocation and inversions in parental chromosomes were particularly important causes of recurrent miscarriage.

Dong et al. reported results of their studies on cases of recurrent miscarriage. In their study peripheral blood samples were obtained from parents. Chromosome analyses with G-banding and low-pass genome sequencing was carried out. Among 1077 individuals 126 had balanced chromosome abnormalities, 78 with balanced translocation, and 48 with inversions. In eight cases translocation was revealed only on low pass genome sequencing. The reciprocal exchanges identified were reported to be below the resolution level for cytogenetic studies.

In 70 cases in which balanced translations were reported on cytogenetic studies, increased precision of breakpoint determination was achieved with low-pass sequencing. Additional findings beyond those found in cytogenetic studies were discovery of a three-way translocation between chromosomes and a rearrangement that involved four different chromosome.

In the cases where inversions were found the inversion sizes ranged between 114.7 kb and 94.5 kb in size.

6.19 Sequencing in prenatal diagnosis: noninvasive prenatal testing

In 2015 a position statement on noninvasive prenatal testing (NIPT) was published jointly by the European Society of Human Genetic and the American Society of Human Genetics (Dondorp et al., 2015).

They noted that NIPT was a promising approach to testing for more prevalent fetal aneuploidies. However, they stressed that NIPT is a screening test and should not be considered to be a diagnostic test.

NIPT refers to a prenatal test based on quantitative and qualitative analyses of cell-free fetal DNA present in maternal blood. The presence of cell-free DNA was reported by Lo et al. (1997). Initial use of NIPT was applied to analysis of the Rhesus D blood group locus. Following that cell-free fetal DNA recovered

6.19 Sequencing in prenatal diagnosis: noninvasive prenatal testing

from maternal blood was used in massive parallel genome sequencing to quantitatively determine the quantity of chromosome 21-derived DNA and to estimate the chromosome 21 copy number in the fetus.

Subsequent reports indicated that NIPT had 99% sensitivity for trisomy 21 detection, a 96% sensitivity for detection of trisomy 18, and 91% sensitivity for detection of trisomy 13 (Gil et al., 2014).

Dondorp et al. (2015) noted that there was growing evidence that good results were achieved in general obstetric population so the NIPT constituted an alternative to other first trimester screening procedures.

It is, however, important to note that a positive NIPT can also be generated from placental mosaicism, vanishing twin occurrence or maternal tumor.

It is also important to note that in maternal blood only 10% of the cell-free DNA is fetal in origin and is primarily derived from the placenta. At 4 weeks of gestation the fraction of cell-free DNA derived from the fetus is insufficient for accurate diagnosis. Another limiting factor is maternal weight and failure to detect fetal DNA is more common in obese pregnant women.

Positive NIPT test for aneuploidy must be followed up by direct testing of fetal tissue, for example, through amniocentesis or chorionic villus sampling.

Wilkins-Haug et al. (2018) noted that discordant results had sometimes been reported between fetal cell-free DNA in maternal blood and fetal karyotype. They undertook a study to establish causes of discordant results.

Causes found included vanishing twin in pregnancy, prior history of maternal transplantation, cancer in the mother. Other important causes of discordance included placental mosaicism and true fetal mosaicism.

Health Quality Ontario reported results of an evaluation of clinical benefits, harms, budgetary impact, and patient preferences for testing of trisomies 21, 18, 13, sex chromosome aneuploidies and chromosomal microdeletions in an average at risk population.

They concluded that NIPT was safe for screening of trisomies 21, 18, and 13. They emphasized that NIPT results indicating any of these trisomies should be confirmed by diagnostic testing, for example, amniocentesis.

NIPT testing for sex chromosome anomalies and microdeletions was not supported.

van der Meij et al. (2019) reported that NIPT was offered in the Netherlands in 2017. The first tests were designed to check for trisomies 21, 18, and 13, The first part of the test evaluated uptake, technical performance, and logistics and the second part of the test evaluated perspectives of pregnant women. These included decision making and satisfaction. The NIPT was considered part of standard prenatal care and costs were covered by health insurance.

Maternal blood draw for testing was at or after 11 weeks gestation. Cell-free DNA was isolated from plasma and DNA sequencing was performed. This was described as genome-wide shallow sequencing on HiSeq400 or on Illumina Nest seq 500. Bioinformatic analyses were performed using WISECONDR with setting designed to identify aneuploid or unbalanced chromosome translocation. For initial studies, analyses were confined to chromosomes 21, 18, and 13, Follow-up of high-risk results involved amniocentesis or in some cases chorionic villus sampling. Positive testing during gestation was also followed up after birth.

The study included 73,239 pregnancies and high-risk cases were identified; 230 were cases of trisomy 21; 49 cases were trisomy 18; and 55 cases were trisomy 13.

Complex abnormalities on NIPT were also detected and these were found to be due to maternal malignancies and in one case due to vitamin B12 deficiency.

References

Abel HJ, Larson DE, Regier AA, et al: Mapping and characterization of structural variation in 17,795 human genomes, *Nature.* 583(7814):83–89, 2020. Available from: https://doi.org/10.1038/s41586-020-2371-0. PMID: 3246030.

ACMG Board of Directors Clinical utility of genetic and genomic services: a position statement of the American College of Medical Genetics and Genomics Genet Med. 2015 June;17(6):505–7. Available from: https://doi.org/10.1038/gim.2015.41. PMID: 25764213.

Aref-Eshghi E, Bend EG, Colaiacovo S, et al: Diagnostic utility of genome-wide DNA methylation testing in genetically unsolved individuals with suspected hereditary conditions, *Am. J. Hum. Genet* 104(4):685–700, 2019. Available from: https://doi.org/10.1016/j.ajhg.2019.03.008. PMID: 30929737.

Aref-Eshghi E, Schenkel LC, Lin H, et al: Clinical validation of a genome-wide DNA methylation assay for molecular diagnosis of imprinting disorders, *Mol. Diagn* 19(6):848–856, 2017. Available from: https://doi.org/10.1016/j.jmoldx.2017.07.002. PMID: 28807811.

Arima T, Drewell RA, Arney KL, et al: A conserved imprinting control region at the HYMAI/ZAC domain is implicated in transient neonatal diabetes mellitus, *Hum. Mol. Genet* 10(14):1475–1483, 2001. Available from: https://doi.org/10.1093/hmg/10.14.1475. PMID: 11448939.

Barbosa M, Joshi RS, Garg P, et al: Identification of rare de novo epigenetic variations in congenital disorders, *Nat. Commun* 9(1):2064, 2018. Available from: https://doi.org/10.1038/s41467-018-04540-x. PMID: 29802345.

Bartha I, di Iulio J, Venter JC, Telenti A: Human gene essentiality, *Nat. Rev. Genet* 19(1):51–62, 2018. Available from: https://doi.org/10.1038/nrg.2017.75. PMID: 29082913.

Bastepe M, Jüppner H: GNAS locus and pseudohypoparathyroidism, *Horm. Res* 63(2):65–74, 2005. Available from: https://doi.org/10.1159/000083895. PMID: 15711092.

Begemann M, Rezwan FI, Beygo J, et al: Maternal variants in NLRP and other maternal effect proteins are associated with multilocus imprinting disturbance in offspring, *J. Med. Genet* 55(7):497–504, 2018. Available from: https://doi.org/10.1136/jmedgenet-2017-105190. PMID: 29574422.

Brioude F, Kalish JM, Mussa A, et al: Expert consensus document: clinical and molecular diagnosis, screening and management of Beckwith–Wiedemann syndrome: an international consensus statement, *Nat. Rev. Endocrinol* 14(4):229−249, 2018. Available from: https://doi.org/10.1038/nrendo.2017.166. PMID: 29377879.

Butler MG: Imprinting disorders in humans: a review, *Curr. Opin. Pediatr* 32(6):719−729, 2020. Available from: https://doi.org/10.1097/MOP.0000000000000965. PMID: 33148967.

Coe BP, Witherspoon K, Rosenfeld JA, et al: Refining analyses of copy number variation identifies specific genes associated with developmental delay, *Nat. Genet* 46(10):1063−1071, 2014. Available from: https://doi.org/10.1038/ng.3092. PMID: 25217958.

Cummings BB, Marshall JL, Tukiainen T, et al: Improving genetic diagnosis in Mendelian disease with transcriptome sequencing, *Sci. Transl. Med* 9(386):eaal5209, 2017. Available from: https://doi.org/10.1126/scitranslmed.aal5209. PMID: 28424332.

Deakin JE, Potter S, O'Neill R, et al: Chromosomics: bridging the gap between genomes and chromosomes, *Genes. (Basel)* 10(8):627, 2019. Available from: https://doi.org/10.3390/genes10080627. PMID: 31434289.

Deciphering Developmental Disorders Study. Prevalence and architecture of de novo mutations in developmental disorders. *Nature*. 2017 February 23;542(7642):433−438. Available from: https://doi.org/10.1038/nature21062. PMID: 28135719

Docherty LE, Rezwan FI, Poole RL, et al: Mutations in NLRP5 are associated with reproductive wastage and multilocus imprinting disorders in humans, *Nat. Commun* 6:8086, 2015. Available from: https://doi.org/10.1038/ncomms9086. PMID: 26323243.

Dondorp W, de Wert G, Bombard Y, et al: Non-invasive prenatal testing for aneuploidy and beyond: challenges of responsible innovation in prenatal screening, *Eur. J. Hum. Genet* 23(11):1438−1450, 2015. Available from: https://doi.org/10.1038/ejhg.2015.57. PMID: 25782669.

Dong Z, Yan J, Xu F, Yuan J, et al: Genome sequencing explores complexity of chromosomal abnormalities in recurrent miscarriage, *Am. J. Hum. Genet* 105(6):1102−1111, 2019. Available from: https://doi.org/10.1016/j.ajhg.2019.10.003. PMID: 31679651.

Evans DGR, van Veen EM, Byers HJ, et al: A dominantly inherited 5' UTR variant causing methylation-associated silencing of BRCA1 as a cause of breast and ovarian cancer, *Am. J. Hum. Genet* 103(2):213−220, 2018. Available from: https://doi.org/10.1016/j.ajhg.2018.07.002. PMID: 30075112.

Gannamani R, van der Veen S, van Egmond M, et al: Challenges in clinicogenetic correlations: one phenotype—many genes, *Mov. Disord. Clin. Pract* 8(3):311−321, 2021. Available from: https://doi.org/10.1002/mdc3.13163. eCollection 2021 Apr. PMID: 33816658.

Garg P, Jadhav B, Rodriguez OL, et al: A survey of rare epigenetic variation in 23,116 human genomes identifies disease-relevant epivariations and CGG expansions, *Am. J. Hum. Genet* 107(4):654−669, 2020. Available from: https://doi.org/10.1016/j.ajhg.2020.08.019. PMID: 32937144.

Ghose S, Ghosh S, Tanwar VS: Investigating coronary artery disease methylome through targeted bisulfite sequencing, *Gene.* 721:144107, 2019. Available from: https://doi.org/10.1016/j.gene.2019.144107. PMID: 31499127.

Gil MM, Akolekar R, Quezada MS, et al: Analysis of cell-free DNA in maternal blood in screening for aneuploidies: *meta*-analysis, *Fetal Diagn. Ther* 35(3):156−173, 2014. Available from: https://doi.org/10.1159/000358326. PMID: 24513694.

Godler DE, Amor DJ: DNA methylation analysis for screening and diagnostic testing in neurodevelopmental disorders, *Essays Biochem* 63(6):785−795, 2019. Available from: https://doi.org/10.1042/EBC20190056. PMID: 31696914.

Grubert F, Srivas R, Spacek DV, et al: Landscape of cohesin-mediated chromatin loops in the human genome, *Nature* 583(7818):737–743, 2020. Available from: https://doi.org/10.1038/s41586-020-2151-x. Epub 2020 Jul 29.PMID: 32728247.

Gurbich TA, Ilinsky VV: ClassifyCNV: a tool for clinical annotation of copy-number variants, *Sci. Rep* 10(1):20375, 2020. Available from: https://doi.org/10.1038/s41598-020-76425-3. PMID: 33230148.

Hayeems RZ, Dimmock D, Bick D, et al: Clinical utility of genomic sequencing: a measurement toolkit, *NPJ Genom. Med.* 5(1):56, 2020. Available from: https://doi.org/10.1038/s41525-020-00164-7. PMID: 33319814.

Iglesias-Platas I, Martin-Trujillo A, Petazzi P, et al: Altered expression of the imprinted transcription factor PLAGL1 deregulates a network of genes in the human IUGR placenta, *Hum. Mol. Genet* 23(23):6275–6285, 2014. Available from: https://doi.org/10.1093/hmg/ddu347. PMID: 24993786.

Johnson AF, Nguyen HT, Veitia RA: Causes and effects of haploinsufficiency, *Biol. Rev. Camb. Philos. Soc* 94(5):1774–1785, 2019. Available from: https://doi.org/10.1111/brv.12527. PMID: 3114978.

Lee H, Huang AY, Wang LK, et al: Diagnostic utility of transcriptome sequencing for rare Mendelian diseases, *Genet. Med* 22(3):490–499, 2020. Available from: https://doi.org/10.1038/s41436-019-0672-1. Epub 2019 Oct 14.PMID: 31607746.

Lee MP, DeBaun MR, Mitsuya K, et al: Loss of imprinting of a paternally expressed transcript, with antisense orientation to KVLQT1, occurs frequently in Beckwith–Wiedemann syndrome and is independent of insulin-like growth factor II imprinting, *Proc. Natl Acad. Sci. U S A* 96(9):5203–5208, 1999. Available from: https://doi.org/10.1073/pnas.96.9.5203. PMID: 10220444.

Lo YM, Corbetta N, Chamberlain PF, et al: Presence of fetal DNA in maternal plasma and serum, *Lancet.* 350(9076):485–487, 1997. Available from: https://doi.org/10.1016/S0140-6736(97)02174-0. PMID: 9274585.

Lord J, McMullan DJ, Eberhardt RY, et al: Prenatal exome sequencing analysis in fetal structural anomalies detected by ultrasonography (PAGE): a cohort study, *Lancet.* 393(10173):747–757, 2019. Available from: https://doi.org/10.1016/S0140-6736(18)31940-8. PMID: 30712880.

Mackay DJ, Temple IK: Transient neonatal diabetes mellitus type 1, *Am. J. Med. Genet. C. Semin. Med Genet* 154C(3):335–342, 2010. Available from: https://doi.org/10.1002/ajmg.c.30272. PMID: 20803656.

Magrinelli F, Balint B, Bhatia KP: Challenges in clinicogenetic correlations: one gene—many phenotypes, *Mov. Disord. Clin. Pract* 8(3):299–310, 2021. Available from: https://doi.org/10.1002/mdc3.13165. eCollection 2021 Apr.PMID: 33816657.

Mitsuya K, Meguro M, Lee MP, et al: LIT1, an imprinted antisense RNA in the human KvLQT1 locus identified by screening for differentially expressed transcripts using monochromosomal hybrids, *Hum. Mol. Genet* 8(7):1209–1217, 1999. Available from: https://doi.org/10.1093/hmg/8.7.1209. PMID: 10369866.

Monk D, Mackay DJG, Eggermann T, et al: Genomic imprinting disorders: lessons on how genome, epigenome and environment interact, *Nat. Rev. Genet* 20(4):235–248, 2019. Available from: https://doi.org/10.1038/s41576-018-0092-0. PMID: 30647469.

Müller V, Dvirnas A, Andersson J, et al: Enzyme-free optical DNA mapping of the human genome using competitive binding, *Nucleic Acids Res* 47(15):e89, 2019. Available from: https://doi.org/10.1093/nar/gkz489. PMID: 31165870.

Neu MB, Bowling KM, Cooper GM: Clinical utility of genomic sequencing, *Curr. Opin. Pediatr* 31(6):732–738, 2019. Available from: https://doi.org/10.1097/MOP.0000000000000815. PMID: 31693580.

Õunap K: Silver − Russell syndrome and Beckwith − Wiedemann syndrome: opposite phenotypes with heterogeneous molecular etiology, *Mol. Syndromol* 7(3):110–121, 2016. Available from: https://doi.org/10.1159/000447413. PMID:. Available from: 27587987.

Rexach J, Lee H, Martinez-Agosto JA, et al: Clinical application of next-generation sequencing to the practice of neurology, *Lancet Neurol* 18(5):492–503, 2019. Available from: https://doi.org/10.1016/S1474-4422(19)30033-X. PMID: 30981321.

Richards S, Aziz N, Bale S, et al: Standards and guidelines for the interpretation of sequence variants: a joint consensus recommendation of the American College of Medical Genetics and Genomics and the Association for Molecular Pathology, *Genet. Med* 17(5):405–424, 2015. Available from: https://doi.org/10.1038/gim.2015.30. PMID: 25741868.

Riggs ER, Andersen EF, Cherry AM, et al: Technical standards for the interpretation and reporting of constitutional copy-number variants: a joint consensus recommendation of the American College of Medical Genetics and Genomics (ACMG) and the Clinical Genome Resource (ClinGen), *Genet. Med* 22(2):245–257, 2020. Available from: https://doi.org/10.1038/s41436-019-0686-8. Epub 2019 Nov 6.PMID: 31690835.

Roca I, González-Castro L, Fernández H, et al: Free-access copy-number variant detection tools for targeted next-generation sequencing data, *Mutat. Res.* 779:114–125, 2019. Available from: https://doi.org/10.1016/j.mrrev.2019.02.005. PMID: 31097148.

Russo S, Calzari L, Mussa A, et al: A multi-method approach to the molecular diagnosis of overt and borderline 11p15.5 defects underlying Silver − Russell and Beckwith − Wiedemann syndromes, *Clin. Epigenetics* 8:23, 2016. Available from: https://doi.org/10.1186/s13148-016-0183-8. eCollection 2016.PMID: 26933465.

Spencer T, Pan KS, Collins MT, Boyce AM: The clinical spectrum of McCune − Albright syndrome and its management, *Horm. Res. Paediatr.* 92(6):347–356, 2019. Available from: https://doi.org/10.1159/000504802. PMID: 31865341.

van der Meij KRM, Sistermans EA, Macville MVE, et al: TRIDENT-2: National implementation of genome-wide non-invasive prenatal testing as a first-tier screening test in the Netherlands, *Am. J. Hum. Genet* 105(6):1091–1101, 2019. Available from: https://doi.org/10.1016/j.ajhg.2019.10.005. PMID: 31708118.

Wilkins-Haug L, Zhang C, Cerveira E, et al: Biological explanations for discordant noninvasive prenatal test results: preliminary data and lessons learned, *Prenat. Diagn* 38(6):445–458, 2018. Available from: https://doi.org/10.1002/pd.5260. PMID: 29633279.

Wright CF, Fitzgerald TW, Jones WD, et al: Genetic diagnosis of developmental disorders in the DDD study: a scalable analysis of genome-wide research data, *Lancet.* 385(9975):1305–1314, 2015. Available from: https://doi.org/10.1016/S0140-6736(14)61705-0. PMID: 25529582.

Further reading

Health Quality Ontario. https://www.hqontario.ca/Portals/0/documents/evidence/open-comment/hta-noninvasive-prenatal-testing.pdf.

CHAPTER 7

Improved analyses of regulatory genome, transcriptome and gene function, mutation penetrance, and clinical applications

7.1 Introduction

In a 2016 review entitled "Beyond Simplicity," Schacherer (2016) noted that potentially continuous levels of underlying genetic complexity of phenotypes were often overlooked.

In a review in 2015 Zhang and Lupski noted that copy number variants and single nucleotide changes in noncoding regions of the genome can play important roles in human traits and in common disorders. They noted that some progress had been made in defining the functional impact of variants in the noncoding genome identified through genome-wide association studies (GWAS).

With respect to variants identified in GWAS studies they drew attention to a locus on chromosome 1q that has been associated with variability in the interval between the Q and T motifs on electrocardiograms. There is evidence for the role of a specific variant rs7539120 in an enhancer and that this impacts the QT (motifs on electrocardiogram) interval through its impact on NOS1AP that encodes nitric oxide synthase adaptor protein.

Variants in a specific 3′ noncoding region have also been shown to impact function through specific effects. The single nucleotide polymorphism (SNP) rs2266788 was shown to be associated with one form of hypertriglyceridemia. Functional studies revealed that the variant alleles generated a microRNA binding site for a liver expressed microRNA miR 485.5p that impacted expression of APOA5 apolipoprotein A5, a component of high-density lipoprotein.

Zhang and Lupski also emphasized the importance of noncoding variants that act as expression quantitative trait loci (eQTL).

With respect to copy number variants in upstream noncoding regions, Zhang and Lupski documented how a number of these had been shown to be implicated in specific diseases. These included a 150-bp dup. downstream of the PLP1 gene proteolipid 1 on the X chromosome, associated with spastic paraplegia and peripheral neuropathy and a 47-kb deletion downstream of SHOX (short stature homeobox gene) involved in a specific form of skeletal dysplasia.

7.1.1 Gene expression

In 2019 Perenthaler et al. reviewed key elements controlling gene expression and emphasized that in patients with suspected genetic diseases in whom no protein defects were found on sequencing could potentially harbor mutation in nonprotein coding genomic elements (Fig. 7.1).

7.2 Regulatory genome, gene expression, phenotype, and variability

Improved analysis of the regulatory genome, transcriptomes, gene function, mutations, and their penetrance has important clinical applications.

Modified penetrance of coding variants may be in part due to variants in regulatory elements.

7.2.1 Long noncoding RNAs (long nonprotein coding RNAs)

McDonel and Guttman (2019) reviewed progress in understanding mechanisms through which long noncoding RNAs (LNC RNAs) influence gene expression and noted that thousands of these have been identified.

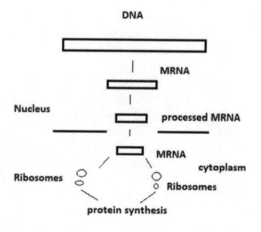

FIGURE 7.1

Diagram intended to indicate pathway from DNA to RNA synthesis and protein synthesis.

7.2 Regulatory genome, gene expression, phenotype, and variability

LNC RNAs were noted to have some properties found in protein coding RNAs, in that they are 5′ capped, spliced, and polyadenylated but they do not encode proteins. Functional mechanisms of LNC RNAs were noted to be poorly documented with the possible exception of XIST, an LNC RNA that is involved in X-linked inactivation.

XIST was noted to bind to the entire X chromosome and to recruit repressive chromatin proteins. Other named LNC RNAs include MALAT1, NORAD, NEAT1.

MALAT1 is an LNC RNA that binds mRNAs in nuclear speckles and was reported to impact splicing of bound mRNAs.

NORAD is an LNC RNA activated by PUMILO RNA binding proteins reported to be a translation regulator of protein coding RNAS through binding to their 3′ regions.

The LNC RNA NEAT1 was reported to be transcribed from the MEN1 locus and to be a component of speckle suborganelles.

One method used to determine LNC RNA function is to disrupt it and then attempt to analyze consequences of the disruption. Disruption of LNC RNAs has in some cases been introduced using antisense mRNAs.

McDonel and Guttman reported that antisense labeling of LNC RNAs has also been used to map LNC RNA localization and its hybridization within cells and organelles. MALAT1 was noted. This transcript is retained in the nucleus where it is thought to form molecular scaffolds for ribonucleoprotein complexes. It may act as a transcriptional regulator for numerous genes, including some genes involved in cancer metastasis and cell migration, and it is involved in cell cycle regulation.

NEAT1 nuclear paraspeckle assembly transcript 1 is reported to be retained in the nucleus It may act as a transcriptional regulator for numerous genes, including some genes involved in cancer progression.

FIRRE LNC RNA is described to have repeat elements. It was shown to be retained within the nucleus and to be important in the maintenance of repressive elements and to influence the innate immune response.

There is thus evidence that LNC RNAs play roles in the local coordination of regulatory responses.

Gil and Ulitsky (2020) emphasized evidence that LNC RNAs contribute to gene regulatory networks and to fine-tuned spatial and temporal gene expression.

7.2.2 Defining functions of specific regulatory elements and their relationship to diseases

Kircher et al. (2019) carried out saturation mutagenesis analyses, sequencing, and reporter assays to investigate nonprotein coding regulatory elements in the genome. They also determined the density of transcription factor binding sites in regulatory elements. They noted that these data help in the interpretation of pathogenicity of observed variants, noting that disruptions in regulatory elements are major drivers of disease.

Kircher et al. documented effects on gene expression of mutations in promoters and enhancers of specific genes. The disease-associated regulatory elements they analyzed included 10 promoters of genes and 10 documented enhancers that were known to harbor mutation, which lead to specific diseases. They specifically examine promoters and enhancers known to have effects on specific cell lines.

One promoter studied was in the low density lipoprotein receptor (LDLR) gene. Defects in this gene lead to lipoprotein-associated metabolic diseases. The LDLR mutations c-152C > T was known to reduce promotor activity by 40%. A mutation c-217C > T increased transcription levels to 160% of normal.

The saturation mutagenesis studies revealed that variants located in close proximity to transcription factor binding sites had deactivating effects.

Kircher et al. noted that the reporter assays enable qualitative ascertainment of the effects of genetic variants in regulatory elements.

Kircher et al. noted that specific mutations in the TERT (telomerase) gene promoter are known to increase promoter activity; these include c-124C > T and c-146C > T that were known to constitute transcription factor binding sites. In their study effects of additional TERT promoter mutations were studied. Specific TERT promoter mutations are associated with specific cancers.

With respect to enhancer elements, Kircher et al. reported that the SORT1 gene-associated enhancer has been found to undergo relatively common mutation that is associated with increased risk of myocardial infarction. This variant is thought to alter expression in SORT1 leading to changes in LDL and VLDL (very low density lipoprotein) levels.

Gasperini et al. (2018) reported a framework for mapping enhancer gene pairs using enhancer element perturbations and single gene RNA sequencing. Their study led to the identification of 470 high confidence cis enhancer gene pairs. They noted that particularly in nonhouse keeping genes related enhancers, the enhancer elements were enriched for transcription factor binding sites.

The median distance of paired enhancers and gene transcription start sites was 24.1 kb.

In reviewing cis and trans factors in gene expression in evolution, Signor and Nuzhdin (2018) noted that regulation of expression in cis often occurs through gene-linked polymorphisms, and eQTLs were noted to primarily be dependent on physical proximity in space or through chromatin looping.

Regulation of gene expression in trans through diffusible factors may occur through metabolites. Metabolites can influence chromatin modifications that lead to alteration in gene expression (Schvartzman et al., 2018).

7.3 Epigenetic factors relevant to gene expression

Key epigenetic mechanisms include DNA methylation, histone modification, and chromatin remodeling. DNA methylation occurs frequently in genomic DNA regions close to promoters, but it also occurs in other genomic regions including regions rich in transposons. DNA methylation occurs primarily but not exclusively at CpG dinucleotides. Extended regions of CpG methylation upstream of promoters occurs in genes where gene expression is silenced and in imprinted gene regions. CpG methylation does not occur upstream of genes being actively expressed. Methylation of DNA is carried out in part through activity of DNA methyl transferases and requires availability of a methyl group; S-adenosyl methionine is an important methyl donor. Demethylation of DNA, 5-methylCytosine demethylation can take place through activity of a specific enzyme TET a DNA demethylase. It can also take place through DNA methyltransferase 1 (Fig. 7.2).

Regions of heterochromatin such as occur in pericentromeric regions of chromosomes are rich in repeat sequences and DNA in those region is inactivated through action of DNA methyltransferase and histone methyltransferase (Fig. 7.3).

Histone modifications occur primarily in the tails of histones that surround the histone core of nucleosomes. In regions where genes are expressed, nucleosomes are widely spaced along the strand and key lysine residues in histones are acetylated.

In regions where genes are not expressed nucleosomes are closer together and more tightly packed and key residues are acetylated.

Histones undergo methylation and the specific histone residues and their degree of methylation vary depending on whether the specific genes are expressed or silenced. Histone methylation involves the activity of histone methyltransferase enzymes or histone demethylases.

Dietary components influence the degree of DNA methylation. It is important to consider metabolism of methionine, homocysteine, folate and vitamin B12

FIGURE 7.2

Dietary components for methylation.

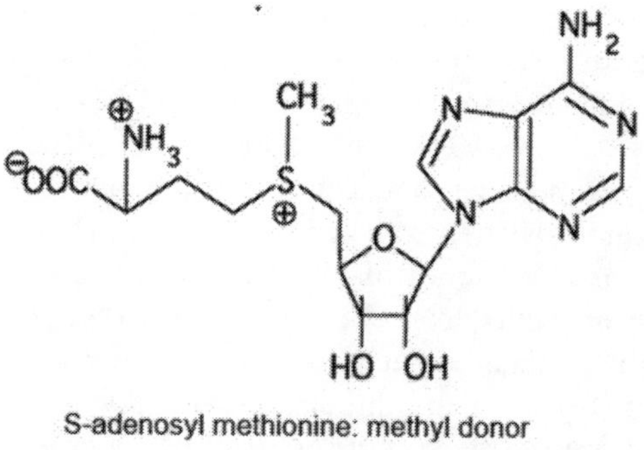

FIGURE 7.3

S-Adenosyl methionine.

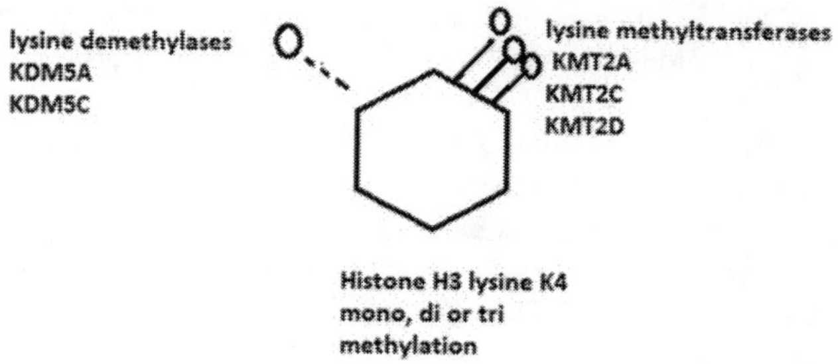

FIGURE 7.4

Histone methylation.

The different factors that lead to histone modifications that impact gene expression are sometimes described as readers, writers, and erasers (Borrelli et al., 2008) (Fig. 7.4).

Main elements of the regulatory genome include gene promoters, enhancers, transcription factor binding sites regions of open chromatin where genes are to be expressed that then constitute regions of DNAse 1 hypersensitivity. Chromatin looping is reported to bring enhancers in close proximity to specific promoters (Bhatia and Kleinjan, 2014).

LNC RNAs are reported to potentially impact gene expression in part through impacting chromatin folding and remodeling.

Chromatin remodeling impacts gene expression particularly through influence on nucleosome positioning and chromatin accessibility. Chromatin remodeling requires functioning of a large number of gene products (Son and Crabtree, 2014). Insights into genes and gene products that play roles in chromatin remodeling have been obtained in part through studies on patients with certain genetic disorders.

Another relatively recent concept in genome organization is the definition of topologically associated domains (TADs) (Lupiáñez et al., 2016). A TAD is defined as a genomic unit that may contain several genes and a TAD has distinct boundaries demarcated by binding of a specific factor CTCF.

Structural chromosome rearrangements that disrupt a TAD can impair gene expression and may lead to malformations.

Mendelian disorders of the epigenetic machinery that impact gene expression are sometimes referred to as transcriptomopathies.

7.3.1 Disorders of the epigenetic machinery leading to neurodevelopmental disorders

In 2015 Bjornsson reported analyses of 44 cases of disorders of the epigenetic machinery. Of the 44 disorders 18 cases involved functional impairments of writers; in 13 cases chromatin remodelers were impacted; seven cases involved erasers of modifications; and six cases involved readers of modifications.

Some degree of overlap of phenotypic manifestations occurred in patients. Common phenotypic features encountered in the 44 patients included intellectual disability, developmental delay, growth abnormalities, facial dysmorphism. In some cases epilepsy occurred.

One case with more unusual features was found to have hypomethylation of centromeric repeats and this patient had evidence of centromeric instability on chromosome studies, and in addition manifested immunodeficiency and facial anomalies, features that together were referred to as ICF syndrome (immuno-deficiency centromeric instability syndrome due to DNA methyltransferase 3 deficiency).

Another related condition with distinct features is ATRX syndrome due to defect in the ATRX chromatin remodeler gene, located on the X chromosome that encodes a chromatin remodeler. ATRX syndrome occurs in males and is associated with craniofacial abnormalities, hypotonia, developmental delay, and alpha thalassemia. Specific laboratory studies recommended to diagnose this syndrome include microarray and bead chip studies that can display altered patterns of CpG methylation across the genome and determine if these are altered in patients relative to controls.

7.3.2 Cohesinopathies

Cohesinopathies are another important category of disorders that can lead to aberrant gene expression. The cohesin complex plays an important role in chromatin looping and in bringing enhancers and transcription factors into close proximity to gene promoters.

Cornelia de Lange syndrome is a cohesinopathy that may arise due to defects in any member of the cohesin complex (Kline et al., 2018). Components of the cohesin complex include NIPBL, a cohesin loading factor, and defects in NIPBL are reported to occur in approximately 70% of cases of Cornelia de Lange syndrome. Other components of the cohesin complex with defects leading to this syndrome include:

SMC1A structural maintenance of chromosomes 1A.
SMC3 structural maintenance of chromosomes 3.
RAD21 cohesin complex component.
BRD4 bromodomain containing 4.
HDAC8 histone deacetylase 8.
ANKRD11 ankyrin repeat domain 11.

7.4 Regulatory circuit: Epimap

Boix et al. (2021) reported generation of Epimap, which they defined as "documentation of chromatin states, high resolution enhancers, upstream regulators and downstream target genes of regulators."

Development of Epimap required assessment of chromatin state and enhancers. It was also necessary to analyze epigenetic modifications, to correlate active enhancers with transcription. Their studies identified 3.3 million tissue-specific enhancers and revealed that each gene was estimated to be linked to 13 enhancers. On average each enhancer was linked to 3.5 genes.

They correlated findings in the Epimap catalog with information gathered from GWAS.

Significant findings from their analyses include revelation that each gene is controlled by a high number of enhancers and that there is tissue specificity for long range enhancer links. They also identified a number of master regulators that were found to interact with diverse partners to establish a tissue-specific gene regulatory program. These master regulators included the following:

RFX2 Regulatory factor X2 encodes transcription factors that contain a highly conserved winged helix DNA binding domain.
RFX4 regulatory factor X4, transcription factor.
GRHL1 grainyhead like transcription factor 1.
HNF1 HNF1 homeobox transcription factor.
AP-1A (AP1AR) adaptor-related protein complex 1 associated regulatory protein.

Boix et al. reported that they established correlation between 540 GWAS traits and 30,000 single nucleotide polymorphisms in tissue-enriched enhancers. They emphasized that in GWAS it is important to take into account that multiple different enhancers can converge on a single target and that a specific enhancer can impact multiple genes.

Development of Epimap was shown to be particularly valuable in studies of complex traits. Boix et al. noted that genomic loci associated with complex disease have, in 90% of cases, been found to map outside the coding genome.

7.4.1 Combinations of variants in different genes and impact of phenotype

Gazzo et al. (2017) carried out studies on disorders where combinations of variants in different genes were found to be responsible for disease phenotype. In some cases the phenotypic variability is due to oligogenic factors. Guzzo et al. drew attention to 44 disorders in which digenic components contributed to disease as documented in the Digenic Diseases Database.

Bardet Biedl syndrome (BBS) is one example. There are 20 different genes described as BBS genes. This syndrome is described as autosomal recessive ciliopathy characterized by retinal degeneration, polydactyly, renal disease, hypogonadism, obesity, dysmorphic features, and variable degrees of cognitive impairment (Zaghloul and Katsanis, 2009).

In some cases defects in two different BBS genes interact leading to disease manifestations. In other cases a BBS gene interacts with another gene, for example, MKKS centrosome shuttling protein.

Gazzo et al. noted digenic effects in oculocutaneous albinism. This disorder was noted to involve variants in genes involved in melanin synthesis. Examples included heterozygous variants in TYR and OCA2.

7.5 Genotype phenotype axis

Baek and Lee (2020) emphasized that understanding the genotype phenotype axis requires information on functional noncoding genomic elements. They noted that development of genome-wide maps of open chromatin regions can provide information on functional cis and trans regulatory elements. Specific sequencing methods have been developed for identification of open chromatin. These include assays of transpose accessible chromatin (ATAC),. For analysis of specific functional sites, single cell ATAC is conducted (Shashikant and Ettensohn, 2019). They reported that ATAC sequencing utilizes hyperactive transposase enzyme to cleave DNA followed by insertion of adaptors. The hyperactive transposase enzyme preferentially cleaves open chromatin elements.

Klemm et al. (2019) reviewed chromatin accessibility and regulation of expression. They emphasized that accessible chromatin creates a network of permissible physical interactions. This network includes chromatin binding factors, enhancers, promoters, and initiators. The network facilitates dynamic responses to external stimuli.

Identification of regulatory elements and their functions have also been facilitated through utilization of reporter assays to analyze gene expression (Elkon and Agami, 2017).

A key factor in gene regulation is for actively expressed genes to be located in open chromatin accessible to transcriptional elements. An early discovery regarding open chromatin is that it manifests DNASE 1 hypersensitivity. Status of chromatin is impacted by epigenetic modifications including histone modification such as acetylation.

Long et al. (2016) described enhancers as sequence elements usually 100–1000 nucleotides in length that are located upstream or downstream of the genes with which they interact.

Gene transcription is also influenced by CTCF (CCCTC binding factor) elements binding to specific regions forming transcription boundaries.

Another important factor in gene transcription has to do with nucleosome positioning and elements that cause nucleosomes to be widely spaced in regions that are to be transcribed.

Transcription factor binding represents a key process in gene transcription.

7.6 Promoters

Perenthaler et al. (2019) noted that a specific promoter may be regulated by more than one enhancer and that a particular enhancer may interact with more than one promoter. Long range interaction between promoter and enhancer were noted to be mediated by chromatin looping. Enhancers include transcription factor binding sites.

Three promoter regions are often distinguished. The core promoter is defined as being proximal to the transcription start codon and to harbor the RNA polymerase binding site, TATA box (Thymine Adenine rich segment in core promoter), and transcription binding sites. The proximal promoter is reported to upstream from the core promoter; general transcription factors bind to this site.

The distal promoter is upstream from the proximal promoter and is reported to contain regulator elements and transcription binding sites.

Haberle and Stark (2018) reviewed core gene promoters and transcription. Core promoters were noted to most commonly be located within nucleotide sequence 50 base pairs upstream and 50 base pairs downstream of transcription start sites. Promoters bind enhancer transcription factors and elements to anchor RNA polymerase II.

Stepwise interaction between promoters and transcription factors was reported to occur particularly when nucleotides in the promoter region have H3 lysine 27 acetylation and H3 lysine 4 methylation H3K4Me3. Stepwise recruitment of transcription factors occurs beginning with recruitment of general transcription factors TFI, TFIID, TFIIA, TFIIB, and then TFIIF.

Karnuta and Scacheri (2018) reviewed enhancers and their relation to gene expression. They noted that active enhancers were acetylated at H3K27 and manifested monomethylated histones H3K4Me1. There is evidence that active enhancers generate RNA referred to as eRNA.

7.7 Transcription initiation and promoters

The mediator complex has been shown to be key to transcription initiation. In 2015 Allen and Taatjes noted that RNA polymerase II is regulated by mediator complex. This complex was reported to be composed of variable subunits. Other specific processes impacted by mediator include chromatin organization, preinitiation, pausing, and elongation.

In mammals, 26 different subunits contribute to mediator and different transcription factors bind to mediator. Allen and Taatjies noted that mediator communicates regulatory signals between DNA bound transcription factors and RNA POLII.

Other processes to which mediator contributes were reported to be enhancer promoter looping. Mediator was shown to contribute not only to transcription initiation but also to transcription elongation.

Mediator mutations have been identified in a number of different diseases including cancer.

Mutations in MED12 and MED13, MED13L were noted to harbor mutations in specific syndromal developmental disorders (Calpena et al., 2019). MED13L, a subunit in the mediator kinase, was reported to lead to transcriptional defects that resulted in intellectual disability (Nizon et al., 2019).

7.7.1 Genes with more than one promoter

It is important to note that some genes have more than one promoter. Kunkel et al. (2020) reported the presence of two upstream promoters in the gene that encodes CHD8, chromatin helicase DNA binding protein that acts as a chromatin remodeler and plays important roles in development. Kunkel et al. reported that the upstream promoters are located thousands of base pairs upstream of the coding sequence. These two promoters were noted to interact with different set of transcription start sites.

They stressed that detailed analyses of promoter usage and transcription sites will be important in analyses of neural development and its defects. CHD8 mutations have been implicated in autism (Hoffmann and Spengler, 2021).

An et al. (2018) reported results of whole genome sequencing in 1902 autism spectrum disorder families. Sequencing results were analyzed in part to determine de novo variant risk scores in noncoding genomic regions.

Prior studies of autism revealing de novo mutations in coding regions had been reported in 2015 by Sanders et al. (2015). These studies had reported that de novo variants occurred particularly in the genes that encoded NRXN1 (neurexin 1) on chromosome 2p16.3; GTF2I general transcription factor 2I on chromosome 7q11.2; GABRA5, GABRA3, GABRG3 on chromosome 15q11.2.

An et al. (2018) carried out a specific type of GWAS study referred to as GWAS category-wide association studies. They assessed variation in 163 categories of noncoding genomic regions. These included evolutionarily conserved genomic regions and promoter regions described as regions 2-kb upstream of transcription start sites. Their analyses revealed that the majority of de novo variants occurred in the region they defined as the promoter region.

Within the promoter region the strongest de novo variants associated with autism were found at regions identified as conserved loci, and particularly at transcription factor binding sites.

An et al. noted further that the strongest signals were not found at the core promoter but at distal promoters.

An et al. concluded that their studies revealed a definitive trend toward promoter-associated risk being implicated in autism.

7.7.2 Ornithine transcarbamylase gene promoters and enhancers

Jang et al. (2018) reported that in 10% − 15% of patients with clinical diagnosis of ornithine transcarbamylase (OTC) deficiency, mutations in the coding regions were not detected and splice site mutations were not found.

OTC is involved in the urea cycle and this enzyme generated citrulline from ornithine and carbamoyl phosphate.

OTC is encoded by a gene on the Xp11.4 and was reported to be 70 kb in length with 10 exons.

Defective OTC manifests in males. Defective activity of OTC resulting in reduced levels of functional enzyme leads to lethargy, vomiting hyperammonemia, and neurological impairments. Biochemical abnormalities resulting from impaired or reduced OTC activity include elevated plasma levels of ammonia, elevated glutamine, increased orotic acid, and low levels of citrulline.

There is evidence that transcription of OTC mRNA can initiate at a number of different start sites and there is evidence that an upstream enhancer plays an important role in transcription initiation.

Jang et al. carried out sequencing analyses in 38 individuals with clinical diagnosed OTC deficiency in whom no coding mutations had been identified. Their results revealed that one of these individuals harbored mutation in an upstream enhancer. In five of these individuals, mutations were identified in transcription factor binding sites and two individuals harbored mutations in evolutionarily conserved elements in upstream sequences.

Detailed analyses of the three conserved regions upstream of the first OTC exon were carried out. These regions were defined as a 600-bp region upstream of the (adenine thymine guanine initiation codon; a second region was 900-kb long and harbored an enhancer. This was also studied in rat and was found to be important for maximal expression of OTC. The third conserved upstream region extended 2-kb upstream of the enhancer bearing region.

In examination of the upstream mutations, Jang et al. used different analysis tools including CADD analysis of the specific mutations to determine their likely functional impact. They also carried out functional analyses and reported assays in culture cells to investigate the impacts of sequence variants on OTC expression. Functional studies were noted to reveal reduced reported expression with all the variants identified on sequence analyses.

7.8 Transcription factors

In reviewing transcription factors Lambert et al. (2018) noted that many questions remained regarding transcription factor binding and control of gene expression. Studies have revealed that more than 1600 transcription factors are required in the human genome.

Mutations in transcription factors and transcription factor binding sites have been implicated in a number of different human diseases.

Transcription factors have been described as factors that can influence gene expression through binding to specific elements in DNA. Each transcription factor can regulate a number of different genes. A number of different databases have been developed that document transcription factors and the DNA elements to which they bind. Transcription factors are classified into distinct families in part on the basis of their structure and on their binding sites.

There is evidence that transcription factors can interact with nucleosomes and they are also reported to recruit RNA polymerase. Some transcription factors are reported to recruit other accessory factors and to influence local chromatin state. The mediator complex was noted to bridge transcription factors and RNA polymerase II.

Lambert et al. reported that of the 1639 known transcription factors 1211 had known binding sites. There is growing evidence for the existence of tissue and cell type-specific expression of transcription factors indicating some specificity of function. One-third of transcription factors were known to have tissue-specific expression. For example, transcription factors expressed particularly in the cerebral cortex included:

SOX2 SRY-box transcription factor 2 is required for stem-cell maintenance in the central nervous system.

POU3F2 (also known as OCT7) POU class 3 homeobox 2 protein is involved in neuronal differentiation and enhances the activation of corticotropin-releasing hormone regulated genes.

OLIG1 oligodendrocyte transcription factor 1.

Transcription factors reported to be expressed primarily in cardiac tissue included:

GATA4 GATA binding protein 4 is thought to regulate genes involved in embryogenesis and in myocardial differentiation and function. Mutations in this gene have been associated with cardiac septal defects.

TBX20 T-box transcription factor 20 essential for heart development. Mutations in this gene are associated with diverse cardiac pathologies, including defects in septation, valvulogenesis, and cardiomyopathy.

Lambert et al. drew attention to a condition associated with growth hormone deficiency, known as anterior pituitary hypoplasia, that can arise due to defects in any one of 15 different genes and 12 of these genes encode transcription factors.

Mutations in the transcription factor HOXD3 homeobox D3 were identified in cases of limb malformations. Homeobox gene products were noted to have important roles in morphogenesis in all multicellular organisms.

Specific transcription factors were noted to be encoded in loci identified through genome wide associated as risk loci for polygenic disease. Lambert et al. noted that these included particularly risk loci for autoimmune diseases.

Lambert et al. also noted that transcription factor binding site disruption has been implicated in specific polygenic diseases. Modulation of the binding site for transcription factor ARID5B (AT-rich interaction domain 5B) occurs in the FTO locus. This secondarily alters adipocyte cell fate and mitochondrial thermogenesis in adipose tissue leading to obesity. ARID5B protein forms a histone H3K9Me2 demethylase complex with PHD2 (plant homeodomainlike finger protein 2) and regulates the transcription of target genes involved in adipogenesis and liver development. FTO is alpha-ketoglutarate dependent dioxygenase.

7.8.1 Transcription elongation and RNA polymerase II

Reines (2020) reviewed recent advances in the analyses of RNA polymerase II structure and function and noted that it is a DNA-dependent DNA nucleotide transferase. Its activity was noted to be highly regulated and to include actions at the transcription initiation site and transcript elongation of RNA and generation of the primary RNA transcript. Polymerase II was also noted to be involved in recognition of the transcription termination site.

Key factors in initiation involved interaction of POLII with transcription initiation factors and subsequently POLII associated with transcription elongation factors.

Reines noted that new information has been gathered on the role of POLII in transcription termination and evidence of interaction with a specific factor SPT5 also known as SUPT5H in humans that travels along with POLII during transcription and is dephosphorylated just after POLII transcribes the polyadenylation site.

RBFOX2 RNA binding protein that is thought to be a key regulator of alternative exon splicing in the nervous system and other cell types. It has broad expression in different tissues.

RBFOX3 has highest expression in the central nervous system and plays a prominent role in neural tissue development and regulation of adult brain function. Primarily expressed in brain.

RBFOX1 is primarily expressed in brain and heart.

NOVA1 a neuron-specific RNA binding protein.

MBNL1 a C3H-type zinc finger protein that modulates alternative splicing of pre-mRNAs.

PTBP1 polypyrimidine tract binding protein 1.

7.9 Transcription termination

Tian and Manley (2017) reviewed aspects of transcription termination and noted that defined nucleotide elements AAUUAAA or AUUAAA and GU elements downstream of these hexamers constitute transcription termination sites. It is known that more than one transcription termination site and polyadenylation may occur in a particular gene.

7.10 Polyadenylation

Polyadenylation requires activity of a complex of at least 20 different factors. The length of the 3' untranslated region upstream of the polyadenylation site was noted to vary in different genes. In some cases transcription termination may occur upstream of the last potential exon present in a gene.

Particular codons act as terminator codons:

DNA codons	RNA codons
TAG	UAG
TAA	UAA
TGA	UGA

7.11 Alternate polyadenylation

In a 2019 review Gruber and Zavalon noted that alternate promoter usage and use of alternative transcription termination determine the boundaries of the 5' and 3' untranslated regions. They emphasized that different isoforms with the same coding regions but with different untranslated regions can have different consequences for function.

At the 3' end most genes were noted to have multiple potential polyadenylation sites and particular regulators determine which polyadenylation site is used. Alternate polyadenylation was reported to impact the cellular location of transcripts. Genes with multiple polyadenylation sites were noted to give rise to changes in isoform generation rates and this also influenced tissue expression.

Specific proteins shown to impact Poly A site use include CSTF2, cleavage stimulation factor subunit 2, this protein binds GU-rich elements within the 3' untranslated region of mRNAs and FIP1L1 Fip1 (factor interacting with PAP (polyA polymerase), binds to U-rich sequences of pre-mRNA and stimulates poly (A) polymerase activity. PAPOLA poly(A) polymerase alpha belongs to the poly (A) polymerase family. It is required for the addition of adenosine residues for the creation of the 3'-poly(A) tail of mRNAs. PABPN1 protein is required for progressive and efficient polymerization of poly(A) tails at the 3' ends of eukaryotic transcripts and controls the size of the poly(A) tail to about 250 nt.

Alternate cleavage and Poly A site usage have been reported in specific gene products involved in hematological and immunological disorders (Gruber and Zavolan, 2019) (Fig. 7.5).

7.11.1 Alternate splicing of transcripts

In 2014 Gamazon and Stranger reviewed transcription and noted that alternative splicing of transcripts occurs in approximately 90% of genes. The complexity of mRNAs that can be derived from a single gene is related to use of alternative transcription start sites, alternative splicing of transcripts, and variations in 3' transcription termination sites. Alternative splicing is associated with use of alternative 5' or 3' splice elements and can lead to exon exclusion or intron inclusion.

Will and Lührmann (2011) reviewed splicing and the spliceosome. Key elements in splicing include the 5' and 3' splice sites that occur at exon − intron junctions. These are nucleotides GU (guanine uridine) in mRNA, (GT in DNA sequence at the start of the intron and the end of the 5' exons) and AG (adenine guanine) at the end of the intron and at the start of the 3' exon. Upstream of the AG acceptor site there is a polypyrimidine tract that binds key proteins. Other key elements in splicing include the splicing branch point that occurs between 18 base pairs and 40 base pairs upstream from the 3' splice site and the spliceosome, a complex multiprotein complex.

Transesterification reactions were shown to be essential for splicing. A key nucleotide is adenosine at the branchpoint. GU from the 5-exon couples to adenosine at the branchpoint and it the branchpoint AG from the 5-end of downstream exon couples with GU.

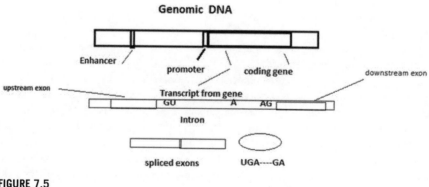

FIGURE 7.5

Exon splicing.

7.12 The spliceosome

The spliceosome is a ribonucleoprotein complex reported to use adenosine triphosphatase (ATP) hydrolysis to promote splicing (Matera and Wang, 2014). Key components of the spliceosome include multiple small nuclear riboproteins SNRPN small nuclear ribonucleoprotein polypeptide N is encoded on chromosome 15q11.2.

Shi (2017) reported on the atomic structure of the spliceosome. They noted that two catalytic metal ions occur at the active site.

Splice regulatory proteins rich in serine and arginine amino acids also impact splicing. Other important proteins in splicing include polypyrimidine tract binding proteins.

In 2019 Montes et al. reported that alternative splicing was responsive to development stimuli and also to environmental stimuli.

The generation of alternative splice forms was reported by Furlanis and Scheiffele (2018) to impact stability of transcripts, transcript localization, and cell type specificity. They specifically noted that alternate splicing impacts axon guidance and also formation of synapses. Examples of the impact of alternate splice forms were derived from studies on calcium channels where specific variants were found to impact cell type specificity of calcium channels.

It is important to note that intronic variants in specific genes can lead to specific disorders. For example, intronic mutations in the SCNA1 were reported to cause a specific form of epilepsy referred to as Dravet syndrome.

Cummings et al. (2020) emphasized the importance of taking into account cell type-specific differences in the splice forms of a specific gene and taking into account specific splice patterns when interpreting the likely impact of genomic variants.

Alternative promoter use and alternative splicing result in a multitude of transcript variants encoding the same protein.

7.12.1 Generation of microexons

Specific genes have been noted to give rise to microexons. Irimia et al. (2014) reported that microexons are particularly included in transcripts of genes related to neurogenesis, vesicle trafficking, calcium signaling, and protein kinase signaling. They reported misregulation of microexon inclusion in autism. Microexon inclusion defects in autism were also reported in 2016 by Quesnel-Vallières et al. (2016), and in 2019 by Parras et al. (2018).

Weyn-Vanhentenryck and Zhang (2016) studied alternative splicing in different categories of neuronal processes. They noted that specific RNA binding proteins were particularly important in splice switching processes in different neuronal processes. These included the RBFOX 1, 2, and 3 proteins.

7.13 MicroRNAs and posttranscriptional regulation

MicroRNAs have been reported to participate in posttranscriptional regulation of gene expression. Bartel (2004) reported that microRNAs act through suppression of translation and that they can also promote RNA cleavage.

7.14 Translation, ribosomes biogenesis, functions, and defects

Ribosome biogenesis is dependent on transcription of ribosomal genes. In 2019 von Walden reviewed key steps in ribosome biogenesis. The ribosome is known to be composed of a 40s subunit with 33 ribosomal proteins and a 60s subunit with 46 ribosomal proteins together with 5s, 5.8s, and 28s ribosomal RNAs.

The smaller ribosomal subunit 40s was known to function in matching the mRNA trinucleotide code to the correct transfer RNA during translation of mRNA. Incorporation of the specific amino acid into the growing polypeptide chain is known to be facilitated by activity of the 60s ribosomal subunit.

Regarding de novo synthesis of ribosomes, von Walden noted that their de novo synthesis requires transcription of RNA genes, synthesis of RNA subunits and ribosomal proteins, and coordination of the processes is reported to be dependent on the phosphatidyl inositol, MTOR, p70S6K1 network, and cMyc transcription factor.

Ribosomal RNA genes in human number on average 300; however, the range is from 61 to 1590; these genes are located on five human chromosomes 13, 14, 15, 21, and 22. The nucleolus within the nucleus is the site of ribosome biogenesis. Ribosomal genes are transcribed by RNA polymerase 1 and requires transcription initiation factor.

Ribosomal genes are described as 43–45 kb in length and the transcribed units are reported to be composed of repetitive units.

7.14.1 Human disorders associated with impaired ribosomal biogenesis or function

A gene positionally cloned on the basis of abnormalities in a genetic disorder, Treacher Collins syndrome, was designated TCOF, Treacher Collins Collaborative Group (1996).

Treacher Collins syndrome is associated with hypoplasia of facial bones including malar bones, zygomatic arch, and mandible, and with abnormalities of the ear leading to conductive hearing loss. In rare cases cleft palate and choanal atresia may be present.

The TCOF1 encoding gene was shown to be targeted to the nucleolus (Winokur and Shiang, 1998). They determined that TCOF1 is a phosphoprotein containing nucleolus targeting elements. Both highly penetrant and reduced penetrant mutations have been identified in TCOF1.

Subsequently, mutations in POLR1C and POLR1D genes were documented to lead to Treacher Collins syndrome. These proteins are subunits of RNA polymerase I and RNA polymerase III.

Ribosomopathies are defined as rare disorders due to defects in ribosome biogenesis of functions. These disorders were reviewed in 2020 by Orgebin et al. (2020). Defects have been identified in the ribosome genes, in the processing of their gene products, in assembly of ribosomal subunits, and in export of ribosomes to the cytoplasm.

Orgebin et al. noted that certain pharmacological substances have been shown to have positive therapeutic effects in these conditions, these include modulators of the MTOR pathway.

One of the first ribosomopathies observed was noted to be X-linked dyskeratosis congenita. Blackfan − Diamond anemia is a ribosomopathy associated with defects in any one of 19 different gene products involved in determination of ribosomal structure or function (Ulirsch et al., 2018). Mutations in ribosomal subunit assembly were reported to occur in Shwachman − Diamond syndrome.

Orgebin et al. reported that in Treacher Collins syndrome transcription of the ribosomal gene transcription is impaired in part due to premature transcription stop codons. Thus ribosomal RNA levels are reduced and the TCOF1 is not adequately generated.

The cartilage hair hypoplasia syndrome was noted to be due to defects in protein RMRP that acts as a component of mitochondria RNA processing endonuclease. RMRP endonuclease also interacts with telomeric reverse transcriptase.

7.15 Translation of mRNA to proteins and associated defects leading to disease

Steps in translation of mRNA to proteins were reviewed by Gabut et al. (2020). Early steps include passage and coupling of the 5′ mRNA regions to the 40s ribosome through action of the eif4 complex that merges with a bound preinitiation complex (PIC). The PIC includes GTP (Guanosine triphosphate), EIF3, EIF5, EIF1 subunits, and EIF1A that together with methionyl initiator TRNA bind to the 40s ribosome unit. The PIC complex was noted to scan the mRNA to identify the AUG initiation codon. Subsequent release of GTP (guanosine triphosphate(and bound initiation factors allow for recruitment and binding of EIF3B and the 80s ribosome unit. The aminoacyl TRNA binding site is exposed, and translation can begin.

7.15.1 Aminoacyl tRNA synthases

Translation elongation requires specific aminoacyl TRNAs being escorted to the ribosome by GTP-coupled elongation factor. Elongation requires movement along the ribosome-coupled mRNA, three nucleotides at a time to add aminoacids that have been bound to TRNAs. A specific aminoacyl tRNA synthetase must select the correct amino acid, and it must select the correct tRNA from the TRNA pool. Fuchs et al. (2019) reviewed aminoacyl tRNA synthetase defects noting that they were particularly encountered in children with psychomotor retardation and seizures.

7.15.2 Noncanonical functions of aminoacyl tRNA synthetases

Musier-Forsyth (2019) reported that the canonical function of aminoacyl tRNA synthetases is well-known and is attachment of cognate tRNAs to specific aminoacids. Additional noncanonical functions have been discovered. Those include regulation or splicing, stimulation of MTOR activity, and DNA repair.

Rubio Gomez and Ibba (2020) reviewed aminoacyl tRNA synthetases and noted that they are not only essential for accurate translation of the genetic code, but they had roles in other processes.

In their primary function, aminoacyl tRNA synthetases were noted to catalyze a two-step reaction. This involved esterification of aminoacid and hydrolysis of ATP to generate aminoacyl tRNA.

Aminoacyl tRNA synthetase defects have been particularly associated with nervous system disorders.

7.16 tRNAs

Rak et al. (2018) reviewed the repertoire of tRNAs and stressed the importance of tRNAs in allowing the ribosome to decode the genetic information. The different TRNAs are named according to the aminoacid anticodon they specify.

In a review of tRNAs Schaffer et al. (2019) noted that in the human genome there are more than 600 potential tRNA encoding segments. Specific maturation steps are required for the tRNA to be functional, and more than 100 modifications were identified. Impaired biogenesis of tRNAs was reported in some cases of leukodystrophy and pontocerebellar hypoplasia.

The mitochondrial genome has 22 tRNA encoding segments. Deletions and mutations in mitochondrial tRNAs have been identified as causes of specific mitochondrial disorders including mitochondrial myopathies and in encephalopathies associated with lactic acidosis.

Defects in translation processes have been reported in specific genetic disorders, particularly in neurodevelopmental disorders (McLachlan et al., 2019).

Kapur and Ackerman (2018) reviewed translation and noted that one technique used to investigate translation errors includes ribosome profiling.

Lant et al. (2019) reported that in humans 600 different genes encode tRNAs and the sequencing studies have revealed variants and mutations in cytoplasmic and mitochondrial tRNAs.

The first disease-associated tRNA mutations were identified in mitochondrial encoded tRNA by Kobayashi et al. (1990). A point mutation in mitochondrial tRNAleu was identified in MELAS syndrome that is characterized by lactic acidosis and stroke-like episodes. Defects in the cytosolic tRNA sec (selenocysteine) was found in a patient with muscle weakness and episodes of abdominal pain.

Lant et al. noted that the tRNA mutations can act synergistically with other mutations to impact protein synthesis and protein homeostasis. They noted that further studies are required to analyze possible deleterious effects of tRNA mutations.

7.17 RNA surveillance

Wolin and Marquat (2019) reported on quality control systems that exist to detect and degrade defective RNAs. Specific surveillance systems were noted to exist to monitor both coding and noncoding RNAs. Importantly, many disease manifestations were noted to result from inappropriate functioning of the RNA surveillance system. One feature potentially defective involved inadequate binding of RNAs to nucleoproteins.

Specific degradation of defective RNAs was carried out by ribonucleases; degradation was also carried out through binding of MTR4 and interactions through nuclear exosome targeting complex NEXT, that is composed of MTR4 (ATP-dependent RNA helicase), RBM7 (RNA binding protein), and zinc finger protein ZCCHC8.

Other alterations were reported to include addition of adenine or uridine nucleotide tails through activity of terminal nucleotidyl transferases.

RNA surveillance processes were noted to take place at the 5′ capping site and also at transcription termination sites. An additional quality control step occurs at the stage of export of the mRNA from the nucleus to the cytoplasm.

Transport out of the nucleus involves the nuclear pore complex, the identification of transport signals and transport receptors in addition the Ran that provides energy for transport (Cautain et al., 2015).

Wolin and Marquat noted evidence that mRNAs that contain an intron may be decayed through nonsense-mediated decay (NMD) mechanisms.

They documented specific diseases that result from inadequate degradation of defective mRNAs. These included pontocerebellar hypoplasia due to defects in EXOSC3 exosome component 3 and EXOSC8. EXOSC3 is a noncatalytic component of the human exosome, a complex with 3′ − 5′ exoribonuclease activity that plays a role in numerous RNA processing and degradation activities. EXOSC8 is a 3′ − 5′ exoribonuclease that specifically interacts with mRNAs containing AUrich elements. The encoded protein is also part of the exosome complex that is important for the degradation of numerous RNA species.

One case of pontocerebellar hypoplasia was found to have a defect in RBM7 (RNA binding protein) that is a component of the RNA exosome (NEXT) complex. Precise mRNA processing was reported to take place in the RNA exosome.

7.17.1 Posttranscriptional control and RNA binding proteins

Corbett (2018) emphasized the importance of posttranscriptional control of gene expression and concentrated particularly on RNA binding proteins and their complexing with mRNA was noted to be particularly important in interactions with the nuclear pore complex and export to the cytoplasm. Within the cytoplasm some forms of mRNA were noted to be stored in cytoplasmic bodies for future translation.

Corbett documented specific diseases due to altered posttranscriptional processing. One disease reviewed was myotonic dystrophy DM1 due to altered posttranslational regulation in DMPK that arises due to expanded trinucleotide repeats in the 3' untranslated region. This altered posttranscriptional regulation results though impaired interactions with an RNA binding protein MBNL1 that binds to expanded repeats.

A condition referred to as oculo-pharyngeal muscular dystrophy was reported to be due to defects in PABPN1, a polyA binding protein reported to control the length of the poly A tail in the transcript.

A specific mRNA export factor was reported to be defective in a condition referred to as Lethal Congenital Contracture syndrome LCCS1 that is associated with loss of motor neurons.

7.17.2 RNA modifications and regulation of gene expression

Frye et al. (2018) reported that RNA modifications act as important posttranscriptional regulators of gene expression. Important modification on mRNA includes methyl adenosine M^6A that is reported to impact the stability of transcripts. More than 170 modifications are known. More than 100 different modifications of tRNAs occur, a particularly important modification involved methylation. Modifications of ribosomes also occur at key sites.

RNA editing was reviewed by Christofi and Zaravinos (2019). They noted that posttranscriptional modifications and RNA editing were carried out primarily by adenosine deaminase and cytidine deaminase. Adenosine to inosine editing was noted to be important in regulation of alternate splicing and transcriptional control. Adenosine to inosine editing was noted to be carried out by ADAR (adenosine deaminase RNA specific) and ADAT (adenosine deaminase tRNA specific).

Cytidine to uridine editing is carried by members of the AID/APOBEC family of cytidine deaminases and was noted to play important roles in immune activity.

These authors noted that analyses of RNA editing might provide insights into new therapies.

AID also known as AICDA, RNA-editing deaminase that is a member of the cytidine deaminase family. AICDA is specifically expressed and active in germinal center-like B-cells. In the germinal center, AICDA is involved in somatic hypermutation, gene conversion, and class-switch recombination of immunoglobulin genes.

ABOBEC1 apolipoprotein B mRNA editing enzyme catalytic subunit 1 member of the cytidine deaminase enzyme family.

7.18 Translation

In a report in 2020 Lombardi et al. reviewed translation. They defined translation as a ribosome catalyzed process to pair specific codon in mRNA with cognate aminoacyl tRNA anticodons in a specific process that includes initiation, elongation, and termination. Termination of translation was known to be triggered by entry of stop codons UAA, UAG, or UGA into the ribosomal A site leading to recruitments of release factors eRF1, eRF3 (in human eukaryotic translation termination factor ETF1).

7.19 Nonsense-mediated decay

The NMD surveillance mechanism reported by Hug and Cáceres (2014) was noted to involve the following:

SMG1 NMD association phosphatidyl inositol related kinase, UPF1 RNA helicase and ATPase and translation termination factors including UPF2 and UPF3 regulators of NMD and DHX34 dead box helicase 34. Together these components were referred to as the DECID complex.

If mRNA with a nonsense codon escapes degradation, translation can be terminated during the translation process. Cognate tRNAs do not exist for termination codons UAA, UAG, UGA. These codons are recognized by release factors, in humans encoded by a class-1 polypeptide chain release factor. ERF1 also known as ETF1 in humans. The encoded protein plays an essential role in directing termination of mRNA translation from the termination codons UAA, UAG, and UGA

Kurosaki et al. (2019) reviewed quality and quantity of control of gene expression through NMD. They noted that NMD occurs primarily in mutated mRNA but that it may also occur as a mechanism of adaptation to cellular or environmental changes.

NMD was noted to be a quality control mechanism at the translation level that degrades mRNA, particularly mRNA with a premature termination codon. NMD can also be activated by mRNA with a particularly long 3′ untranslated region.

Kurosaki et al. also reviewed mRNA inspection processes to detect defects.

Karousis and Mühlemann (2019) reviewed NMD and noted that it was originally interpreted as a mechanism involved in degradation of mRNA in which premature termination codons were introduced. Currently, however, NMD is considered to be more broadly involved in posttranscriptional gene regulation.

They noted that there is evidence that translation termination selection processes occur regarding whether an mRNA should remain is use in additional rounds of translation or whether they should be degraded.

Recognition of translation signals results in release of the peptide from the ribosome and possibly also leads to dissociation of the 60s and 40s ribosome.

Kurosaki et al. (2019) noted that NMD processes are triggered by premature termination codons and other features including an abnormally long untranslated region.

Key to the NMD quality control pathways is UPF1, UPF1 RNA helicase, and ATPase multiprotein complex involved in both mRNA nuclear export and mRNA surveillance.

7.19.1 Suppression of nonsense mutations

Morais et al. (2020) reviewed new technologies designed to suppress nonsense mutations. These were defined as mutations leading to change of a sense code to a nonsense code or mutations that lead to a translation stop codon.

Morais et al. initiated their review by noting the 64 different nucleotide codons that exist that include 61 codons for amino acids and three codons that are designated as nonsense or stop codons that are reported to terminate generation of proteins, UAA, UAG, and UGA. These three codons were noted to promote mRNA decay or to lead to translation termination and termination of synthesis of the protein. They noted that in some cases a small section of the protein is synthesized generating a shortened protein that can potentially be harmful. They described details of the NMD process that is initiated when a premature termination codon is encountered.

Morais et al. noted that a large number of human genetic diseases result from nonsense mutations and that several approaches have been adopted to suppress nonsense mutation, including use of chemical compound gentamycin and ataluren that were noted to promote readthrough of premature termination codons. They also noted that new nucleic acid-based approaches are being explored to impact nonsense mutations. These include antisense oligonucleotides to downregulate components of the NMD.

Cripr-Cas gene editing was noted to being investigated to suppress NMD in specific genetic diseases involving the eye. One trial reported involved treatment of Leber's congenital amaurosis LCA10 through suppression of a mutation in CEP290 centrosomal protein 290.

7.20 Nonsense mutations and human disease

Nonsense mutations were reported in Morais et al., a review to account for more that 11% of inherited human diseases. Nonsense mutations included single nucleotide mutations that converted a sense mutation to stop codon. Premature termination of translation was noted to lead to decreased stability and degradation of the partially translated protein and its loss of function. In 5.25% of cases NMD did not occur leading to production of a truncated protein. Importantly, in autosomal dominant disorders the truncated protein was noted to potentially interfere with the function of the normal protein, referred to as dominant negative effects.

Lombard et al. described the use of medications to induce readthrough in cases with stop codon mutations leading to lysosomal storage diseases.

Translation terminating mutations have also been found in some cases with coagulation disorders. Compounds being investigated to suppress NMD decay include aminoglysosides PTC 124, ataluren, geneticin newer compounds gentamycin B1, and amlexanox.

Crispr-Cas technologies are being evaluated for treatment of disorders due to premature termination codons. Borgatti et al. (2020) reviewed screening of compounds to promote readthrough of premature termination codons for treatment of specific forms of thalassemia. They noted that PTC or nonsense mutations are generally associated with a dramatic decrease in gene expression.

Borgatti et al. noted other genetic disorders due to effects of premature termination codons. These include cystic fibrosis, Duchenne muscular dystrophy, spinal muscular atrophy, Usher syndrome with risk for deafness and visual impairment, ataxia telangiectasia. They noted that gene therapy therapeutic studies were primarily at the level of investigation in model organism but that there was some evidence of progression to clinical applications.

7.21 Approved RNA targeted therapeutics

Crooke et al. (2018) reviewed RNA targeted therapeutics. They noted that by 2018 four RNA therapeutics were approved for clinical use by the FDA. They had different approved administration routes including subcutaneous, intravitreal, or intrathecal. Medicinal chemistry adaptations of oligonucleotides to be used in therapy included promoting of affinity of nucleotides to target sequence, and resistance of oligonucleotides to degradation. Crooke et al. noted that more recent studies aimed at focusing RNA therapies to specific cell types, for example, through coupling to receptor moieties, for example, asialoglycoprotein receptors on hepatocytes for therapies targeted to liver.

Some RNA target nucleotide therapies were produced as single-stranded entities while others were double stranded. Double-stranded RNAs were noted to be more readily excreted and were therefore coupled to lipids or nanoparticles.

Other modifications also changed the properties of oligonucleotides. Crooke et al. documented oligonucleotide modification used to generate therapeutic agents. These included base modification, sugar modification, internucleotide linkage modification, and conjugates to dinucleotides.

Lipid nanoparticle modifications were noted to protect nucleic acids. N-acetyl galactosamine modifications were often used for liver targeted oligonucleotides. Single-stranded oligonucleotides were noted to preferentially be taken up by cells.

In general, single-stranded molecules with hydrophobic groups exposed and with phosphothiolates were noted to have good protein binding and tissue distribution.

Crooke et al. provided details on antisense mechanisms used to modulate gene expression. These included use of antisense RNAs that modulate gene expression by occupancy only methods or by triggering RNA cleavage by RNAse H. Antisense RNAs that function by the occupancy only process include splice modifiers and translation modifiers. Specific modification of antisense RNAs were found to be necessary to avoid cleavage, for example, through addition of morpholinos. For some antisense therapies oligonucleotides were designed to simulate cleavage by RNAse H.

Another therapeutic pathway includes inhibition of microRNAs by base pairing with microRNAs or by inhibiting the microRNA binding to mRNAs.

7.22 Therapy with short inhibitory RNAs

Other forms of RNA therapy investigated include use of short inhibitory RNAs also known as short interfering RNAs.

Small interfering RNAs are double-stranded RNAs 20 − 27 nucleotides in length that function within a specific pathway in the RNA interference pathway.

The RNA interference pathway was described by Fire et al. (1998). It occurs in multiple organisms and is thought to play an important role in regulation of gene expression. Both microRNAs and short inhibitory RNAs are involved. In eukaryotes this pathway includes long double-stranded RNAs with a guide strand and a passenger strand.

Hannon and Rossi (2004) published on the applications of RNA interference. They reported that RNAi had been adopted as a standard methodology for silencing gene expression. They also predicted that RNAi would become a potent therapeutic tool.

Long dsRNA and microRNAs were shown to be processed by Dicer an RNase II enzyme to generate shorter double-stranded RNAs that were then subsequently unwound and assembled into effector complexes designated RISCs. RISC complexes were shown to direct RNA cleavage and could bring about translational repression and induce chromatin modifications.

The RISC RNA-induced silencing complex has many components.

In RNA interference therapy effective delivery is a problem to overcome. The first FDA approval for use of an siRNA product was for Patirisan used to treat transthyretin-associated amyloidosis.

A number of investigators are considering prospect for RNAi therapy in cancer.

The allele-specific RNA interference approaches have been proposed to treat autosomal dominant disorders (Giorgio et al., 2019).

Trochet et al. (2015, 2018) reported the use of allele-specific silencing by RNA interference to treat diseases including autosomal dominant centronuclear myopathy caused by mutation in the DNM2 (dynamin2). They demonstrated successful use of this therapy in a mouse model of the disease and in human cells.

7.23 MicroRNAs in therapeutic use

Miller et al. (2020) reported promising results of phase 1 and phase 2 clinical trials of use of an antisense molecule to target superoxide dismutase (SOD1) microRNA to treat patients with amyotrophic lateral sclerosis.

Specific microRNAs can be targeted with blockers or with antibodies.

RNAi therapy was reviewed by Roberts et al. (2020). They noted that this therapy utilizes short inhibitory RNA modules with a duplex structure. One strand is referred to as the guide or antisense strand and is designed to couple with the target transcript. The other strand is referred to as the passenger strand. The siRNAs then guide a silencing complex, the RISC complex to the target transcript. The RISC complex can target siRNAs of microRNAs to complementary mRNA sequences. A key component of the multiprotein RISC complex is argonaute that then cleaves the mRNA.

7.24 RNA sequencing in diagnosis of genetic diseases

Byron et al. (2016) reviewed clinical applications of RNA sequencing and particularly focused on detection of rare transcripts, fusion transcripts, and allele-specific expression.

They noted that RNA sequencing can detect nucleotide changes and expression levels, and can be used to quantify expression and detect alternative transcripts and can reveal expression levels of different transcripts.

Byron et al. also noted challenges in translating RNA analysis into clinical practice.

Lee et al. (2020) reported investigations on the utility of RNA sequencing and transcriptome analysis in undiagnosed diseases thought to be Mendelian in origin. They specifically reviewed 234 cases referred to the undiagnosed Disease Network in Los Angeles.

They reported that in 30% of undiagnosed cases likely to be of Mendelian origin diagnosis was achieved based on exome sequencing. They noted further that genome sequencing achieved diagnosis in a further 3%–7% of cases through detection of structural genomic abnormalities.

Lee et al. reported results of RNA sequencing in 48 families. They also noted integration of RNA sequencing with genomic sequencing in some of these cases. Through these studies the diagnostic rate was increased to 38%. In seven cases there were splice region variants and in four cases deep intronic variants caused alteration in splicing, exon skipping, or intron inclusion.

7.25 Penetrance of mutations and modified penetrance

Modified penetrance of coding variants may be in part due to variants in regulatory elements.

Castel et al. (2018) noted that variable penetrance was a feature of many Mendelian disorders. They utilized the term variable penetrance to describe variable expressivity of the phenotype and noted that the term penetrance can also refer to the proportion of individuals with a particular pathogenic variant who manifest a specific phenotype.

Evidence for the role of the regulatory genome in penetrance comes in part from evidence of sequence variants or structural variants in the regulatory region linked to genes involved in Mendelian disorders.

Castel et al. noted that even significant pathogenic variants can be associated with intraindividual differences in phenotype severity, referred to as variable penetrance.

They undertook analyses of variant penetrance in the context of specific haplotype configurations. Their studies revealed definitive evidence of specific haplotype configuration that impact phenotype penetrance. In addition, they carried out gene editing to determine the impact of nucleotide changes in haplotypes.

Included in their analyses were probands in the Deciphering Developmental Disorders Study in the United Kingdom. Their study also involved analyses in individuals in this cohort who were not found to have abnormalities on exome sequencing; in these individuals, specific regulatory elements were analyzed. They included evolutionarily conserved noncoding elements (CNEs), validated enhancers, putative heart enhancers, control intronic elements.

CADD metrics were utilized to assess the likely pathogenic impact of pathogenic variants in coding sequences cases where these were identified.

The investigators also analyzed DNAse 1 hypersensitivity sites.

Regarding evolutionarily CNEs, 93% of these were shown to be active in at least 1 of the 111 tissues analyzed. In cases studied versus controls the CNEs manifested an increased number of de novo mutations. An excess of variants defined as fetal active CNEs were identified in the exome cases with neurodevelopmental disorders.

Between 50% and 70% of brain active CNEs were found to act as enhancers, and target genes were identified for 28% of CNEs.

Castel et al. also analyzed the impact of de novo mutation in 45 transcription factor binding sites that were associated with CNE. They also carried out analyses of the chromatin state of predicted evolutionarily CNEs.

Castel et al. emphasized that both coding and regulatory elements can influence penetrance. They studied haplotypes associated with specific genes in which mutations occurred. Their studies revealed that deleterious coding mutations association with specific disease phenotype were often found to be associated with aberrant regulatory haplotypes.

Castel et al. concluded that de novo mutations in regulatory elements contribute to severe neurodevelopmental disorders.

7.26 Variable penetrance of disease due to polymorphisms in regulatory factors

In 2007 Wilkins et al. noted that differential expression of alleles could arise due to polymorphisms in cis regulatory elements. Lee et al. (2019) reported that in some cases preferential decreased expression of a mutant allele through action of cis acting regulatory elements could influence degree of phenotype abnormality.

Cooper et al. (2013) reviewed the molecular basis for reduced penetrance in human inherited diseases. They noted that a combination of genetic and environmental factors influence penetrance. Penetrance measures involved assessment of the measure of number of individuals with a specific disease genotype that express the disease phenotype.

They noted that reduced penetrance was most frequently seen in autosomal dominant diseases.

Difference in penetrance of a specific gene mutation was noted to present significant problems in genetic counseling.

Cooper et al. noted that although penetrance and expressivity are defined as distinct terms, they are related. Within the context of penetrance, modifier genes must also be considered.

They noted that more widespread genome sequencing had revealed that some individuals harbor disadvantageous variants and do not have phenotypic features associated with those specific genomic variants.

Reduced or incomplete penetrance was also noted in specific autosomal recessive conditions, for example, in HFE (human homeostatic iron regulator) mutations leads to hemochromatosis. Clearly gene environment interactions in the form of iron intake versus iron loss come into play here.

In considering variable penetrance it is now more common to take into account the phenotype resulting from a specific sequence variant in a particular gene.

Variable penetrance in Gaucher disease-associated mutations was noted with respect to specific mutations in the glucocerebrosidase gene GBA. There is evidence that modifier genes impact the penetrance of specific GBA mutations (Schierding et al., 2020).

In addition, some variants lead to disease symptoms only later in life and present with a Parkinson disease-like phenotype, for example, GBA p.E326K (Mata et al., 2016).

Cooper et al. emphasized that modifying genomic elements can influence the effects of a specific variant in a particular gene and these modifiers may be in cis act by influencing gene expression.

Differential allelic expression may also impact the disease phenotype. In dominant disorders increased expression of the normal gene relevant to the mutant gene will mitigate effects of the disease-causing variants.

Cooper et al. emphasized that modifying genomic elements can influence the effects of a specific variant in a particular gene and these modifiers may be in cis acting processes by influencing gene expression.

Clinical penetrance can also be influenced by copy number variants or either the disease allele or the normal allele.

7.27 Penetrance in inherited eye diseases

Green et al. (2020) reported incomplete penetrance in specific inherited eye disease and noted their association with variable gene expression. In their analyses they utilized genomic and transcriptomic data sets and information on 340 genes implicated in inherited eye diseases. Results of their analyses revealed that eye diseases with variable penetrance tended to have more variable levels of expressivity than is usually found. These results led them to propose that cis or trans acting elements that impact levels of gene expression influence penetrance in these diseases.

In the study by Green et al., pathogenic variants with CADD scores greater than 15 were identified in genome aggregation data in the human gene mutation database. Data on variability in levels of gene expression and mRNA data were examined for inherited eye disease-related genes.

Together these data analyses revealed that in general the genes associated with inherited eye disease and that manifest variable penetrance have variable expression in the general population. They noted that this variable expressivity could be due to cis or trans acting factors or due to modifier genes.

7.28 Primary immunodeficiency and incomplete penetrance

In a 2020 review Gruber and Bogunovic (2020) noted that there are more than 400 genetic disorders that manifest the phenotypes of immunodeficiency. Evidence for important factors that altered penetrance included epigenetic modifications, environmental influences, and genetic mosaicism.

They noted that much progress has been made in understanding the molecular basis of immunodeficiency. In 2020 Tangye et al. (2020), in reviewing inborn errors of immunity, documented different classes of these disorders.

1. Combined immunodeficiency with cellular and humoral impairments.
2. Predominantly antibody deficiency.
3. Immune deregulation diseases.
4. Congenital defects of phagocytes.
5. Defects in intrinsic and innate immunity.
6. Auto-inflammatory disorders.
7. Complement deficiency.
8. Bone marrow failure.

In their review of immunodeficiencies, Gruber and Bogunovic noted that an important distinction should be made between clinical penetrance and cellular penetrance where cellular penetrance more directly reflects impact on cellular or biological processes.

In one form of immunodeficiency due to pathogenic mutation in the gene IFNGR1 that encodes interferon gamma receptor 1, susceptibility to mycobacterial infection is impaired. In autosomal recessive pathogenic mutation mycobacterial infection due to environmental exposure of BCG vaccination can develop in early childhood. In the presence of autosomal dominant forms mycobacterial infections may occur later in life or may never develop.

In exploring altered penetrance in immunodeficiency disorders genetic modifiers and epigenetic factors were noted to often be involved. In common, variable immunodeficiency differences in clinical manifestations in monozygotic twins were found and these twins differed in the degree of methylation of the genes BCL2L1, a pro-apoptotic regulator, PIKCD phosphatidylinositol-4,5-bisphosphate 3-kinase catalytic subunit delta that is involved in immune response, TCF3 a transcription factor that plays a role in lymphopoiesis, and KCNN4 potassium calcium-activated channel subfamily N member 4 active in lymphocytes.

Gruber and Bogunovic also noted that some forms of combined variable immunodeficiency are reported to likely involve several genes that potentially have epistatic interactions.

Differences in environments and in the exposome were also noted to impact clinical presentations in specific immunodeficiencies. Somatic mutations have also been determined to influence severity in severe combined immunodeficiency due to adenosine deaminase deficiency.

References

Allen BL, Taatjes DJ: The mediator complex: a central integrator of transcription, *Nat Rev Mol Cell Biol* 16(3):155−166. Available from: https://doi.org/10.1038/nrm3951, 2015.

An JY, Lin K, Zhu L, Werling DM, et al: Genome-wide de novo risk score implicates promoter variation in autism spectrum disorder, *Science* 362(6420):eaat6576, 2018. Available from: https://doi.org/10.1126/science.aat6576. PMID: 30545852.

Baek S, Lee I: Single-cell ATAC sequencing analysis: from data preprocessing to hypothesis generation, *Comput Struct Biotechnol J* 18:1429−1439, 2020. Available from: https://doi.org/10.1016/j.csbj.2020.06.012. eCollection 2020. PMID: 32637041.

Bartel DP: MicroRNAs: genomics, biogenesis, mechanism, and function, *Cell* 116 (2):281−297, 2004. Available from: https://doi.org/10.1016/s0092-8674(04)00045-5. PMID: 14744438.

Bhatia S, Kleinjan DA: Disruption of long-range gene regulation in human genetic disease: a kaleidoscope of general principles, diverse mechanisms and unique phenotypic consequences, *Hum Genet* 133(7):815−845, 2014. Available from: https://doi.org/10.1007/s00439-014-1424-6. PMID: 24496500.

Bjornsson HT: The Mendelian disorders of the epigenetic machinery, *Genome Res* 25 (10):1473−1481, 2015. Available from: https://doi.org/10.1101/gr.190629.115. PMID: 26430157.

Boix CA, James BT, Park YP, et al: Regulatory genomic circuitry of human disease loci by integrative epigenomics, *Nature* 590(7845):300−307. Available from: https://doi.org/10.1038/s41586-020-03145-z, 2021.

Borgatti M, Altamura E, Salvatori F, et al: Screening readthrough compounds to suppress nonsense mutations: possible application to β-thalassemia, *J Clin Med* 9(2):289, 2020. Available from: https://doi.org/10.3390/jcm9020289. PMID: 31972957.

Borrelli E, Nestler EJ, Allis CD, Sassone-Corsi P: Decoding the epigenetic language of neuronal plasticity, *Neuron* 60(6):961−974, 2008. Available from: https://doi.org/10.1016/j.neuron.2008.10.012. PMID: 19109904.

Byron SA, Van Keuren-Jensen KR, Engelthaler DM, et al: Translating RNA sequencing into clinical diagnostics: opportunities and challenges, *Nat Rev Genet* 17(5):257−271, 2016. Available from: https://doi.org/10.1038/nrg.2016.10. Epub 2016 Mar 21. PMID: 26996076.

Calpena E, Hervieu A, Kaserer T, et al: De novo missense substitutions in the gene encoding CDK8, a regulator of the mediator complex, cause a syndromic developmental disorder, *Am J Hum Genet* 104(4):709−720, 2019. Available from: https://doi.org/10.1016/j.ajhg.2019.02.006. PMID: 30905399.

Castel SE, Cervera A, Mohammadi P, et al: Modified penetrance of coding variants by cis-regulatory variation contributes to disease risk, *Nat Genet* 50(9):1327−1334. Available from: https://doi.org/10.1038/s41588-018-0192-y, 2018.

Cautain B, Hill R, de Pedro N, Link W: Components and regulation of nuclear transport processes, *FEBS J* 282(3):445–462, 2015. Available from: https://doi.org/10.1111/febs.13163. Epub 2014 Dec 22. PMID: 25429850.

Christofi T, Zaravinos A: RNA editing in the forefront of epitranscriptomics and human health, *J Transl Med* 17(1):319, 2019. Available from: https://doi.org/10.1186/s12967-019-2071-4. PMID: 31547885.

Cooper DN, Krawczak M, Polychronakos C, et al: Where genotype is not predictive of phenotype: towards an understanding of the molecular basis of reduced penetrance in human inherited disease, *Hum Genet* 132(10):1077–1130. Available from: https://doi.org/10.1007/s00439-013-1331-2, 2013.

Corbett AH: Post-transcriptional regulation of gene expression and human disease, *Curr Opin Cell Biol* 52:96–104, 2018. Available from: https://doi.org/10.1016/j.ceb.2018.02.011. Epub March 6, 2018. PMID: 29518673.

Crooke ST, Witztum JL, Bennett CF, Baker BF: RNA-targeted therapeutics, *Cell Metab* 27(4):714–739, 2018. Available from: https://doi.org/10.1016/j.cmet.2018.03.004. PMID: 29617640.

Cummings BB, Karczewski KJ, Kosmicki JA, et al: Transcript expression-aware annotation improves rare variant interpretation, *Nature* 581(7809):452–458. Available from: https://doi.org/10.1038/s41586-020-2329-2, 2020.

Elkon R, Agami R: Characterization of noncoding regulatory DNA in the human genome, *Nat Biotechnol* 35(8):732–746, 2017. Available from: https://doi.org/10.1038/nbt.3863. PMID: 28787426.

Fire A, Xu S, Montgomery MK, Kostas SA, Driver SE, Mello CC: Potent and specific genetic interference by double-stranded RNA in *Caenorhabditis elegans*, *Nature* 391(6669):806–811, 1998. Available from: https://doi.org/10.1038/35888. PMID: 9486653.

Frye M, Harada BT, Behm M, He C: RNA modifications modulate gene expression during development, *Science* 361(6409):1346–1349, 2018. Available from: https://doi.org/10.1126/science.aau1646. PMID: 30262497.

Fuchs SA, Schene IF, Kok G: Aminoacyl-tRNA synthetase deficiencies in search of common themes, *Genet Med* 21(2):319–330, 2019. Available from: https://doi.org/10.1038/s41436-018-0048-y. Epub June 6, 2018. PMID: 29875423.

Furlanis E, Scheiffele P: Regulation of neuronal differentiation, function, and plasticity by alternative splicing, *Annu Rev Cell Dev Biol* 34:451–469. Available from: https://doi.org/10.1146/annurev-cellbio-100617-062826, 2018.

Gabut M, Bourdelais F, Durand S: Ribosome and translational control in stem cells, *Cells* 9(2):497, 2020. Available from: https://doi.org/10.3390/cells9020497. PMID: 32098201.

Gamazon ER, Stranger BE: Genomics of alternative splicing: evolution, development and pathophysiology, *Hum Genet* 133(6):679–687, 2014. Available from: https://doi.org/10.1007/s00439-013-1411-3. PMID: 24378600.

Gasperini M, Hill AJ, McFaline-Figueroa J, et al: A genome-wide framework for mapping gene regulation via cellular genetic screens, *Nat Methods* 15(4):271–274. Available from: https://doi.org/10.1038/nmeth.4604, 2018.

Gazzo A, Raimondi D, Daneels D, et al: Understanding mutational effects in digenic diseases, *Nucleic Acids Res* 45(15):e140, 2017. Available from: https://doi.org/10.1093/nar/gkx557. PMID: 28911095.

Gil N, Ulitsky I: Regulation of gene expression by cis-acting long non-coding RNAs, *Nat Rev Genet* 21(2):102–117, 2020. Available from: https://doi.org/10.1038/s41576-019-0184-5. PMID: 31729473.

Giorgio E, Lorenzati M, Rivetti di Val Cervo P, et al: Allele-specific silencing as treatment for gene duplication disorders: proof-of-principle in autosomal dominant leukodystrophy, *Brain* 142(7):1905–1920, 2019. Available from: https://doi.org/10.1093/brain/awz139. PMID: 31143934.

Green DJ, Sallah SR, Ellingford JM, et al: Variability in gene expression is associated with incomplete penetrance in inherited eye disorders, *Genes (Basel)* 11(2):179, 2020. Available from: https://doi.org/10.3390/genes11020179. PMID: 32050448.

Gruber AJ, Zavolan M: Alternative cleavage and polyadenylation in health and disease, *Nat Rev Genet* 20(10):599–614, 2019. Available from: https://doi.org/10.1038/s41576-019-0145-z. PMID: 31267064.

Gruber C, Bogunovic D: Incomplete penetrance in primary immunodeficiency: a skeleton in the closet, *Hum Genet* 139(6–7):745–757, 2020. Available from: https://doi.org/10.1007/s00439-020-02131-9. Epub February 17, 2020. PMID: 32067110.

Haberle V, Stark A: Eukaryotic core promoters and the functional basis of transcription initiation, *Nat Rev Mol Cell Biol* 19(10):621–637, 2018. Available from: https://doi.org/10.1038/s41580-018-0028-8. PMID: 29946135.

Hannon GJ, Rossi JJ: Unlocking the potential of the human genome with RNA interference, *Nature* 431(7006):371–378, 2004. Available from: https://doi.org/10.1038/nature02870. PMID: 15372045.

Hoffmann A, Spengler D: Chromatin remodeler CHD8 in autism and brain development, *J Clin Med* 10(2):366, 2021. Available from: https://doi.org/10.3390/jcm10020366. PMID: 33477995.

Hug N, Cáceres JF: The RNA helicase DHX34 activates NMD by promoting a transition from the surveillance to the decay-inducing complex, *Cell Rep* 8(6):1845–1856, 2014. Available from: https://doi.org/10.1016/j.celrep.2014.08.020. PMID: 25220460.

Irimia M, Weatheritt RJ, Ellis JD, Parikshak NN, et al: A highly conserved program of neuronal microexons is misregulated in autistic brains, *Cell* 159(7):1511–1523, 2014. Available from: https://doi.org/10.1016/j.cell.2014.11.035. PMID: 25525873.

Jang YJ, LaBella AL, Feeney TP, et al: Disease-causing mutations in the promoter and enhancer of the ornithine transcarbamylase gene, *Hum Mutat* 39(4):527–536, 2018. Available from: https://doi.org/10.1002/humu.23394. PMID: 2928279.

Kapur M, Ackerman SL: mRNA translation gone awry: translation fidelity and neurological disease, *Trends Genet* 34(3):218–231. Available from: https://doi.org/10.1016/j.tig.2017.12.007, 2018.

Karnuta JM, Scacheri PC: Enhancers: bridging the gap between gene control and human disease, *Hum Mol Genet* 27(R2):R219–R227, 2018. Available from: https://doi.org/10.1093/hmg/ddy167. PMID: 29726898.

Karousis ED, Mühlemann O: Nonsense-mediated mRNA decay begins where translation ends, *Cold Spring Harb Perspect Biol* 11(2):a032862, 2019. Available from: https://doi.org/10.1101/cshperspect.a032862. PMID: 29891560.

Kircher M, Xiong C, Martin B, et al: Saturation mutagenesis of twenty disease-associated regulatory elements at single base-pair resolution, *Nat Commun* 10(1):3583, 2019. Available from: https://doi.org/10.1038/s41467-019-11526-w. PMID: 31395865.

Klemm SL, Shipony Z, Greenleaf WJ: Chromatin accessibility and the regulatory epigenome, *Nat Rev Genet* 20(4):207–220, 2019. Available from: https://doi.org/10.1038/s41576-018-0089-8. PMID: 30675018.

Kline AD, Moss JF, Selicorni A, et al: Diagnosis and management of Cornelia de Lange syndrome: first international consensus statement, *Nat Rev Genet* 19(10):649–666, 2018. Available from: https://doi.org/10.1038/s41576-018-0031-0. PMID: 29995837.

Kobayashi Y, Momoi MY, Tominaga K, Momoi T, et al: A point mutation in the mitochondrial tRNA(Leu)(UUR) gene in MELAS (mitochondrial myopathy, encephalopathy, lactic acidosis and stroke-like episodes), *Biochem Biophys Res Commun* 173(3):816–822, 1990. Available from: https://doi.org/10.1016/s0006-291x(05)80860-5. PMID: 2268345.

Kunkel GR, Lisciandro HG, Winter HL: The human chd8 gene is transcribed from two distant upstream promoters, *Biochem Biophys Res Commun* 532(2):190–194. Available from: https://doi.org/10.1016/j.bbrc.2020.08.051, 2020.

Kurosaki T, Popp MW, Maquat LE: Quality and quantity control of gene expression by nonsense-mediated mRNA decay, *Nat Rev Mol Cell Biol* 20(7):406–420, 2019. Available from: https://doi.org/10.1038/s41580-019-0126-2. PMID: 30992545.

Lambert SA, Jolma A, Campitelli LF, et al: The human transcription factors, *Cell* 172(4):650–665, 2018. Available from: https://doi.org/10.1016/j.cell.2018.01.029. PMID: 29425488.

Lant JT, Berg MD, Heinemann IU, et al: Pathways to disease from natural variations in human cytoplasmic tRNAs, *J Biol Chem* 294(14):5294–5308, 2019. Available from: https://doi.org/10.1074/jbc.REV118.002982. Epub January 14, 2019. PMID: 30643023.

Lee C, Kang EY, Gandal MJ, et al: Profiling allele-specific gene expression in brains from individuals with autism spectrum disorder reveals preferential minor allele usage, *Nat Neurosci* 22(9):1521–1532. Available from: https://doi.org/10.1038/s41593-019-0461-9, 2019.

Lee H, Huang AY, Wang LK, et al: Diagnostic utility of transcriptome sequencing for rare Mendelian diseases, *Genet Med* 22(3):490–499. Available from: https://doi.org/10.1038/s41436-019-0672-1, 2020.

Lombardi S, Testa MF, Pinotti M, Branchini A: Molecular insights into determinants of translational readthrough and implications for nonsense suppression approaches, *Int J Mol Sci* 21(24):9449, 2020. Available from: https://doi.org/10.3390/ijms21249449. PMID: 33322589.

Long HK, Prescott SL, Wysocka J: Ever-changing landscapes: transcriptional enhancers in development and evolution, *Cell* 167(5):1170–1187, 2016. Available from: https://doi.org/10.1016/j.cell.2016.09.018. PMID: 27863239.

Lupiáñez DG, Spielmann M, Mundlos S: Breaking TADs: how alterations of chromatin domains result in disease, *Trends Genet* 32(4):225–237, 2016. Available from: https://doi.org/10.1016/j.tig.2016.01.003. Epub February 7, 2016. PMID: 26862051.

Mata IF, Leverenz JB, Weintraub D, et al: GBA variants are associated with a distinct pattern of cognitive deficits in Parkinson's disease, *Mov Disord* 31(1):95–102. Available from: https://doi.org/10.1002/mds.26359, 2016.

Matera AG, Wang Z: A day in the life of the spliceosome, *Nat Rev Mol Cell Biol* 15(2):108–121, 2014. Available from: https://doi.org/10.1038/nrm3742. PMID: 24452469.

McDonel P, Guttman M: Approaches for understanding the mechanisms of long noncoding RNA regulation of gene expression, *Cold Spring Harb Perspect Biol* 11(12):a032151, 2019. Available from: https://doi.org/10.1101/cshperspect.a032151. PMID: 31791999.

McLachlan F, Sires AM, Abbott CM: The role of translation elongation factor eEF1 subunits in neurodevelopmental disorders, *Hum Mutat* 40(2):131−141, 2019. Available from: https://doi.org/10.1002/humu.23677. PMID: 30370994.

Miller T, Cudkowicz M, Shaw PJ, et al: Phase 1−2 trial of antisense oligonucleotide tofersen for SOD1 ALS, *N Engl J Med* 383(2):109−119, 2020. Available from: https://doi.org/10.1056/NEJMoa2003715. PMID: 32640130.

Montes M, Sanford BL, Comiskey DF, Chandler DS: RNA splicing and disease: animal models to therapies, *Trends Genet* 35(1):68−87, 2019. Available from: https://doi.org/10.1016/j.tig.2018.10.002. Epub November 19, 2018. PMID: 30466729.

Morais P, Adachi H, Yu YT: Suppression of nonsense mutations by new emerging technologies, *Int J Mol Sci* 21(12):4394, 2020. Available from: https://doi.org/10.3390/ijms21124394. PMID: 32575694.

Musier-Forsyth K: Aminoacyl-tRNA synthetases and tRNAs in human disease: an introduction to the JBC reviews thematic series, *J Biol Chem* 294(14):5292−5293. Available from: https://doi.org/10.1074/jbc.REV119.007721, 2019.

Nizon M, Laugel V, Flanigan KM, et al: Variants in MED12L, encoding a subunit of the mediator kinase module, are responsible for intellectual disability associated with transcriptional defect, *Genet Med* 21(12):2713−2722. Available from: https://doi.org/10.1038/s41436-019-0557-3, 2019.

Orgebin E, Lamoureux F, Isidor B, et al: Ribosomopathies: new therapeutic perspectives, *Cells* 9(9):2080, 2020. Available from: https://doi.org/10.3390/cells9092080. PMID: 32932838.

Parras A, Anta H, Santos-Galindo M, et al: Autism-like phenotype and risk gene mRNA deadenylation by CPEB4 mis-splicing, *Nature* 560(7719):441−446, 2018. Available from: https://doi.org/10.1038/s41586-018-0423-5. Epub August 15, 2018. PMID: 30111840.

Perenthaler E, Yousefi S, Niggl E, Barakat TS: Beyond the exome: the non-coding genome and enhancers in neurodevelopmental disorders and malformations of cortical development, *Front Cell Neurosci* 13:352, 2019. Available from: https://doi.org/10.3389/fncel.2019.00352. eCollection 2019. PMID: 31417368.

Quesnel-Vallières M, Dargaei Z, Irimia M, et al: Misregulation of an activity-dependent splicing network as a common mechanism underlying autism spectrum disorders, *Mol Cell* 64(6):1023−1034, 2016. Available from: https://doi.org/10.1016/j.molcel.2016.11.033. PMID: 27984743.

Rak R, Dahan O, Pilpel Y: Repertoires of tRNAs: the couplers of genomics and proteomics, *Annu Rev Cell Dev Biol* 34:239−264. Available from: https://doi.org/10.1146/annurev-cellbio-100617-062754, 2018.

Reines D: Recent advances in understanding RNA polymerase II structure and function, *Fac Rev* 9:11, 2020. Available from: https://doi.org/10.12703/b/9-11. eCollection 2020. PMID: 33659943.

Roberts TC, Langer R, Wood MJA: Advances in oligonucleotide drug delivery, *Nat Rev Drug Discov* 19(10):673−694, 2020. Available from: https://doi.org/10.1038/s41573-020-0075-7. PMID: 32782413.

Rubio Gomez MA, Ibba M: Aminoacyl-tRNA synthetases, *RNA* 26(8):910–936, 2020. Available from: https://doi.org/10.1261/rna.071720.119. Epub April 17, 2020. PMID: 32303649.

Sanders SJ, He X, Willsey AJ, et al: Insights into autism spectrum disorder genomic architecture and biology from 71 risk loci, *Neuron* 87(6):1215–1233, 2015. Available from: https://doi.org/10.1016/j.neuron.2015.09.016. PMID: 26402605.

Schacherer J: Beyond the simplicity of Mendelian inheritance, *C R Biol* 339(7–8):284–288. Available from: https://doi.org/10.1016/j.crvi.2016.04.006, 2016.

Schaffer AE, Pinkard O, Coller JM: tRNA metabolism and neurodevelopmental disorders, *Annu Rev Genomics Hum Genet* 20:359–387, 2019. Available from: https://doi.org/10.1146/annurev-genom-083118-015334. Epub May 13, 2019. PMID: 31082281.

Schierding W, Farrow S, Fadason T, et al.: Common variants coregulate expression of GBA and modifier genes to delay Parkinson's disease onset, *Mov Disord* 35(8):1346-1356, 2020. doi: 10.1002/mds.28144. PMID: 32557794.

Schvartzman JM, Thompson CB, Finley LWS: Metabolic regulation of chromatin modifications and gene expression, *J Cell Biol* 217(7):2247–2259. Available from: https://doi.org/10.1083/jcb.201803061, 2018.

Shashikant T, Ettensohn CA: Genome-wide analysis of chromatin accessibility using ATAC-seq, *Methods Cell Biol* 151:219–235, 2019. Available from: https://doi.org/10.1016/bs.mcb.2018.11.002. PMID: 30948010.

Shi Y: The spliceosome: a protein-directed metalloribozyme, *J Mol Biol* 429(17):2640–2653. Available from: https://doi.org/10.1016/j.jmb.2017.07.010, 2017.

Signor SA, Nuzhdin SV: The evolution of gene expression in cis and trans, *Trends Genet* 34(7):532–544. Available from: https://doi.org/10.1016/j.tig.2018.03.007, 2018.

Son EY, Crabtree GR: The role of BAF (mSWI/SNF) complexes in mammalian neural development, *Am J Med Genet C Semin Med* 166C(3):333–349. Available from: https://doi.org/10.1002/ajmg.c.31416, 2014.

Tangye SG, Al-Herz W, Bousfiha A, et al: Human inborn errors of immunity: 2019 update on the classification from the International Union of Immunological Societies Expert Committee, *J Clin Immunol* 40(1):24–64, 2020. Available from: https://doi.org/10.1007/s10875-019-00737-x. Epub January 17, 2020. PMID: 31953710.

The Treacher Collins Syndrome Collaborative Group: Positional cloning of a gene involved in the pathogenesis of Treacher Collins syndrome., *Nat Genet* 12(2):130–136. Available from: https://doi.org/10.1038/ng0296-130, 2022.

Tian B, Manley JL: Alternative polyadenylation of mRNA precursors, *Nat Rev Mol Cell Biol* 18(1):18–30, 2017. Available from: https://doi.org/10.1038/nrm.2016.116. Epub September 28, 2016. PMID: 27677860.

Trochet D, Prudhon B, Beuvin M, et al: Allele-specific silencing therapy for Dynamin 2-related dominant centronuclear myopathy, *EMBO Mol Med* 10(2):239–253, 2018. Available from: https://doi.org/10.15252/emmm.201707988. PMID: 29246969.

Trochet D, Prudhon B, Vassilopoulos S, Bitoun M: Therapy for dominant inherited diseases by allele-specific RNA interference: successes and pitfalls, *Curr Gene Ther* 15(5):503–510, 2015. Available from: https://doi.org/10.2174/1566523215666150812115730. PMID: 26264709.

Ulirsch JC, Verboon JM, Kazerounian S, et al: The genetic landscape of diamond-blackfan anemia, *Am J Hum Genet* 103(6):930–947, 2018. Available from: https://doi.org/10.1016/j.ajhg.2018.10.027. Epub November 29, 2018. PMID: 30503522.

von Walden F: Ribosome biogenesis in skeletal muscle: coordination of transcription and translation, *J Appl Physiol* 127(2):591–598, 2019. Available from: https://doi.org/10.1152/japplphysiol.00963.2018. (1985).

Weyn-Vanhentenryck SM, Zhang C: mCarts: genome-wide prediction of clustered sequence motifs as binding sites for RNA-binding proteins, *Methods Mol Biol* 1421:215–226, 2016. Available from: https://doi.org/10.1007/978-1-4939-3591-8_17. PMID: 26965268.

Wilkins JM, Southam L, Price AJ, et al: Extreme context specificity in differential allelic expression, *Hum Mol Genet* 16(5):537–546, 2007. Available from: https://doi.org/10.1093/hmg/ddl488. Epub January 12, 2007. PMID: 17220169.

Will CL, Lührmann R: Spliceosome structure and function, *Cold Spring Harb Perspect Biol* 3(7):a003707, 2011. Available from: https://doi.org/10.1101/cshperspect.a003707. PMID: 21441581.

Winokur ST, Shiang R: The Treacher Collins syndrome (TCOF1) gene product, treacle, is targeted to the nucleolus by signals in its C-terminus, *Hum Mol Genet* 7(12):1947–1952, 1998. Available from: https://doi.org/10.1093/hmg/7.12.1947. PMID: 9811939.

Wolin SL, Marquat LE: Cellular RNA surveillance in health and disease, *Science* 366 (6467):822–827. Available from: https://doi.org/10.1126/science.aax2957, 2019.

Zaghloul NA, Katsanis N: Mechanistic insights into Bardet-Biedl syndrome, a model ciliopathy, *J Clin Invest* 119(3):428–437. Available from: https://doi.org/10.1172/JCI37041, 2009.

Zhang F, Lupski JR: Non-coding genetic variants in human disease, *Hum Mol Genet* 24(R1):R102–R110. Available from: https://doi.org/10.1093/hmg/ddv259, 2015.

CHAPTER 8

Standardized phenotype documentation, documentation of genotype phenotype correlations

In 2003 Freimer and Sabati (Freimer and Sabati, 2003) drew attention to the limitations of the then current descriptions of genetic disease phenotypes and proposed tasks for a potential phenome project. They noted that phenotype descriptions could be included at different levels: organism, tissue, cellular, and molecular, and they noted further that inclusion of quantitative measures would be useful.

Freimer and Sabati also proposed gathering of comparative phenomics that would include relevant information from other species and potential model organisms.

They also noted that information on the impacts of environmental factors on specific phenotypes would be useful.

Ambitions for the Human Phenome Project were later included in the Human Variome Project (Oettin et al., 2013). Goals included establishment of standardized terms for phenotypic abnormalities and attention was also paid to making data and computational resources and analyses accessible. The goals of the Variome Project were stated to also include gathering information on the natural history of genetic diseases and the spectrum of complications.

8.1 Phenotype and clinical genetics

Victor McKusick (1993) defined three cardinal principals of clinical genetics: pleiotropism (pleiotropy), genetic heterogeneity, and variability.

Pleiotropism was noted to refer to the multiple phenotypic manifestations that can result from a single gene defect.

Genetic heterogeneity refers to a single phenotype resulting from abnormalities in different genes.

Variability referred to differences in phenotype in different individuals with the same genetic changes.

McKusick noted that linkage analysis could lead to identification of genetic heterogeneity. He noted that variability in the phenotypic features and expressivity of a specific mutant gene could sometimes be related to the impact of modifying genes.

Clinical syndromes can be defined as disorders with specific combination of phenotypic features and/or biochemical and metabolic alterations due to specific gene defects.

Over the years much information has been gathered on phenotypic features associated with specific inborn errors of metabolism. These include documentation of systems associated with lysosomal storage disease, peroxisomal disorders, and mitochondrial disorders.

In addition to documentation of phenotypic features assessed on physical examinations and on imaging for example, it is important to note that blood and or urine chemical studies can also provide important clues into disease diagnosis. Key expansions of knowledge in concerning the range of manifestations of metabolic disorders came in part through several editions of the "Metabolic Basis of Inherited Disease" initially by Stanbury, Wyngaarden, and Fredrickson, first published in 1972 (Stanbury et al., 1972).

Expertise has also expanded on the range of manifestations of neurological diseases and increased technology to measure abnormal muscle tone abnormal movements and seizure disorders. Specific neurological disorders of genetic origin are sometimes suspected based on the presence of specific neurologic manifestations.

8.2 Congenital malformations and syndromes

Congenital malformations can be defined as phenotypic manifestation of abnormal gene functions during development.

Clinical syndromes can be defined as disorders with specific combination of phenotypic features and/or biochemical and metabolic alterations due to specific gene defects.

Specific syndromes may include constellation of manifestations due to defects in the functions of components in a specific pathway. Early links were also postulated between disorders associated with connective tissue abnormalities.

8.2.1 Inborn errors of development

In "Inborn errors of development" Epstein and Erickson (2004) core developmental pathways associated with malformations were reviewed. These pathways included specific signaling pathways, sonic hedgehog signaling, WNT signaling, transforming growth factor beta (TGFbeta), fibroblast growth factor pathway (FGF/FGFR, RAS/MAP/ERK signaling pathway). Current information on defects in these signaling pathways will be discussed later in this chapter.

In addition, defects in genes linked to specific physiological processes were noted to give rise to congenital defects with overlapping phenotypic manifestations, for example, genes encoding products involved in ciliary and microtubule function, genes encoding products involved in vesicle trafficking and endocytosis.

8.2.2 Twin studies and analysis of gene effects on phenotype

Traditional suppositions were that monozygotic twins should have identical phenotypes and be closely similar in behaviors. Cardno and Gottesman (2000) reported that in monozygotic twins the concordance rates for schizophrenia across different studies ranged between 41% and 65%.

Studies on the concordance rates for autism in twins were reported by Rosenberg et al. (2009) and were reported to be 88% in monozygotic twins and 31% in dizygotic twins.

The differences observed, lower concordance rates in monozygotic twins with schizophrenia than in monozygotic twins with autism, were considered to be likely due to the fact that autism was due to factors acting early in development while causative factors in schizophrenia may arise incrementally over longer time periods.

8.2.3 Accounting for phenotypic differences in individuals with the same genetic defect

Increasing insights into factors involved in regulation of gene expression provide possible explanations for phenotypic differences that occur in individuals with the same genetic defect.

In 2009 (Wang et al., 2009) defined goals of transcription profiling noting that they included the following:

1. Development of catalogs of RNA transcripts including knowledge of functions of coding and noncoding RNA transcripts.
2. Determination of factors that influence generation of all alternate splice forms.
3. Elucidation of structure of 5′ and 3′ gene regions.
4. Increase in understanding of range and effects of posttranscription modification of mRNA. There is evidence that posttranslational modification of mRNA influences mRNA functions. One form of RNA editing involves conversion of adenosine to inosine.
5. Wang et al. also noted that increased information on environmental factors that influence gene expression would be valuable.

Significant advances in understanding of factors involved in regulation of gene expressions have been made through activities of the Encode project. The Encyclopedia of DNA Elements (ENCODE) Project was launched in 2003 and updated in 2020. The goal of this project was to develop a comprehensive map of functional elements in the human genome. A report on this project from the Encode Consortium in 2020 (ENCODE Project Consortium, 2020) documented significant advances in developing a comprehensive map of functional elements in the human genome.

8.3 Variable phenotypes associated with specific mitochondrial mutations

One example of this was reported by Pierron in 2008 (Pierron, 2008). They noted that a specific mutation A3243 in the gene that encodes mitochondrial tRNAleu occurred in individuals with a range of phenotypes and there were also variations in the severity of diseases in different individuals with the same mutation. Some individuals were reported to suffer stroke like episodes while other individuals manifested diabetes and deafness.

Differences in the severity of disease could potentially be due to mitochondrial heteroplasmy, that is the occurrence in any one individual of both normal mitochondria with the disease carrying mutation.

However, heteroplasmy is unlikely to be responsible for striking differences in the type of phenotypic defects encountered.

There are also reports of differences in the range of phenotypic manifestations in different individuals in one family where all affected individuals have a homoplasmic mutation, a T to C mutation in nucleotide mt DNA 9185 (Childs et al., 2007). This mutation occurs in the gene that encodes ATP6, a subunit of the ATP synthase complex. Childs reported that five affected individuals in the family had neurological defects due to basal ganglia damage as shown on MRI. Some family members had neurogenic muscle weaknesses. Other affected family members had gastro-intestinal symptoms and facial dysmorphology.

8.4 Variable genomic abnormalities in individuals with the same phenotype

Defects in different genes have sometimes been encountered in patients with a defined phenotypic syndrome. One example is CHARGE syndrome characterized by choanal atresia, heart defects retardation of growth and development, genital anomalies, and eye abnormalities. One gene responsible for this disorder was mapped to chromosome 8q12 and was found to encode chromodomain helicase CHD7. Further studies revealed that mutations throughout CHD7 could lead to CHARGE syndrome (Sanlaville and Verloes, 2007).

Legendre et al. (2017) reported results of studies on 119 patients diagnosed with CHARGE syndrome. They reported finding CHD7 mutation in 87% of patients with typical CHARGE syndrome.

Moccia et al. (2018) reported results of exome sequencing studies in 28 patients diagnosed clinically with CHARGE syndrome; 15 of the 28 individuals (53.6%) were reported to have pathogenic mutations in CHD7 and four patients (14.3%) had pathogenic mutation in other genes including KMT2D, RERE, EP300, PUF60. They concluded that the phenotypic features in CHARGE syndrome overlap with features of other single gene disorders.

Genes with defects in CHARGE syndrome in different patients include:

1. CHD7 chromodomain helicase DNA binding protein 7.
2. KMT2D lysine methyltransferase 2D, histone methyltransferase t transcription regulator.
3. RERE arginine-glutamic acid dipeptide repeats thought to function as a transcriptional corepressor.
4. EP300 E1A binding protein p300, histone acetyltransferase regulates transcription via chromatin remodeling.
5. PUF60 poly(U) binding splicing factor 60 involved in premRNA splicing and transcriptional regulation.

8.5 Standardized phenotype documentation, documentation of genotype phenotype correlations databases

Human Phenotype Ontology (HPO) is defined as, "a resource to describe and to computationally analyze phenotype abnormalities found in human disease" (Köhler et al., 2021). In this resource, attempts are made to systematically define and logically organize phenotypes.

Other online resources documenting phenotypic features include Online Mendelian Inheritance in Man (OMIM) https://www.omim.org/, Decipher (https://www.deciphergenomics.org/) designed to share and compare phenotypic and genotypic data, and ORPHANET (http://www.orpha.net) designed to document primarily information related to rare diseases.

In 2020 HPO databases were noted to include 15,247 terms; examples presented included Marfan syndrome, noted to have 50 phenotypic abnormalities. Specific phenotypic features known to occur across disorders were also documented, for example, brachydactyly is noted to be present in 484 different disorders.

HPO documents disorders that impact specific systems, for example, head and neck, skeletal, eye, genitourinary, integument, muscle, cardiovascular, metabolism, digestion, ear, growth, blood respiratory system, limbs, endocrine, immunology, prenatal, connective tissue, breast, neoplasms, voice, cellular abnormalities, thoracic cavity.

Other terms introduced into HPO relate to pharmacology, abnormal drug concentration levels, altered drug efficiency, and adverse drug response.

HPO database can also be searched by specific manifestations, for example., hydrocephalus and the search will then link to specific databases with description of conditions with those disorders and gene links. Useful search terms in HPO database include phenotypic abnormality, mode of inheritance, clinical modifiers, for example, triggering factors, clinical course, frequency.

8.5.1 Clinical genetics and genomics databases

The clinical genome database https://research.nhgri.nih.gov/CGD/ is currently confined to data on single gene disorders. It is defined as "a manually curated database for condition with known genetic causes focusing on medically significant genetic data with available intervention." Clinical characterization categories include manifestations, organ systems involved. Interventions also include optimal supplemental care, prognosis, and reproductive information.

From the clinical genome database (CGD) there are links to OMIM, HGMD (Human Gene Mutation Database), ClinVar (HGMD), and Gene Reviews https://www.ncbi.nlm.nih.gov/books/NBK1116/.

The HGMD database is defined as "an attempt to collect information on gene and gene lesions responsible for human inherited disease."

Another database often referred to is the Decipher database with information utilized to compare phenotypic and genotypic data (https://www.deciphergenomics.org/).

The gnomAD database (https://gnomad.broadinstitute.org/) presents aggregated and harmonized exome and genome sequencing data for the human genome with data on population frequency of variants and data on significant pathogenic variants,

The OMIM databases (https://www.omim.org/# and https://www.ncbi.nlm.nih.gov/omim/) can be searched by specific symptom, by genes, by named Mendelian conditions. Information includes clinical manifestations, gene loci and map positions, and allelic variants for specific conditions.

The database https://www.omim.org/# can also be searched by chromosome and by map position on a specific chromosome to identify genes at a specific chromosome position.

Yang et al. (2015) described development of Phenolyzer http://phenolyzer.usc.edu, a machine learning tool that integrates disorder features and prioritizes candidate genes. Components of Phenolyzer were reported to include: a test to map phenotypic features to a disease; a resource that interprets knowledge about disease genes; a tool to predict previously unassociated genes for a particular disease; a tool to examine disease genes and gene − gene relationships.

PhenCards (https://phencard.org) was described by Havrilla et al. (2021) as a resource that integrates biomedical knowledge related to human clinical phenotypes. This database can be interrogated using phenotype terms or clinical notes. PhenCards was noted to also include information on medications relevant to specific conditions.

8.6 Phenome-wide association studies

Studies have been carried out to determine if constellations of manifestations as collected in the medical records can be utilized to compute a risk score for certain genetic conditions (Bastarache et al., 2018). They also proposed that data obtained from phenome-wide association studies could be used in conjunction with sequencing data to determine the pathological significance of specific sequence variants.

8.7 Dysmorphology syndromes with overlapping features due to defect in gene products that function in a specific pathway

It is interesting to note that some syndromes with overlapping features may be due to different genes that encode products that function in a specific pathway.

Rauen (2013) reviewed phenotypic manifestations in syndromes described as Rasopathies due to defects in genes that encode products that function in the RAS MAP kinase signaling pathways. The disorders included in the Rasopathy category include Noonan syndrome, Costello syndrome, Legius syndrome, Neurofibromatosis type 1, cardio-faciocutaneous syndrome, capillary malformation-arterio-venous malformation. Phenotypic manifestations that occur in these conditions include growth abnormalities, developmental delay, cutaneous pigmentary changes, capillary vascular changes, cardiac abnormalities, unusual facial features like hypertelorism, Down slanting palpebral fissures, anteverted nares, deep philtrum.

8.7.1 Gene products involved in the RAS/MAP signal transduction pathway and chromosomal map positions

1. CBL proto-oncogene encodes a RING finger E3 ubiquitin ligase 11q23.3.
2. (SHP2) PTPN11 member of the protein tyrosine phosphatase (PTP) family.
3. SOS1 SOS Ras/Rac guanine nucleotide exchange factor 2p22.1
4. NF1 Neurofibromin 1 negative regulator of the Ras signal transduction pathway. 17q11.2.
5. NRAS proto-oncogene, GTPase 1p13.2.
6. HRAS proto-oncogene, GTPase 11p15.5.
7. KRAS proto-oncogene, GTPase 12p12.1.
8. SHOC2 leucine-rich repeat scaffold protein links RAS to downstream signal transducers 10q25.2.
9. CRAF Raf-1 proto-oncogene, serine/threonine kinase can phosphorylate MEK1 and MEK2, 3p25.2.
10. BRAF B-Raf proto-oncogene, serine/threonine kinase 7q34.

11. SPRED sprouty-related EVH1 domain containing 1 regulates activation of the MAP kinase cascade 15q14.
12. MEK1 (ERK2) MAP2K1 essential component of MAP kinase signal transduction pathway 15q22.31.
13. MEK2 MAP2K2 It phosphorylates activates MAP2K1/ERK2 19p13.

It is important to note that some of syndromes are known to be caused by defects in one gene, for example, Costello syndrome is caused by mutation in HRAS; Legius syndrome is caused by mutation in SPRED1; neurofibromatosis is caused by defects in NF1. Other syndromes in this group may be caused by mutations in any one of several genes. Cardio-faciocutaneous syndrome can be caused by defects in BRAF, MAP2K1, MAP2K2, or KRAS.

Noonan syndrome can be caused by mutations in PTPN11, or SOS1, RAF1, KRAS, NRAS SHOC2. CBL (Fig. 8.1).

8.8 Phenotypic defects due to defects in sonic hedgehog signaling pathway

The SHH pathway was implicated in holoprosencephaly. In addition, the SHH pathway is now known to be implicated in causation of neoplasms. Sasai et al. (2019) reviewed the SHH signaling pathway in the context of genetic disorders. They noted that there are three different hedgehog genes and their products to consider.

SHH (7q36.3) protein is reported to be synthesized as a precursor that is then autocatalytically cleaved; the N-terminal portion is soluble. The C-terminal product covalently attaches a cholesterol moiety to the N-terminal product, restricting

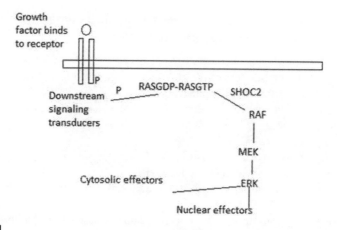

FIGURE 8.1

Ras signaling pathway activated by binding of growth factor to receptor.

8.8 Phenotypic defects

the N-terminal product to the cell surface and preventing it from freely diffusing throughout the developing embryo (Fig. 8.2).

Indian hedgehog IHH (2q35) reported to have properties similar to SHH defects involved in specific bone disorders including acrocapitofemoral dysplasia, an autosomal recessive disorder with cone-shaped epiphyses in hands and hips and brachydactyly.

Desert hedgehog DHH (12q13.12) has defects in gonadal dysgenesis with minifascicular neuropathy and 46 XY sex reversal.

Sasai et al. also noted the importance of a number of mediators in the hedgehog signaling system and emphasized the importance of the cleavage that generated an amino-terminal product that served as a signaling molecule that required cholesterol modification before becoming functional. The modified functional molecule was required to bind to the protein PTCH (patched) encoded on 9q22.32 prior to being taken up by the molecule Smoothened (SMO), a G-coupled receptor encoded on chromosome 7q32.1. Following this, a specific transcription factor, GLI, travels into the nucleus to promote transcription.

GLI stability and activity were noted to be impacted by KIF7 and SUFU. KIF7 encode on 15q26.1 is noted to be a cilia-coupled protein that functions as a negative regulator of the SHH pathway. SUFU is also defined as a negative regulator of the SHH signaling pathway; it is encoded on 10q24.32.

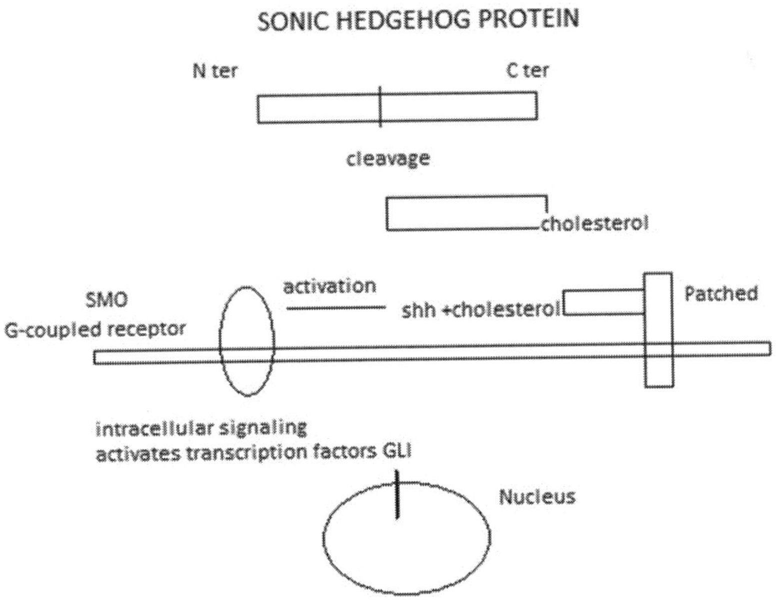

FIGURE 8.2

Sonic hedgehog protein processing, downstream signaling and transcription activation.

The hedgehog signaling pathway was noted to be involved in many processes in development and growth. SHH was described as a morphogen that determined cell fate in a concentration-dependent manner. It was noted to be particularly expressed in the central nervous system and in limb buds from early embryonic life. It was also shown to be expressed in some epithelial cells.

SHH expression was noted to be repressed postnatally in cells, particularly in brain region. IHH was noted to be expressed primarily in the developing skeletal system, particularly in chondrocytes. Studies in mice revealed that it played a role in osteoarthritis. DHH was shown to be expressed in testis, particularly in Sertoli cells and there is evidence that it may also be expressed in perineural cells there.

De Mori et al. (2017) reported that SUFU is the main negative regulator of the SHH pathway.

Specific mutations in SUFU were reported by Schröder et al. (2021) who noted that defects in SUFU led to increased levels of hedgehog signaling, increased levels of GL1, GLI2, GLI3, and Patched 1.

Individuals with defective SUFU function were reported to have congenital, ocular, motor apraxia, and impairment of horizontal gaze.

PTCH1 Mutations of this gene have been associated with basal cell nevus syndrome and holoprosencephaly.

GLI1 mutations have been associated with cases of polydactyly, including preaxial (Ullah et al., 2019) and postaxial polydactyly (Yousaf et al., 2020).

GLI3 mutations have been reported in several patients with Pallister – Hall (PH) syndrome. Roscioli et al. (2005) reported two patients with GLI 3 mutations and PH syndrome manifestations including upper and lower acromesomelic limb shortening and the previously unreported fibular hypoplasia, radio-ulnar bowing, and proximal epiphyseal hypoplasia.

Genes and their products with defects leading to holoprosencephaly as documented in the CGD (https://research.nhgri.nih.gov/CGD/):

1. SHH, PTCH, GLI2 sonic hedgehog pathway.
2. FGF8 Fibroblast growth factor 8.
3. TGIF2 DNA-binding homeobox protein and a transcriptional repressor.
4. DLL1 delta-like canonical Notch ligand 1.
5. CNOT1 CCR4-NOT transcription complex subunit 1.
6. CDON cell surface receptor that is a member of the immunoglobulin superfamily.
7. SIX3 member of the sine oculis homeobox transcription factor family.
8. STAG2 a subunit of the cohesin complex.

8.9 Fibroblast growth factor signaling pathway

Ornitz and Itoh (2015) reviewed the FGF signaling pathway and noted that it included 18 secreted peptides that interacted with the FGF receptors. When activated these receptors signal to cytosolic adapters and signaling pathways including RAS-MAP, phosphatidyl inositol PI3K and AKT, and to phospholipase C.

FGFs were also noted to regulate function of voltage gate sodium channels.

They reported that different subfamilies of FGFs exist, and FGFs differ with respect to the cofactors they require. The FGF7 subfamily was reported to require heparan. FGF23 subfamily requires Klotho (sometimes referred to as alpha klotho). Klotho was reported to activate FGF23. Klotho levels decline in chronic kidney disease and this decline is reported to be an early marker of kidney dysfunction.

FGFs are reported to have mitogenic stimulatory activities, to be involved in cell growth, to be important in organogenesis and tissue repair. Some FGFs have oncogenic activity.

8.10 Fibroblast growth factor receptor defects

FGF receptors play important roles in appropriate developments of bone and cartilage, and genetic defects in these receptors cause a number of disorders associated with abnormalities of skull, including craniosynostosis, while other FGF receptors impact development of long bones. Different FGF receptors differ with respect to chromosomes to which their encoding genes are assigned and differ in diseases that arise from defects in their function. Some forms of FGFR defects manifest dermatological abnormalities, some forms manifest genital anomalies.

1. FGFR1 8p11.23: Trigonocephaly, Pfeiffer syndrome, Jackson – Weiss syndrome, Hartfield syndrome, holoprosencephaly.
2. FGFR2 10q26.13: Apert syndrome, Crouzon syndrome, Jackson – Weiss syndrome, Pfeiffer syndrome, Antley – Bixler syndrome.
3. FGFR3 4p16.3: Achondroplasia, Crouzon syndrome, Muenke syndrome, SADDAN syndrome, thanatophoric dysplasia I, thanatophoric dysplasia II.

Pfeiffer syndrome is also described as craniofacial skeletal dermatologic dysplasia autosomal dominant (AD) disorder that can be caused by pathologic mutations in FGFR1, FGFR2, or FGFR3. Manifestations include craniosynostosis, midface deficiency, defects in hands and feet.

It is also interesting to note that some pathogenic mutations in FGFR3 are associated with a specific condition, for example., mutation NM_000142.5 (FGFR3):c.791 C > T (p.Thr264Met) with hypochondroplasia, while other pathogenic mutations in FGFR3 are described in a number of different conditions, for example, NM_000142.5(FGFR3):c.251 C > T (p.Ser84Leu) pathogenic mutation described in thanatophoric dysplasia, Crouzon disease, and hypochondroplasia.

Specific FGFR3 mutations are also associated with bladder cancer.

FGFR2 pathogenic mutations occur in a number of conditions associated with skull and skeletal abnormalities and FGFR2 mutations are also reported to occur in intestinal cancers and endometrial cancers.

The pathogenic mutation NM_000141.5(FGFR2):c.1150 G > A (p.Gly384Arg) was found in cranial and skeletal malformation syndromes and in stomach neoplasms.

Achondroplasia is reported to be associated with FGFR3 mutations; one unusual form of achondroplasia is also associated with pigmentary abnormalities.

Two specific pathogenic FGFR3 mutations NM_000142.5(FGFR3):c.251 C > T (p.Ser84Leu) and NM_000142.5(FGFR3):c.742 C > T (p.Arg248Cys) have each been reported in ClinVar to be have been identified in cases of achondroplasia, hypochondroplasia, thanatophoric dwarfism, cases with Crouzon syndrome, and cases with Muenke syndrome. The p.Arg248Cys mutation has also been reported in squamous carcinoma.

8.10.1 Mutations reported as pathogenic in achondroplasia multiple submitters

1. NM_000142.5(FGFR3):c.1138 G > C (p.Gly380Arg).
2. NM_000142.5(FGFR3):c.1620C > A (p.Asn540Lys) achondroplasia, also in hypochondroplasia.
3. NM_000142.5(FGFR3):c.1620C > A (p.Asn540Lys) achondroplasia, hypochondroplasia, craniosynostosis.

FGFR1 pathogenic mutations have been found in cases with dwarfism and in cases with skull defects including craniosynostosis. Hartfield syndrome is an FHFR1 defect with holoprosencephaly and cleft palate. Specific FGFR1 mutations may also be associated with hypogonadism, for example, NM_023110.2 (FGFR1):c.1825C > T (p.Arg609Ter).

8.11 Transforming growth factor beta signaling pathway

Massagué (2012) provided insights into processes through which TGFbeta signaling crossed from the cell membrane and to the nucleus and evidence for the multifunctional nature of TGFbeta signaling. It was noted to drive developmental programs and to impact cell proliferation and differentiations and tissue homeostasis. Key factors in this signaling system involve more than 30 different proteins.

Key factors include TGFB1 encoded on chromosomes 19q13.2, TGFB2 encoded on chromosome 1q4.1, and TGB3 encoded on 14q24.3. In addition, orthologs of TGFbeta exist.

TGFbeta binds to TGFbeta receptors and intracellularly this binding leads to activation of SMAD family of proteins (Fig. 8.3).

8.11 Transforming growth factor beta signaling pathway

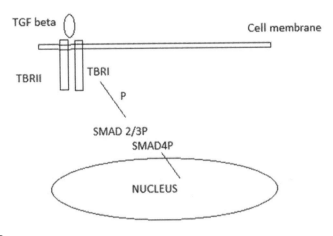

FIGURE 8.3

TGF binding to receptor and downstream signaling.

MacFarlane et al. (2019) reviewed the TGFbeta signaling pathway and they stressed the ligand diversity of receptors in this pathway. Signaling molecules include not only TGFbeta but also activins, inhibin, bone morphogenetic proteins (BMPs), growth and differentiation factors.

TGFbeta signaling was noted to play important roles pre and postnatally in blood vessel wall and homeostasis.

Both TGFbeta and BMP were shown to be involved in limb and digital development, in elongations of digits, in formation of joint cartilage, and in bone remodeling. TGFbeta and BMP play roles in endochondral ossification, chondrogenic differentiation, maturation of cartilaginous elements, and in palate development.

Key to functioning of the pathway is binding to TGFbeta receptors TGFBR1 encoded on 9q22.33 and TGFR2 encoded on 3p24.1 and downstream activation of SMAD2, SMAD3, and SMAD4. The SMAD family proteins are activated by serine receptor kinases in response to TGFbeta receptor signaling.

Fibrillin protein encoded on chromosome 15q21.1 is reported to have a number of functions that include modulation of TGFbeta signaling. Defects in fibrillin as seen in Marfan syndrome, can dysregulate TGFbeta receptor signaling.

1. TGFBR1 9q22.23 mutations are implicated in Loeys – Dietz syndrome type 1.
2. TGFBR2 3p24.1 mutations are implicated in Loeys – Dietz syndrome type 2.
3. SMAD3 15q22.2 mutations implicated in Loeys – Dietz syndrome type 3.
4. TGFB2 1.q41 mutations implicated in Loeys – Dietz syndrome type 4.
5. TGFB3 14q24.3 mutations implicated in Loeys – Dietz syndrome type 5.

8.11.1 Phenotypic features Loeys – Dietz syndromes

1. LDS1 arterial tortuosity, cranio-facial involvements including craniosynostosis, hypertelorism, cleft palate, bifid uvula.
2. LDS2 skeletal and cardiovascular manifestations similar to Marfan syndrome.
3. LDS3 aneurysms of aorta and/or pulmonary, splenic, iliac arteries.
4. LDS4 mitral prolapse. Aortic aneurysm, aortic dissection, arterial tortuosity, cerebrovascular aneurysms, hypertelorism, bifid uvula, high-arched palate.
5. LDS5 thoracic and/or abdominal aneurysms, mitral valve disease, cleft palate, bifid uvula, skeletal overgrowth, club feet, cervical spine instability.
6. TGFBR1 pathogenic mutation NM_004612.4(TGFBR1):c.700 T > C (p.Phe234Leu) implicated in fatal familial aortic aneurysm, aortic dissections (multiple submissions).
7. TGFBR1 pathogenic mutations NM_004612.4(TGFBR1):c.722 C > T (p.Ser241Leu) implicated in Loeys – Dietz syndrome, aortic aneurysm multiple submissions.
8. TGFBR2 deletion in the chromosome 3p26.3p22 region leads LDS2.
9. Pathogenic mutation multiple submitters NM_003242.6(TGFBR2): c.1067 G > C (p.Arg356Pro).
10. Pathogenic mutation multiple submitters NM_003242.6(TGFBR2): c.1570 G > A (p.Asp524Asn).
11. Pathogenic mutation multiple submitters NM_003242.6(TGFBR2): c.1583 G > A (p.Arg528His).
12. Genomic deletions in SMAD3 have been documented as causes of Loeys – Dietz syndrome type 3.
13. SMAD3 pathogenic mutations have been documented in Loeys – Dietz syndrome type 3.
14. Likely pathogenic mutation in SMAD 3 leading to Loeys – Dietz 3 syndrome multiple submitters.
15. NM_005902.4(SMAD3):c.1 A > T (p.Met1Leu).
16. Likely pathogenic mutation, familial aortic aneurysm, aortic dissection, submitted by multiple submitters NM_005902.4(SMAD3):c.5 C > A (p.Ser2Ter); NM_005902.4(SMAD3):c.138del (p.Gln47fs)

8.11.2 LDS type 4 TGFB2 mutations

Likely pathogenic multiple submitters:

1. NM_003238.5(TGFB2):c.896 G > A (p.Arg299Gln).
2. NM_003238.5(TGFB2):c.904 C > T (p.Arg302Cys).
3. NM_003238.5(TGFB2):c.958 C > T (p.Arg320Cys) pathogenic in LDS4 and in Holt – Oram syndrome.
4. NM_001135599.3(TGFB2):c.979 C > T (p.Arg327Trp).

8.11.3 LDS type 5 TGFB3 mutations

Pathogenic/likely pathogenic multiple submitters

1. NM_003239.4(TGFB3):c.1034 C > G (p.Ser345Ter).
2. NM_003239.4(TGFB3):c.927−1 G > C.
3. NM_003239.4(TGFB3):c.883_884del (p.Gly295fs).
4. NM_003239.4(TGFB3):c.899 G > A (p.Arg300Gln).
5. NM_003239.4(TGFB3):c.898 C > T (p.Arg300Trp).

8.12 Marfan syndrome 15q21.1 FBN1

Skeletal, cardiovascular, and ocular systems are impacted in Marfan syndrome due to specific mutations in FBN1. Marked pleiotropism was noted. Marfan syndrome was noted to share some phenotypic features with arachnodactyly syndrome due to defects in AD defects in FBN2.

Skeletal features in Marfan syndrome were reported to include unusually long limbs and digits; spinal lordosis and scoliosis high arched palate can occur with teeth crowding (Pyeritz and McKusick, 1979). Ocular findings can include myopia and lens subluxation. Cardiovascular findings can include mitral valve prolapse, mitral regurgitation, dilatation of the aortic root, and aortic regurgitation. In some patients, pneumothorax may occur.

Heterozygous mutations in Fibrillin2 (FBN2) on 5q23 were reported to cause contractural arachnodactyly (CCA1). Features of this disorder include kyphoscoliosis, osteopenia, finger flexion contractures, and abnormally shaped ears (Mirise and Shear, 1979) However, no definitive pathological mutations in FBN2 seen by multiple investigators were found in ClinVar.

8.12.1 FBN1 mutations in Marfan syndrome

It is important to note that defects in FBN1 have also been reported to be associated with other syndromes: Geleophysic dysplasia, Weill − Marchesani syndrome, and stiff skin mutations.

It is also important to note that some of these pathogenic FBN1 mutations have been identified in patients with Loeys − Dietz syndrome manifestations, while other mutations associated with Marfan syndrome are also found in Weil − Marchesani syndrome. This syndrome associated with Weill − Marchesani syndrome is a rare systemic connective tissue disorder consisting of brachydactyly, ectopia lentis, spherophakia, and glaucoma (Evereklioglu et al., 1999).

In looking at data in ClinVar there are indications that the same mutation has, in different cases, been considered typical of both syndromes. In a published statement on distinguishing features of the two syndromes Meester et al. (2017) wrote:

"LDS can be distinguished from MFS by the unique presence of hypertelorism, bifid uvula or cleft palate, and widespread aortic and arterial aneurysm and tortuosity."

8.13 FBN1 mutations, pathogenic, likely pathogenic, Marfan syndrome multiple submitters, without conflicts identified in Clin Var searches

There are mutations throughout the FBN1 gene defined as pathogenic, likely pathogenic, multiple submitters, no conflict. It is interesting to note that a number of these are chain terminating mutations.

There are also mutations in FBN1 where different diseases are diagnosed in cases with the same mutation. Examples are presented below.

1. NM_000138.4(FBN1):c.7916 A > G (p.Tyr2639Cys) Marfan syndrome, Loeys − Dietz, familial aortic thoracic aneurysms.
2. *NM_000138.5(FBN1):c.7754 T > C (p.Ile2585Thr) note this mutation reported pathogenic in different syndromes. Marfan, Loeys − Dietz, Geleophysic dysplasia 2, stiff skin syndrome, Weill − Marchesani syndrome.
3. +NM_000138.4(FBN1):c.7708 G > A (p.Glu2570Lys) reported pathogenic in Marfan syndrome, Loeys − Dietz syndrome, familial thoracic aortic aneurysm syndrome.
4. +NM_000138.4(FBN1):c.7663 G > A (p.Gly2555Arg) reported pathogenic in Marfan syndrome, Loeys − Dietz syndrome, familial thoracic aortic aneurysm syndrome.
5. NM_000138.4(FBN1):c.6388 G > A (p.Glu2130Lys) Geleophysic dysplasia 2, stiff skin syndrome, Weill − Marchesani syndrome 2, ectopia lentis, isolated, AD, acromicric dysplasia, Marfan syndrome.

It thus becomes clear that the sequence data alone are usually not sufficient to make a diagnosis.

8.14 Connective tissue disorder Ehlers − Danlos syndrome disorders

These are connective tissue disorders that share common features including joint hypermobility, skin hyperextensibility, and tissue fragility. Of the 11 defined types, four are due to defects in different forms of collagen. Different Ehlers − Danlos syndrome (EDS) types, their chromosome map positions, gene product defects, mode of inheritance, AD or autosomal recessive (AR) are listed below.

8.14 Connective tissue disorder Ehlers − Danlos syndrome disorders

1. Vascular type 2q33.2 COL3A1 AD collagen type III alpha 1 chain.
2. Classic type ` 9q34.3 COL5A1 AD collagen type V alpha 1 chain.
3. Kyphoscoliosis type 1p36.22 PLOD AR (procollagen-lysine,2-oxoglutarate 5-dioxygenase 1).
4. Periodontal type 12p13.31 C1S AD serine protease, major constituent of the human complement.
5. Classic type 2 2q32.1 AD COL5A2 collagen type 5A2.
6. Spondylodysplastic type 5q35.3 AR B4GALT7 B4 galactosyltransferase T7.
7. Musculocontractural type 15q15.1 AR CHST14 carbohydrate sulfotransferase 14 also type II DSEdermata sulfo-epimerase 6Q22.1.
8. EDS with platelet dysfunction and fibronectin deficiency.
9. Arthrochalasia type 1 7q21.3 AD COL1A1 7q2133, type 2 collagen type I alpha 2 chain 7q21.3.
10. EDS Classic like 6p21.3 AR TNXB tenascin XB, extracellular matrix glycoprotein.
11. EDS kyphoscoliosis type hydroxy lysine deficient PLOD1.
12. Pathogenic, likely pathogenic, mutations, multiple submitters, no conflict.
13. NM_000302.4(PLOD1):c.327del (p.Arg111fs).
14. NM_000302.4(PLOD1):c.404_423del (p.Asp135fs).
15. NM_000302.4(PLOD1):c.1533 C > G (p.Tyr511Ter).
16. NM_000302.4(PLOD1):c.1651−2 A > G.

EDS hypermobility type, multigenic, environmental factors.

8.14.1 COL3A, vascular EDS, pathogenic/likely pathogenic, mutations, multiple submitters, no conflicts

1. NM_000090.3(COL3A1):c.555del (p.Gly186fs).
2. NM_000090.3(COL3A1):c.674 G > C (p.Gly225Ala).
3. NM_000090.3(COL3A1):c.712 C > T (p.Arg238Ter) familial aortic aneurysm and dissecting aneurysm.
4. NM_000090.3(COL3A1):c.962 G > A (p.Gly321Asp).
5. NM_000090.3(COL3A1):c.1618G > A (p.Gly540Arg).
6. NM_000090.3(COL3A1):c.1662 + 1 G > A (p.Gly537_Pro554del).
7. NM_000090.3(COL3A1):c.1763G > A (p.Gly588Asp).
8. NM_000090.3(COL3A1):c.2194 G > A (p.Gly732Arg).
9. NM_000090.3(COL3A1):c.2221 G > T (p.Gly741Cys).
10. NM_000090.3(COL3A1):c.2356 G > A (p.Gly786Arg).
11. NM_000090.3(COL3A1):c.2534dup (p.Gly846fs).
12. NM_000090.3(COL3A1):c.2689 G > A (p.Gly897Ser).
13. NM_000090.3(COL3A1):c.3500 G > A (p.Gly1167Asp).
14. NM_000090.3(COL3A1):c.4087 C > T (p.Arg1363Ter).

8.14.2 Classic type 9q34.3 COL5A AD collagen type V alpha 1 chain

Clinical features observed include carotid-cavernous fistula, complications included rupture of large vessels, hiatus hernia, spontaneous rupture of the bowel, and diverticula of the bowel. Retinal detachment was sometimes reported.

Pathogenic/likely pathogenic mutations multiple submitters no conflicts:

1. NM_001278074.1(COL5A1):c.2034 + 1 G > A.
2. NM_001278074.1(COL5A1):c.2140 C > T (p.Gln714Ter).
3. NM_001278074.1(COL5A1):c.2734 C > T (p.Arg912Ter).
4. NM_001278074.1(COL5A1):c.3069dup (p.Gly1024fs).
5. NM_001278074.1(COL5A1):c.3397 C > T (p.Arg1133Ter).
6. NM_001278074.1(COL5A1):c.4126dup (p.Ser1376fs).
7. NM_001278074.1(COL5A1):c.4474 G > A (p.Gly1492Ser).

Arthrochalasia EDS associated with frequent congenital hip dislocation and extreme joint laxity with recurrent joint subluxations and minimal skin involvement Type 1 COL1A1, Type 2 COL1A2.

COL1A1 note defects in this gene are also associated with osteogenesis imperfecta. No single mutations with definitive evidence for EDS assignment from multiple submitters.

COL1A2

1. NM_000089.3(COL1A2):c.1009 G > A (p.Gly337Ser).
2. NM_000089.3(COL1A2):c.1127 G > T (p.Gly376Val).
3. NM_000089.3(COL1A2):c.1342 G > C (p.Gly448Arg).
4. NM_000089.3(COL1A2):c.2133 + 6 T > A.

Musculocontractural EDS, craniofacial dysmorphism, contractures of thumbs and fingers, clubfeet, kyphoscoliosis, muscular hypotonia, hyperextensible thin skin browsability and atrophic scarring, joint hypermobility, and ocular involvement (Malfait et al., 2017).

Type 1 carbohydrate sulfotransferase; Type II DSE dermatan epimerase: no definitive mutations found in ClinVar.

8.14.3 Hypermobile Ehlers − Danlos syndrome

Chiarelli et al. (2021) reported results of studies on hypermobile EDS. In this disorder joint hypermobility occurs and skin changes, increased elasticity and fragility may occur. There are questions to whether this disorder is genetic or environmental in origin. They reported results of studies and dermal myofibroblasts and reported evidence of dysregulated protein networks and altered protein expression and altered energy metabolism with altered redox balance.

Riley (2020) noted the hypermobile EDS may also manifest postural orthostatic hypotension, an orthostatic tachycardia sometimes referred to as POTS syndrome.

Patients may present with additional symptoms including chronic pain, and some patients have signs of mast cell activation.

8.15 DNA methylation episignatures and phenotypic correlations

Aref-Eshghi et al. (2019), (2020) reported that specific genetic syndromes that include neurodevelopmental defects frequently have overlapping phenotypic manifestations that impede accurate diagnoses. They provided evidence that for a number of these disorders it is possible to determine specific diagnoses through evaluation of DNA methylation patterns.

Their studies involved analyses on DNA derived from peripheral blood from patients and controls. DNA samples were bisulfite treated and were then analyzed using Illumina Infinium 450 or EPIC bead chip methylation detection arrays.

Between 450,000 and 800,000 human genome CpG sites were reported to be detected with these arrays. Parameters to be determined include the level of methylation and the map positions of methylation sites.

They reported results of studies on subjects with 42 different genetic syndromes with neurodevelopmental manifestations. Results revealed that in specific diseases there were specific patterns of altered methylation. In some of the disorders there was some degree of overlap in altered methylation patterns. However, data analyses revealed specific peripheral blood episignatures in individuals with particular neurodevelopmental disorders.

They concluded that episignature analyses were particularly useful in syndromes where phenotypic features partially overlap and when diagnoses cannot be made on the basis of phenotypic features alone.

Syndromes characterized by development and neurodevelopmental defects that were shown to manifest specific methylation episignatures reported in the Aref-Eshghi et al. (2019) study included the following:

1. ADCADN Cerebellar ataxia with deafness and narcolepsy.
2. ADNP1 activity dependent neuroprotector homeobox 1, mutations lead to chromatin remodeling defect.
3. ADNP2 activity dependent neuroprotector homeobox 2, mutations lead to chromatin remodeling defect.
4. ATRX chromatin remodeler, alpha-thalassemia/mental retardation syndrome.
5. BAFOPATHIES Defects in genes leading to chromatin remodeling defects.
6. CDLS Cornelia de Lange syndrome 1, dominant.
7. CHARGE syndrome congenital anomalies including choanal atresia and malformations of the heart, inner ear, and retina.

8. DOWN syndrome Trisomy 21.
9. DUP7 duplication of genes lying within the critical region for Williams − Beuren syndrome.
10. FHS Fetal hydantoin syndrome phenytoin toxicity syndrome.
11. GTPTS microcephaly, psychomotor retardation, facial dysmorphology, absent patella, urogenital anomalies, KAT6B defect.
12. KABUKI syndrome KMT2D defect developmental delay, unusual facial features.
13. SOTOS1 impaired neurodevelopment overgrowth, NSD1 gene defect.
14. Williams − Beuren syndrome 7q11.23 deletion.
15. William's syndrome.

References

Aref-Eshghi E, Bend EG, Colaiacovo S, et al: Diagnostic utility of genome-wide DNA methylation testing in genetically unsolved individuals with suspected hereditary conditions, *Am J Hum Genet* 104(4):685−700, 2019. Available from: https://doi.org/10.1016/j.ajhg.2019.03.008. Epub 2019 Mar 28. PMID: 30929737.

Aref-Eshghi E, Kerkhof J, Pedro VP, et al: Evaluation of DNA methylation episignatures for diagnosis and phenotype correlations in 42 Mendelian neurodevelopmental disorders, *Am J Hum Genet* 106(3):356−370, 2020. Available from: https://doi.org/10.1016/j.ajhg.2020.01.019. PMID: 32109418.

Bastarache L, Hughey JJ, Hebbring S, et al: Phenotype risk scores identify patients with unrecognized Mendelian disease patterns, *Science* 359(6381):1233−1239, 2018. Available from: https://doi.org/10.1126/science.aal4043. PMID: 29590070.

Cardno AG, Gottesman II: Twin studies of schizophrenia: from bow-and-arrow concordances to star- wars Mx and functional genomics, *Am J Med Genet* 97(1):12−17, 2000. PMID: 10813800.

Chiarelli N, Zoppi N, Ritelli M, et al: Biological insights in the pathogenesis of hypermobile Ehlers − Danlos syndrome from proteome profiling of patients' dermal myofibroblasts, *Biochim Biophys Acta Mol Basis Dis* 1867(4):166051, 2021. Available from: https://doi.org/10.1016/j.bbadis.2020.166051. PMID: 33383104.

Childs AM, Hutchin T, Pysden K, et al: Variable phenotype including Leigh syndrome with a 9185T>C mutation in the MTATP6 gene, *Neuropediatrics.* 38(6):313−316, 2007. Available from: https://doi.org/10.1055/s-2008-1065355. PMID: 18461509.

De Mori R, Romani M, D'Arrigo S, et al: Hypomorphic recessive variants in SUFU impair the sonic hedgehog pathway and cause Joubert syndrome with cranio-facial and skeletal defects, *Am J Hum Genet* 101(4):552−563, 2017. Available from: https://doi.org/10.1016/j.ajhg.2017.08.017. PMID: 28965847.

ENCODE Project Consortium, Snyder MP, Gingeras TR, Moore JE, et al: Perspectives on ENCODE, *Nature.* 583(7818):693−698, 2020. Available from: https://doi.org/10.1038/s41586-020-2449-8. PMID: 32728248.

Erickson RP, Wynshaw-Boris AJ: *Epstein's inborn errors of development*, New York, NY, 2004, Oxford University Press.

Evereklioglu C, Hepsen IF, Er H: Weill − Marchesani syndrome in three generations, *Eye (Lond)* 13(Pt 6):773−777, 1999. Available from: https://doi.org/10.1038/eye.1999.226. PMID: 10707143.

Freimer N, Sabatti C: The human phenome project, *Nat Genet* 34(1):15−21, 2003. Available from: https://doi.org/10.1038/ng0503-15. PMID: 12721547.

Havrilla JM, Liu C, Dong X, et al: PhenCards: a data resource linking human phenotype information to biomedical knowledge, *Genome Med* 13(1):91, 2021. Available from: https://doi.org/10.1186/s13073-021-00909-8. PMID: 34034817.

Köhler S, Gargano M, Matentzoglu N, et al: The human phenotype ontology in 2021, *Nucleic Acids Res* 49(D1):D1207−D1217, 2021. Available from: https://doi.org/10.1093/nar/gkaa1043. PMID: 33264411.

Legendre M, Abadie V, Attié-Bitach T, et al: Phenotype and genotype analysis of a French cohort of 119 patients with CHARGE syndrome, *Am J Med Genet C Semin Med Genet* 175(4):417−430, 2017. Available from: https://doi.org/10.1002/ajmg.c.31591. PMID: 29178447.

MacFarlane EG, Parker SJ, Shin JY, et al: Lineage-specific events underlie aortic root aneurysm pathogenesis in Loeys − Dietz syndrome, *J Clin Invest* 129(2):659−675, 2019. Available from: https://doi.org/10.1172/JCI123547. PMID: 30614814.

Malfait F, Francomano C, Byers P, et al: The 2017 international classification of the Ehlers − Danlos syndromes, *Am J Med Genet C Semin Med Genet* 175(1):8−26, 2017. Available from: https://doi.org/10.1002/ajmg.c.31552. PMID: 28306229.

Massagué J: TGFβ signalling in context, *Nat Rev Mol Cell Biol* 13(10):616−630, 2012. Available from: https://doi.org/10.1038/nrm3434. PMID: 22992590.

McKusick VA: *Medical genetics*, Baltimore, MD and London, 1993, The Johns Hopkins University Press.

Meester JAN, Verstraeten A, Schepers D, et al: Differences in manifestations of Marfan syndrome, Ehlers − Danlos syndrome, and Loeys − Dietz syndrome, *Ann Cardiothorac Surg* 6(6):582−594, 2017. Available from: https://doi.org/10.21037/acs.2017.11.03. PMID: 29270370.

Mirise RT, Shear S: Congenital contractural arachnodactyly: description of a new kindred, *Arthritis Rheum* 22(5):542−546, 1979. Available from: https://doi.org/10.1002/art.1780220516. PMID: 444317.

Moccia A, Srivastava A, Skidmore JM, et al: Genetic analysis of CHARGE syndrome identifies overlapping molecular biology, *Genet Med* 20(9):1022−1029, 2018. Available from: https://doi.org/10.1038/gim.2017.233. PMID: 29300383.

Oetting WS, Robinson PN, Greenblatt MS, et al: Getting ready for the Human Phenome Project: the 2012 forum of the Human Variome Project, *Hum Mutat* 34(4):661−666. Available from: https://doi.org/10.1002/humu.22293, 2013.

Ornitz DM, Itoh N: The fibroblast growth factor signaling pathway, *Wiley Interdiscip Rev Dev Biol* 4(3):215−266, 2015. Available from: https://doi.org/10.1002/wdev.176. PMID: 25772309.

Pierron D, Rocher C, Amati-Bonneau P, et al: New evidence of a mitochondrial genetic background paradox: impact of the J haplogroup on the A3243G mutation, *BMC Med Genet* 9(41):. Available from: https://doi.org/10.1186/1471-2350-9-41. PMID: 18462486.

Pyeritz RE, McKusick VA: The Marfan syndrome: diagnosis and management, *N Engl J Med* 300(14):772−777, 1979. Available from: https://doi.org/10.1056/NEJM197904053001406. PMID: 370588.

Rauen KA: The RASopathies, *Annu Rev Genomics Hum Genet* 14:355−369, 2013. Available from: https://doi.org/10.1146/annurev-genom-091212-153523. PMID: 23875798.

Riley B: The many facets of hypermobile Ehlers−Danlos syndrome, *J Am Osteopath Assoc* 120(1):30−32, 2020. Available from: https://doi.org/10.7556/jaoa.2020.012. PMID: 31904772.

Roscioli T, Kennedy D, Cui J, et al: Pallister−Hall syndrome: unreported skeletal features of a GLI3 mutation, *Am J Med Genet A.* 136A(4):390−394, 2005. Available from: https://doi.org/10.1002/ajmg.a.30818. PMID: 16007608.

Rosenberg RE, Law JK, Yenokyan G: Characteristics and concordance of autism spectrum disorders among 277 twin pairs, *Arch Pediatr Adolesc Med* 163(10):907−914, 2009. Available from: https://doi.org/10.1001/archpediatrics.2009.98. PMID: 19805709.

Sanlaville D, Verloes A: CHARGE syndrome: an update, *Eur J Hum Genet* 15(4):389−399, 2007. Available from: https://doi.org/10.1038/sj.ejhg.5201778. PMID: 17299439.

Sasai N, Toriyama M, Kondo T: Hedgehog signal and genetic disorders, *Front Genet.* 10:1103, 2019. Available from: https://doi.org/10.3389/fgene.2019.01103. eCollection2019. PMID: 31781166.

Schröder S, Li Y, Yigit G, Altmüller J, et al: Heterozygous truncating variants in SUFU cause congenital ocular motor apraxia, *Genet Med* 23(2):341−351, 2021. Available from: https://doi.org/10.1038/s41436-020-00979-w. PMID: 33024317.

Stanbury JB, Wyngaarden JB, Fredrickson DS, editors: *The metabolic basis of Inherited disease*, New York, NY, 1972, McGraw Hill.

Ullah A, Umair M, Majeed AI, et al: A novel homozygous sequence variant in GLI1 underlies first case of autosomal recessive pre-axial polydactyly, *Clin Genet* 95(4):540−541, 2019. Available from: https://doi.org/10.1111/cge.13495. PMID: 30620395.

Wang Z, Gerstein M, Snyder M: RNA-Seq: a revolutionary tool for transcriptomics, *Nat Rev Genet.* 10(1):57−63, 2009. Available from: https://doi.org/10.1038/nrg2484. PMID: 19015660.

Yang H, Robinson PN, Wang K: Phenolyzer: phenotype-based prioritization of candidate genes for human diseases, *Nat Methods* 12(9):841−843, 2015. Available from: https://doi.org/10.1038/nmeth.3484. PMID: 26192085.

Yousaf M, Ullah A, Azeem Z, et al: Novel heterozygous sequence variant in the GLI1 underlies postaxial polydactyly, *Congenit Anom (Kyoto)* 60(4):115−119, 2020. Available from: https://doi.org/10.1111/cga.12361. PMID: 31621941.

CHAPTER 9

Expansion of methods of gene editing therapy and analysis of safety and efficacy

9.1 Introduction

Gene therapy involves a number of different approaches. These can include use of factors that block elements within the genes, for example, nucleotides in genes or within immediate product of genes, mRNA. Approaches can also be directed at increasing expression of specific genes.

This chapter will also include discussion of methods to add a normal version of a gene in cases where disease arises due to absent or inadequate function of a specific gene, and more recent efforts directed at editing somatic genes through newly developed methods. In addition, gene therapy approaches to treatment of specific diseases will be presented.

9.2 Therapies designed to block nucleotides or RNA derived from a specific gene

Antisense oligonucleotides were reviewed by Bennett (2019) and Roberts et al. (2020). They are defined as small 18–30 nucleotide single stranded polymers and they were subdivided into steric block polymers and RNAse H1 competent polymers. RNAse H1 enzyme was reported to recognize RNA − DNA heteroduplex substrates that result from the binding of DNA oligonucleotides and matching RNA. RNAse H1 enzyme then leads to destruction of RNA. Specific antisense oligonucleotides referred to as gapmers bind to RNAs and can serve to activate RNAse H enzyme leading to mRNA degradation.

9.3 Oligonucleotide therapies

Roberts et al. (2020) summarized aspects of oligonucleotide therapy and specific goals of this therapy; they included the following.

1. Modulation of gene expression, RNA inhibition, target RNA degradation through RNASe H mediated cleavage, splicing modulation, noncoding RNA inhibition, gene activation, gene silencing.
2. Oligonucleotides were noted to interact with their targets via Watson – Crick pairing and based on the sequence to be targeted highly specific lead compounds can be designed; thus patient-specific sequences can be targeted.

However, one important difficulty involves delivering therapies to specific organs with the exception of liver. Another important factor to consider is off-target effects of therapy.

9.3.1 Steric block oligonucleotides

These were reported to block RNA transcripts with high affinity but not to lead to degradation of the bound transcript, as these oligonucleotides are not RNAseH1 competent. Steric block oligonucleotides can also be designed to modulate splicing, and lead to exclusion of a specific exon and to inclusion of a specific exon. These oligonucleotides can therefore influence which isoforms are generated from a specific transcript. Roberts et al. (2020) noted that by three splice switching oligonucleotides were FDA (Food and Drug Administration) approved. These include:

1. Eteplirsen that impacts dystrophin exon 51; the target tissue skeletal muscle, administration intravenous.
2. Golodirsen impacts dystrophin exon 53; the target tissue is skeletal muscle, administration intravenous.
3. Nusinersen used for treatment of spinal muscular atrophy, promotes expression of SMN2 exon 7 by blocking inhibitory sequences, target tissue includes spinal cord and nerves, administration is intrathecal.

Roberts et al. (2020) noted that by three RNAseH1 competent allele-specific oligonucleotides had been FDA approved. These included:

1. Fomivirsen for treatment of cytomegalovirus retinitis, the target organ is the eye and administration is intravitreal.
2. Mipomersen for treatment of ApoB-related hypercholesterolemia, the target organ is the liver and administration is subcutaneous.
3. Inotersen for treatment of hereditary transthyretin amyloidosis, the target organ is liver and administration is subcutaneous.

9.3.2 RNA inhibition in therapies

Crooke et al. (2018) and Roberts et al. (2020) reviewed RNA target therapeutics. Short inhibitory RNAs (siRNAs) can guide a specific enzyme complex to degrade a specific mRNA. The siRNA has a guide strand that matches specific mRNA sequence and a passenger strand. Within the cell the siRNA binds to RNA inducing silencing complex (RISC); the passenger strand is then discarded. The guide strand then binds to the complementary mRNA sequence and the enzyme Argonaute within RISC then cleaves the mRNA.

Two siRNAs have received FDA approval for therapy. These include Patisiran that targets TTR transthyretin that is defective in transthyretin amyloidosis. Patisiran, an interfering RNA that was developed for the treatment of hereditary transthyretin-mediated amyloidosis, was delivered encased in a lipid particle (Yang, 2019). Givosiran, a small interfering RNA directed toward ALAS1 5′-aminolevulinate synthase 1, that is mutated in acute hepatic porphyria (Sardh et al., 2019).

9.3.3 MicroRNAs as mRNA inhibitors

Roberts et al. noted that microRNAs are endogenous RNA inhibitors with physiological roles. MicroRNAs function in posttranscriptional regulation of gene expression. Defects in microRNA function can lead to pathologies and microRNAs are considered to be important drug targets. Roberts et al. noted that steric block antisense oligonucleotides have been developed to impact specific microRNAs.

A specific microRNA mir21 was reported to play a role in the pathology of Alport syndrome (Gomez et al., 2015). Roberts et al. reported that clinical trials were ongoing to target this disorder with an antisense oligonucleotide.

MicroRNAs were shown to play roles in polycystic kidney disease PKD1 and PKD2 (Li and Sun, 2020). A specific antisense microRNA is in clinical trials to target microRNA miR17 that influences pathology in polycystic kidney disease types PKD1 and PKD2.

It is thus important to note that specific therapeutic oligonucleotides can be directed not only against the primary defective gene transcripts but also against ancillary gene encoded transcripts that impact the pathophysiology in a specific disease.

9.3.4 Long noncoding RNAs, small RNAs, endogenous antisense RNAs

Long noncoding RNAs (lncRNAs) have increasingly been found to be involved in the regulation of gene expression. In some cases, endogenous antisense RNAs and small RNAs are being identified and found to influence gene expression. Roberts et al. noted that specific disorders in humans have been attributed to aberrant gene silencing events and there is evidence that oligonucleotides may constitute target for therapies in these disorders. lncRNAs have been reported to be associated with specific disorders including cancer. Gupta et al. (2020) reported the NF-κB signaling was also reported to be regulated by several long non lncRNAs. The lncRNA and NF-κB signaling were reported to crosstalk during cancer and they discussed possibilities for therapies based on these associations.

A lncRNA is associated with the neurodevelopmental disorder Angelman syndrome (Meng et al., 2015). Use of antisense oligonucleotides to downregulate this long RNA is in clinical trials.

9.4 Delivery challenges in oligonucleotide therapies

Roberts et al. noted that oligonucleotides are large hydrophobic polyanions that do not traverse plasma membranes of cells easily. Within the extracellular space they can be degraded by nucleases. In addition, they may be lost through renal clearance. If they do traverse cell membranes, they may be taken up by endosomes and degraded. For these reasons, a number of therapeutic oligonucleotides have been administered locally, for example, in the intravitreal space in the eye or in the intrathecal space.

The liver was noted to have high concentrations of receptors on cells and that oligonucleotides can be modified to bind to these receptors to expedite uptake.

Additional modifications have been designed to enhance uptake of therapeutic oligonucleotides. These include alterations of ribose or phosphate components of oligonucleotides. Roberts et al. noted that lipid covalent conjugation was being explored to promote delivery of siRNAs.

Conjugation of antisense oligonucleotides with N-acetyl galactosamine (GalNac) was being used to promote oligonucleotide uptake into liver cells through specific receptors.

Antibody coupling of oligonucleotides has been proposed to promote binding to specific cellular proteins. Nanoparticles and particularly cationic polymers are also being explored as carriers for oligonucleotides.

Hammond et al. (2021) reported that a main challenge in nucleic acid-based therapeutic is delivery to the target tissue. They noted that use of conjugates and nanoparticles has proven useful in some cases. Conjugates used included fatty acids, N-acetyl-galactosamine (GalNac), and cell penetrating peptides. Antibodies to specific ligands including cellular receptors are also used.

9.5 Splice mutations and diseases

Aberrant splicing occurs in a number of different diseases. Montes et al. (2019) reviewed splicing defects and development of animal models to help develop therapies. They noted that mistakes in splicing can lead to alterations in transcript code and also to frame shift mutations.

Accurate splicing was noted to be dependent on factors within transcript splice sites and on associated factors and in cofactors involved in splicing. Montes et al.

emphasized that transcripts of a specific gene may undergo splicing at different sites and that there is evidence that 95% of genes are alternatively spliced. Alternative splicing of a specific gene may occur at different stages of development, or in different tissues.

Important elements in splicing include splice regulatory elements (SREs) that occur on primary transcripts; other important elements include trans factors, RNA binding proteins such as serine-arginine rich proteins, and heterogeneous ribonucleoproteins.

Montes et al. noted that 10% of all mutations documented in human disease mutation databases impacted splice sites. In the Human Mutation database (Stenson et al., 2020), mutations that impact trans splicing factors are also documented.

Montes et al. noted the importance that animal models can play in studies of splicing and on the impact of its alterations and that studies of such models are important for development of therapies directed at aberrant splicing.

Montes et al. noted that specific splice switching oligonucleotides have been developed to bind to transcripts of SREs to prevent binding of important proteins. Splice switching oligonucleotides have also been developed to alter activity of splice enhancer or splice inhibitory factors.

For analysis of splicing mutations and development of possible therapies, mouse models of specific mutations have often been developed. Examples include the mdx mouse for studies of DMD mutations leading to Duchenne Muscular Dystrophy. A Duchenne-like phenotype in Golden Retriever dogs arises due to mutation and these dogs also serve as a model for therapeutic applications.

In addition to splice switching oligonucleotides, small molecules have been identified that target splicing factors. Examples include small molecules that impact spliceosomes, such as Spliceostatin. Questions arise regarding the targeting of splice modulators to specific cell types.

Montes et al. reviewed splicing alterations in specific diseases including myelodysplasia syndromes characterized by proliferation of specific subtypes of hematopoietic cells. In myelodysplasia and in acute myeloid leukemia, mutations were reported in splice factors SRF2, U2AF, SFBP1. Splice modulators were developed to target these mutations.

9.6 Antisense therapies under investigation

Bennett et al. (2019) reviewed use of antisense nucleotides under investigation for therapies in specific neurodegenerative diseases, including Huntington's diseases, Alzheimer disease, and amyotrophic lateral sclerosis. They emphasized that beneficial medical therapies were not available for many neurodegenerative disorders including those cited above.

They noted that mechanisms though which antisense RNAs function must initially be tested in cell culture system. Modifications of antisense RNAs were noted to impact their pharmacological properties, tissue distribution, and clearance. Specific modifications include morpholino modifications and phosphorothioate modifications.

It is important to note that after intrathecal antisense oligonucleotides administration, a high percentage of the compound (up to 80%) appeared in the circulation. Studies on the half-life of administered antisense oligonucleotides indicate that this may be 6 weeks — 6 months following injection.

Several conditions where specific strategies to downregulate mutant RNAs are under investigation. These include autosomal dominant diseases due to nucleotide triplet repeat expansions. One example is myotonic dystrophy type 1 due to CUG repeat expansion in the DM1 gene that encodes a protein kinase. A chimeric antisense oligomer was designed to induce degradation of the mutant RNA.

C9ORF 72 mutations that comprise expanded hexanucleotide repeat lead to amyotrophic lateral sclerosis. Investigations to silence this repeat expansion with a gapmer are underway. These investigations are being carried out in neuronal cells. In animal models of the disorder, studies are being carried out using intraventricular administration of the antisense gapmer.

Other therapeutic approaches in specific neurological diseases use silencing to regulate expression of a specific gene. Examples include downregulation of DNM2 dynamin 2, a microtubule associated protein in centronuclear myopathy and downregulation of SOD1 superoxide dismutase 1 in ALS (Amyotrophic lateral sclerosis).

9.7 Genomic data leading to therapeutics

Kaczmarek et al. (2017) noted that the two major classes of compounds included in FDA-approved therapeutics were noted to be small molecules and proteins. Small molecules were noted to be hydrophobic organic compounds that bind to specific sites on proteins. However, such binding sites were noted to be relatively sparse. Antibodies may bind with high specificity to specific proteins; however, antibodies are of large size and are of limited solubility.

Kaczmarek et al. emphasized the therapeutic potential of RNA and DNA because of their high specificity. They noted, however, that therapeutic single stranded RNA is prone to nuclease degradation. It cannot easily cross cell membranes and must be protected from degradation in endosomes.

Engineered adenoviral vectors have been particularly used for RNA delivery. However, nonviral vehicles involving nanoparticle encapsulation are being increasingly explored. Polymers were noted to be used for nanoparticle formation.

Cationic polymers were used for condensation. Amine containing polymers were reported to be used for RNA delivery; these included synthetic poly-L-lysine, a polyaminoamine, and polyethylene-amine.

Howard et al. (2006) reported use of a naturally occurring polymer chitosan for RNA delivery. In delivery of DNA polyaminoesters were reported to be useful.

Kaczmarek et al. reported that lipid-like nanoparticles were also being used as nucleic acid delivery vehicles. Ionizable lipids were thought to escape endosomal uptake and degradation more easily. Semiautomated synthesis of lipid-like molecule synthesis for RNA delivery was being explored.

Kaczamarek et al. noted that an additional approach involved conjugation of N-acylgalactosamine to RNA as this could target RNA to specific receptors on liver cells.

9.7.1 Additional RNA modifications to further improve use in therapy

Specific chemical modifications are also being investigated to identify strategies to improve RNA therapy. Strategies explored include modification of the ribose sugars and modifications of phosphate linkages.

9.8 Pluripotent stem cells for investigation of disease manifestations and effects of therapies

Neural stem cells derived from pluripotent stem cells of patients with lysosomal storage diseases have been developed to obtain further insight into stages of pathology in these disorders. Studies are also undertaken on these cells to investigate the impact of certain therapeutic agents.

Luciani and Freedman (2020) noted that pluripotent stem cells can also be developed into brain organoids; neural stem cells can also be developed to radial glial cells. They specifically considered utilization of pluripotent cells differentiated to neural stem cells in the investigation of certain therapeutic agents proposed to treat lysosomal storage disorders. In Niemann − Pick disease Type 1, patient's neural cells were developed from patient pluripotent stem cells to study effects of certain pharmacologic agents including cyclodextrins, tocopherols, reported promoter autophagy, and reduced accumulation of sphingomyelins.

Effects of gene therapy have been studied in neural stem cells differentiated from patients' cells in cases of mucopolysaccharide storage disease, Sandhoff disease, and Tay − Sachs disease.

Hematopoietic stem cells (HSCs) have also been investigated to determine the effects of gene therapy.

9.9 Gene therapy by adding genes

Specific forms of gene therapy in current use were noted to include viral-based delivery of a specific protein coding gene and evidence that the introduced gene may integrate within the host genome or remain extrachromosomal as an episome. Problems encountered with this strategy involved lack of information on where the introduced gene would integrate in the host genome and potential insertional mutagenesis. Nonintegrating genes that remain as episomes in the host were noted to potentially be safer.

There is evidence that episomes may be lost during cell division; therefore episomes for gene therapy are used primarily for editing nondividing cells.

9.10 Gene therapy

9.10.1 Early gene therapy applications

In a review of gene therapy Zittersteijn et al. (2021) noted that early therapies included treatment of adenosine deaminase (ADA)-related immunodeficiency. ADA encoding sequences were cloned into retroviral sequences for therapy of patent T-cells. Rescue of the ADA deficiency in patients was reported. Another form of immunodeficiency, treated around the same time, involved use of sequences encoding the IL2RG receptor cloned into retroviral vector. It subsequently became clear that the retroviral vector bearing the IL2RG receptor had integrated into the patient genome close to an oncogene and integration was also associated with deletion of a tumor-suppressor gene CDKN2A leading to leukemia in 5 of 19 treated patients. Fortunately, in these cases, leukemia was reported to respond to chemotherapy. None of the ADA-deficient patient treated developed leukemia.

Studies revealed that gamma retroviruses tended to integrate into transcription start sites of active genes.

Shirley et al. (2020) reported that gene therapy with viral introduction is used to treat a number of inherited diseases and also a number of acquired diseases. They noted that innate and adaptive immune responses to viral vectors constitute problems to therapy.

They noted that the viral vector must be capable of carrying and delivering the therapeutic gene and the viral vector must be stripped of pathogenic and replicative elements. Viral vectors in use on gene therapy were noted to include adenovirus, adenoassociated virus, lentivirus, and herpes simplex virus.

Specific viral vectors and gene therapies that have been approved for use in clinical trials include adenoassociated vectors carrying genes for therapy, which are present in LUXturna to treat some forms of congenital blindness and in Zolgensma to treat spinal muscular atrophy.

Shirley et al. outlined additional factors that influence selection of a viral vector for gene therapy in given conditions. They include specific cells or tissue to be targeted, the packaging capacity of the vector, and potential for genome integration of the virus. The adenoassociated viral (AAV) vector has frequently been the most favored vector used in gene therapy. Specific host factors to take into account include pattern receptors that recognize foreign proteins on vectors.

Adenoassociated viruses were reported to elicit a weak inflammatory response and to therefore have more favorable safety. However, CD8 T-cell responses to the vector have been reported in some studies (Nidetz et al., 2020).

Shirley et al. also noted that immunotoxicities can result if high doses of vector are infused in gene therapy. These can impact CNS, liver, skeletal, lung, cardiac muscles, and eye. These responses were considered to involve responses to capsid antigens. An additional concern that has emerged is the risk of antigen response to the transgene product. This was noted in some cases where large doses of the therapeutic product were administered.

Transient therapy with immune-suppressive agents was noted to be potentially valuable in these cases.

Lentiviral vectors were reported to be useful in therapy of both dividing and nondividing cells. Lentiviral vectors derived from human immunodeficiency virus have been used in cancer therapy. The incidence of preexisting immunity against lentiviral vectors was reported to be low.

Shirley et al. concluded that immune response to vectors or transgenes remains as hurdles to gene therapy.

Safe harbor gene therapy involves introduction of new DNA into a safe location, into a suitable position or into a location with suitable epigenetic modifications (Pellenz et al., 2019). This study identified 35 new sites for targeted transgene insertion that have the potentially to serve as new human genomic "safe harbor" sites (SHS). SHS include potenetially 35 sites located on 16 different chromosomes, including both arms of the human X chromosome.

In a report of gene therapies approved for clinical treatment in the United States and Europe, High and Roncarolo (2019) reported that since 2016 six different gene therapies were approved. Two of these were therapies for B-cell cancers. Four were approved for treatment of monogenic disease, these included beta thalassemia, a form of vision loss, spinal muscular atrophy, and a form of immunodeficiency.

They also noted that clinical developments involving cell and gene therapies were underway for more than 800 different diseases for which therapies were not available in 2019. The goals of therapies were to integrate therapeutic genes into precursor or stem cells, also to ensure that the integrated gene was stable, and extrachromosomal location of the introduced gene was favored. The procedures thus involved transduction of patient stem cells and then transplantation of treated stem cells.

It was noted that specific viral vectors used in gene therapy could be associated with significant immune responses in patients. High and Roncarolo noted that lentiviral and adenoassociated viruses were less likely associated to be associated with immune response and were thus used in gene therapy.

High and Roncarolo noted that lentiviral vectors were suitable for transferring genes into hematopoietic cells. However, large-scale production of lentiviral vectors was problematic. Lentiviral vectors were reported to have been used to treat certain blood disorders and metabolic disorders. They were used in treatment of Wiskott − Aldrich syndrome, adrenoleukodystrophy, and metachromatic leukodystrophy.

Gene therapy using lentiviral vectors for gene transfer into HSC was used in treatments of adrenoleukodystrophy and metachromatic leukodystrophy. There were claims that gene treated bone-marrow stem cells gave rise to cells that migrated to the brain. However, high viral copy number was noted to be necessary in these cases.

Some studies have reported immune response problems when AAV vectors were used in therapy. However, AAV vectors have been approved for treatment of autosomal recessive blindness with RPE65 gene therapy. The retina is known to be an immune-privileged site.

9.11 Stem cells and importance in gene therapy

HSCs and stem cell-derived neurons have also been investigated to determine the effects of gene therapy. Effects of gene therapy have been studied in neural stem cells differentiated from patients' cells in cases of mucopolysaccharide storage disease, Sandhoff disease, and Tay − Sachs disease

9.11.1 Hematopoietic stem cells (HSC) and therapies

Ferrari et al. (2020) reviewed aspects of gene therapy in HSCs. They reported that hematopoietic cell transplants have been used to treat specific genetic diseases for more than 50 years. They noted that in (1968) there were reports of treatment of patients with specific types of immunotherapy using HSC transplants.

Ferrari et al. noted that improvements have been made in donor matching, as incomplete matching of donor and recipient resulted in graft versus host reactions in some cases.

Autologous HSC therapy together with gene transfer has been used for treatment of a number of genetic disorders. In successful cases, the treated autologous stem cells were reported to undergo further division so that the benefits of therapy extended over time. Improvements have been made in gene therapy and viral vectors such as gamma retroviral vectors that presented problems are no longer used. In addition to their use in gene therapy, Ferrari et al. noted that HSCs can potentially be used as delivery vehicles for therapeutic proteins.

9.11.2 Collection of hematopoietic stem cells for therapy

This was reported to require multiple collections of bone marrow samples from iliac crest In some cases this collection is followed by leukapheresis to specifically collect CD34 + cells that represent cells that include HSCs and their progenitors.

In some situations specific agents are administered to patients to promote hematopoiesis and if stimulation is significant it is sometimes possible to collect adequate number of stem cells by leukapheresis of peripheral blood. One pretreatment agent used is Plexifor that promotes release of stem cells into peripheral blood. However, plexifor cannot be used in all patients.

Ferrari et al. noted that lentiviral vectors were often used for gene transfer. Modifications to vectors were made to inactivate viral promoter regions, to limit transcription of viruses following introduction, and to produce replication defective viruses.

Ferrari et al. explored evidence of utility of HSC as delivery vehicles for certain enzymes or proteins. Questions arise regarding the efficacy of modified HSC in treating conditions that impacted brain, bone, or other organs.

Ferrari et al. reported that there is some evidence that enzyme proteins released from bone marrow stem cells could be advantageous in reducing enzyme deficiency. There is evidence that HSCs give rise to monocytes and microglia that pass through capillaries and paravascular channels in the interior of the brain.

9.12 Gene editing

9.12.1 Early discoveries

In 1994 Choo and Klug (1994) reported on the capacity of certain zinc finger proteins to bind to DNA. Specific zinc finger units bind to single stranded DNA. Each specific zinc finger binds to three nucleotides and zinc fingers can be linked. The linked zinc fingers can bind to a DNA target of 18 nucleotides. The zinc fingers can also be fused to an endonuclease Fok1. Zinc fingers are designed to bind to two opposite strands of DNA and the endonuclease can then cleave DNA between the bound zinc fingers thus creating a double stranded break. If repair by homologous recombination is required, repair sequence can be included. Specific vectors, for example, adenoviral vectors can be utilized to transport the zinc fingers, the nuclease, and the repair sequence.

TALENs are related to plant transcription factors. Each TALEN is composed of 33–35 amino acids; the different TALENs differ from each other by two amino acids. The two amino acids, one at position 13 and one at position 12, distinguish a TALEN that binds to a specific nucleotide. The TALEN HD binds to nucleotide **C**, TALEN NI binds to nucleotide **A**, TALEN HG or NG binds to nucleotide **T**, TALEN NK binds to nucleotide **G**; TALEN NN can recognize G or A nucleotide. In addition, the TALEN editing system has an endonuclease position between the left and right arm that targets opposite sides of the DNA strand.

TALENs were noted to be easier to generate than zinc fingers but they are very large (Chandrasegaran and Carroll, 2016).

In 2020 Ernst et al. (2020) reviewed aspects of gene editing and its entry into the clinic. They noted use of zinc finger nuclease (ZFN), TALENs, and aspects of

CRISPR-Cas editing. ZFN and TALENs were noted to involve induction of double stranded DNA breaks mediated by FOK1 nuclease. In CRISPR-CAS technologies the CAS nuclease leads to double stranded breaks.

Specific clinical applications of gene editing included cancer therapy, HIV infections to disrupt the CCR5 receptor, and particular genetic disorders including hemoglobinopathies, hemophilia, mucopolysaccharidoses, and Leber's congenital amaurosis.

Gene editing methods were reviewed with CRISPR-Cas 9 by Broeders et al. (2020). In the CRISPR-CAS9 systems, nucleotide sequences in the guide RNA can be changed to match other DNA regions and the guide RNA can be coupled to the CAS cleavage system. Guide RNA takes CAS to regions that need to be edited. However, a specific sequence, the PAM sequence, to determine binding of CAS9 must flank the site to be edited in the genome. The cleaved DNA can be repaired using the body's own repair systems or DNA elements can be provided to be inserted following cleavage. Repair pathways following gene editing can include homologous directed gene repair, nonhomologous gene repair, and end joining. Repair by end joining can be used to remove a DNA segment.

Gene editing in patients has thus far involved bone marrow HSCs or T-lymphocytes. Following the editing procedure, cells are cultivated in vitro to determine accuracy or editing; following this they may be introduced into patients.

Ernst et al. noted that off-targeted editing remains a key problem. On-target editing can also be followed by impaired introduction of elements. There are challenges in delivery of gene editing systems to human cells. Different delivery systems are being devised including liposomes and additional vectors and nanotubes.

Ex vivo approaches involve modifying cells outside the patients. Cells used in these approaches include hematopoietic cells, stem cells, or T-cells

9.13 Delivery of reagents for editing

This sometimes involves use of viral vectors, for example, AAV (adenoassociated vectors), problems with developments of abnormal immune reactions following editing to adenoviral vectors have been described.

There are challenges in delivery of gene editing systems to human cells. Different delivery systems are being devised including liposomes, lipid nanoparticles, and additional vectors and nanotubes.

9.14 Preclinical and clinical trials

Ernst et al. reviewed a number of preclinical and clinical trials in place that utilize gene editing with adenoviral vectors. US trials completed include HIV therapy with ZFN.

Ongoing trials include:

1. Gene therapy in hemoglobinopathies, beta thalassemia, and sickle cell disease.
2. Cancer immunotherapy trials and strategies directed toward immune checkpoints. Cancer-related gene therapy investigations were reported to include deletion of HPV (human papilloma virus) in cervical cancer.
3. MPS1 (mucopolysaccharidosis type 1) ZFN therapies.
4. Leber's congenital amaurosis CRISPR-Cas.
5. Deletion of CCR5 (C-C motif chemokine receptor 5) receptor gene in HIV infections.

Ferrari et al. emphasized that targeted gene editing could ensure that the targeted gene remained under control of endogenous regulatory elements. They noted that gene editing proof of concept studies had been carried out in hemoglobinopathies, in severe combined immunodeficiency, and in Wiskott – Aldrich syndrome.

It is known that double stranded DNA breaks induced by gene editing nucleases can be repaired by homologous recombination if donor homologous sequence is available. In some cases, donor sequence elements need to be introduced. There is evidence that introduction of donor template sequence using adenoassociated virus has yielded encouraging results.

DNA double stranded breaks can also be repaired by nonhomologous end joining. This mechanism could be useful to induce knock-out of damaging sequences.

Ferrari et al. noted that chemical modifications in CRISPR-CAS editing reagents were being investigated. These include chemical modifications of the single strand guide RNA and adaptation of the CAS9 ribonucleoproteins. Modifications in agents used to transfer sequences were also being investigated such as use of adenovirus type 6.

Evaluation of the proportion of treated cells correctly modified prior to transfer were necessary. They noted that high doses of modified HSCs were often required.

In some cases, preconditioning of patients was required to reduce the number of cells within the bone marrow so that the transplanted cells could gain a foothold. This preconditioning required immunosuppression and was not recommended in some patients.

Ferrari et al. explored evidence of utility of HSCs as delivery vehicles for certain enzymes or proteins. Questions arise regarding the efficacy of modified HSC in treating conditions that impacted brain, bone, or other organs.

There is some evidence that enzyme proteins released from bone marrow stem cells could be advantageous in reducing enzyme deficiency. There is evidence that HSCs give rise to monocytes and microglia can pass through capillaries and paravascular channels in the interior of the brain.

9.14.1 Delivery of agents for gene editing

Ates et al. (2020) reported that a major barrier to therapeutic gene editing that remained, included scarcity of ideal methods for delivery of reagents. They also drew attention to liver-directed gene editing for mucopolysaccharidosis and hemophilia B. CRISPR-CAS gene editing tools were noted to have been undertaken for treatments of hemoglobinopathies. They emphasized that in certain diseases, for example, muscular dystrophies, editing needed to occur in widespread cells.

In considering delivery systems they noted use of adenoassociated viruses. They also mentioned hydrodynamic delivery, electroporation, and the use of lipid nanoparticles and cell penetrating peptides.

Ates et al. noted that hydrodynamic delivery was developed primarily for delivery of naked DNA. It has also been used for the delivery of siRNAs and small molecule chemical agents.

9.15 NIH (National Institutes of Health) somatic cell gene editing program

Saha et al. (2021) reviewed the NIH somatic gene editing program designed to investigate and establish safer gene editing methods. The plan includes investigations of new genome editors and effective delivery technologies and also to track editing cells and to assess their effects.

They noted that in some disorders addition of a functioning gene can lead to death, while in other cases there are benefits to editing a defective gene within a patient. Gene editing was noted to include diverse technologies including use of ZFN (zinc finger nucleases, TALENs, (plant transcription factors) or CRISPR-Cas. However, they also noted that challenges remain, including adequate repair of nuclease-induced edits. Other gene editing processes involve nucleotide base editing. Many gene editing studies involve utilization of animal models of disease.

Genome editing processes have been carried out in vivo in some cases while in other cases they are carried out in vitro on patients' cells and treated cells are subsequently administered to patients.

One consideration is that the patient may develop immunity to the delivery cells or to the delivery product.

There has also been progress in use of gene editing of patient T-cells for subsequent immunotherapy in cancer.

Development of safe and effective delivery systems is a key goal. Many of the currently used delivery vectors were not considered to be optimal.

Risks of off-target effects including delivery to inappropriate tissues remained a concern. In addition, currently there are limits in the number and types of tissues to which agents for editing can be delivered.

Editing platforms currently available were noted to include CRISPR system with Cas9 and Cas 12 homologs. Other systems being investigated included a CasX system. Liu et al. (2019) reported a new form of Cas protein referred to

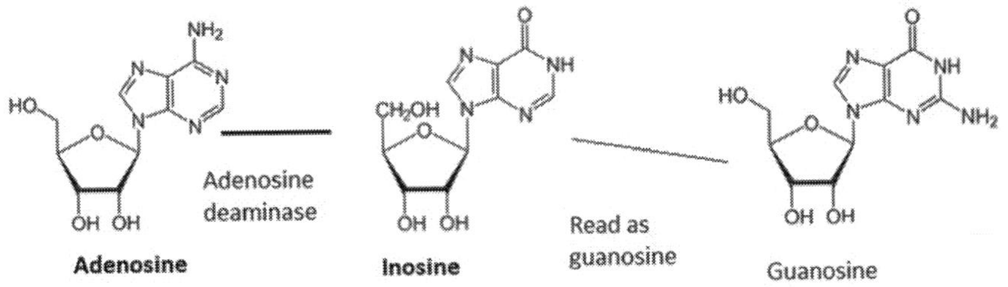

FIGURE 9.1

Example of Base editing.

as CasX and sometimes referred to as Cas12e. It was determined to be an RNA-guided DNA endonuclease that generated staggered double stranded DNA breaks with 20 nucleotides complementary to the guide RNA present in CasX. A particular advantage was the small size of CasX protein, 1000 nucleotoide approximately, its cleavage characteristics, and the fact that it was derived from ground water organisms rather than from an infectious agent. The guide RNA could constitute 26% of the mass. CasX was shown to be capable of cleaving human DNA.

Other editing systems being considered included enzymes DNA helicases, transposase, and epigenetic editors being considered (Fig. 9.1).

9.16 Base editing

(Porto et al., 2020) emphasized that many genetic disorders are known to be caused by single base mutations. They explored possibilities for gene editing of these and emphasized the need for editing tools with high on-target efficiency and minimal off-target effects. Tools initially used included ZFN or TALEN nucleases that required protein synthesis. The CRISPR-CAS9 system was noted to represent a more usable tool. Newer versions of CAS, including CAS12 and CAS13, were noted to be more efficient.

9.17 Programmable base editing

Lewis et al. (2016) reported that deamination of cytosine and of methyl-cytosine occurred frequently in microorganisms. Also this change was frequently observed in mutations in human disease. They examined the mechanism of cytosine deamination. Hydrolytic deamination of cytosine was reported to yield uracil. Uracil may be removed by uracil DNA glycolase.

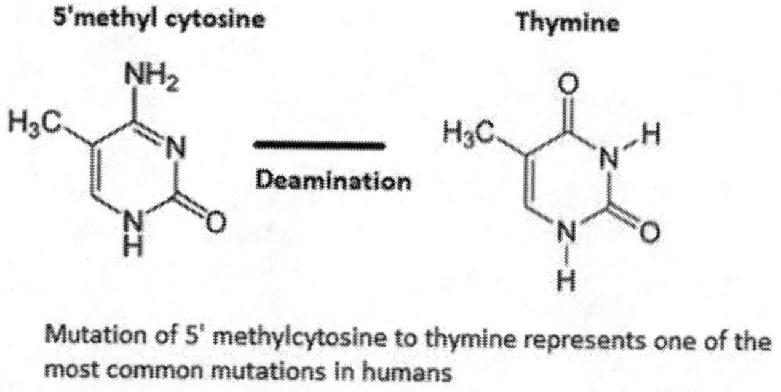

FIGURE 9.2

Example of Base editing.

Mutation of 5' methylcytosine to thymine represents one of the most common mutations in humans.

5-methyl cytosine can undergo deamination to yield thymine and ammonia. The 5-methyl cytosine to thymine conversion was reported to be the most common nucleotide conversion and cytosine to thymine conversion (C to T mutations) was reported to represent common mutations.

Adenine undergoes conversion to inosine, which is read as guanine by polymerase.

Gaudelli et al. (2017) reported that AT to CG mutations represented half of the pathogenic mutations in the human genome.

The development of adenine editors and cytosine editors therefore became a priority.

Base editors were sometimes referred to as Cas9 nickases; Cas 9 was linked to enzymes that included cytidine deaminase or adenine deaminase.

In a 2020 review Kantor et al. noted that base editing could correct point mutations without first inducing double stranded DNA breaks (Fig. 9.2).

9.18 Prime editing

Anzalone et al. (2019) noted that as of July 2019, 75,122 known pathogenic human gene variants were documented in ClinVar.

They described a process that they described as prime editing. This involved use of a modified catalytically impaired Cas9 fused to a reverse transcriptase, and to a prime editing RNA (peg RNA).

The peg RNA specified the target site correction and also the edit. This method enabled editing of insertions and deletions and point mutation. Using prime editing systems they targeted insertions, deletions, and 12 point mutations.

In 2021 Zittersteijn noted that prime editing included a Cas form that nicks DNA and has a guide RNA (pegRNA) that is attached to a reverse transcriptase to synthesize a new DNA from the guide RNA.

9.19 CRISPR-Cas theta

In 2020 Pausch et al. (2020) reported use of a CRISPR-Cas specific system derived from large bacteriophage. The system includes a CRISPR array and Cas theta. It was noted to be a compact system with Cas theta being a 70 kilodalton protein. The molecular size of Cas theta was noted to be half that of Cas9 or Cas12a. The new system was noted to have significant advantages in cellular delivery.

9.20 RNA editing

RNA editing involves modification at the transcript levels and has a lower risk of off-target genomic modifications. Fry et al. (2020) noted that specific RNA editing technologies utilize enzymes that induce adenosine to inosine conversion A to I or cytosine to uracil conversion. Site-directed reagents are used to deliver RNA editing enzymes to targets in RNA that correspond to coding sequence mutations G to A or T to C.

Specific local sequence context can influence the ability of ADAR1 ADA RNA-specific, to deaminate adenosine. There is also some evidence that specifically induced mutations in ADAR1 can influence the editing capability.

ADAR2 has a more limited tissue distribution. However, specific mutations have been introduced to ADAR2 to improve the utility.

With respect to RNA editing and CRISPR-CAS, Fry et al. noted that type VI CRISPR-Cas nuclease CAS13 under direction of a guide was noted not to require PAM, protospacer adjacent motif.

Fry et al. noted that nucleotide editing can be used to treat retinal degeneration due to G to A or T to C mutation. RNA editing may be feasible. However, in dominant disease editing of specific alleles may be necessary.

However, they did note that delivery of DNA and RNA base editing reagents is an ongoing challenge.

9.21 Gene therapy in specific diseases

9.21.1 Eye diseases and retinal degeneration

Fry et al. (2021) reported utilization of RNA editing therapy in treatment of genetic disorders associated with retinal degeneration. Examples of large genes with single point mutations leading to specific forms of retinal degeneration include *USH2A* and *ABCA4*.

USH2A encodes usherin, a laminin-like protein found in the basement membrane and may be important in development and homeostasis of the inner ear and retina. Mutations within this gene have been associated with Usher syndrome type IIa and retinitis pigmentosa.

ABCA4 ATP-binding cassette (ABC) transporter. ABC proteins transport various molecules across extra and intracellular membranes. Mutations in this gene are also associated with retinitis pigmentosa-19, cone-rod dystrophy type 3, early-onset severe retinal dystrophy, fundus flavimaculatus.

Defects in these genes were noted to be present in 25% of cases of genetic retinal degeneration in the United States.

Fry et al. noted that genetic therapies are particularly applicable in the retina because of easy access for delivery, and further because of immune privilege of the retina and because functional improvement can be readily assessed. Thus the FDA approval was obtained for gene therapy for gene defects that lead to Leber congenital amaurosis and occur in REP65. This gene encodes retinoid isomerohydrolase. RPE65 is a component of the vitamin A visual cycle of the retina, which supplies the 11-cis retinal chromophore of the photoreceptors, opsin visual pigments (Russell et al., 2017).

Fry et al. noted that AAV vectors were most commonly used in gene therapy for retinal disease, as they were demonstrated to have tropism for retinal pigment epithelium and photo receptors. These structures are most frequently damaged in retinal degeneration. Also overexpression of transgenes can lead to cell damage.

The packaging capacity of AAV was noted to be 4–7 kb. Importantly, given their size the ABCA$ and USH2A could not be accommodated in the AAV vectors. They noted that CRISPR-Cas9 has been used to reprogram photoreceptors in mouse models of retinal degeneration. They emphasized that gene editing approaches that require a donor template for homology-dependent correction of DNA cleavage have a lower success rate. Approaches that involve base editing without double stranded DNA breaks were reported to be more successful.

Fry et al. emphasized that defects in large genes may be beyond the possibilities of gene replacement therapy based on their size. The large genes cannot be accommodated in AAV vectors. They listed the six large genes most commonly implicated in retinal degeneration and the size of these genes.

These genes, their size in kilobases, and their functions are listed below.

1. ABCA4 6.81Kb ATP-binding cassette (ABC) transporter.
2. USH2A 15.6 Usherin a basement membrane protein.
3. CEP290 7,4 centrosomal protein.
4. MYO7A 6.65 myosin 7a a mechanochemical protein, serves as an anchor.
5. EYS 9.43 eyes shut, protein contain multiple epidermal growth factor and lamin domains.
6. CDH23 10.1 cadherin-related 23, encodes calcium dependent cell – cell adhesion glycoproteins.

9.22 Molecular analyses and therapies relevant to hearing loss

Appler and Goodrich (2011) reported new techniques and improved inner ear access have revealed details regarding connections of the peripheral auditory systems and connection with the central auditory circuit. Further information has been gathered on the arrangement of the inner and outer hair cells, the sound and fluid movements that lead to hair cell movement, and to signaling to the spiral ganglion that is connected via axons to the hair cells and that also sends signals to the 8th cranial nerve. There is evidence that the different spiral ganglion neurons differ in the sounds to which they respond.

Delmaghani and El-Amraoui (2020) reviewed progress in development of inner ear therapies for hearing loss. Disabling hearing loss was reported by the World Health Organization to impact 5% of the world's population. Congenital hearing impairment was reported to occur in 1 in 500 newborns.

The authors of this review noted that increases in knowledge of the molecular mechanism of the auditory and vestibular systems could be applied to analysis of hearing loss. Improved therapy was dependent upon determination of the nature of factors that damage hearing and information on target cells. Cells that were impaired included auditory hair cells, supporting cells, or neurons.

Delmaghani and El-Amraoui noted that development of therapies must take into account auditory hair cells that react to and amplify sound stimuli and inner hair cells that transmit stimulation to the nervous system. Outer hair includes between 9000 and 12,000 cells arranged in three rows. Hair cells were reported to carry actin-rich stereocilia on their apical surfaces. In response to stimuli, inner hair transmit stimuli that ultimately lead to depolarization of cells and release of neurotransmitters.

Early onset of prelingual deafness includes syndromic hearing loss. when hearing loss and other physical or functional abnormalities occur and nonsyndromic hearing loss when hearing loss occurs as apparently the only abnormality. Nonsyndromic hearing loss due to genetic factors includes autosomal dominant, autosomal recessive, X-linked forms, and can arise as a result of mitochondrial defects.

Delmaghani and El-Amraoui noted that defects in more than 140 genes have been reported to lead to hearing loss. The first genetic linkage of hearing loss was reported in 1994 when Guilford et al. (1994) reported mapping of nonsyndromic form of neurosensory, recessive deafness to the pericentromeric region of chromosome 13q. The map region was refined to be 13q12 and the relevant gene was found to encode a gap junction protein connexin 26.

Delmaghani and El-Amraoui documented the different functions impacted by genes with defects that lead to hearing loss. They included genes involved in hair bundle formation and function, genes involved in hair cell adhesion and maintenances. Other genes implicated in hearing loss included gene involved in cochlear ion homeostasis, transmembrane secreted proteins and extracellular matrix components, genes encoding products involved in control of oxidative stress, metabolism, and mitochondrial functions, and genes involved in transcription regulation.

Delmaghani and El-Amraoui noted that the inner ears represent an available site for gene therapy in part because it is relatively isolated, and second because the fluid circulation in ear can disseminate therapeutic agents.

In focusing on specific approaches in gene therapy they noted that using direct administration reagents could be delivered to vestibule structures including the semicircular canal, utricle, and cochlear round window. Injection in the posterior semicircular canal could be used to treat auditory and vestibular dysfunction.

Early in gene therapy, approaches were carried out in mouse models of genetic deafness. Both viral vectors and nonviral vectors were used; the latter included nanoparticles, microsphere of biodegradable polymers, including liposomes.

The most promising viral vectors were reported to be adenoviral vectors and at least 12 different serotypes and variants of these have been identified. Different serotypes of adenovirus were noted to have different tropisms.

Specific serotypes were identified to administer therapy. The adenoassociated virus AAV2 serotypes 2/1, 2/2, 2/9, 26/27 were found to have tropism for hair cells. V9 serotypes were found to be useful in transmitting gene therapies to the cochlea.

However, one disadvantage of the AAV vectors was their limited capacity to accommodate genes lager than 4.8 kb.

Nonviral delivery was shown to be less efficient but more stable and flexible.

9.22.1 RNA-based therapies in deafness

Delmaghani and El-Amraoui noted that antisense oligonucleotides and short interfering RNAs or microRNAs provided possibilities for treating dominant negative deafness. The harmonin encoding gene implicated in USHER syndrome C has a specific mutation c.216 G > A, which leads to generation of a cryptic spice site that in turn leads to a truncated harmonin protein. This defect is associated with early onset profound deafness. Specific antisense oligonucleotides were generated to block the cryptic splice site.

Delmaghani and El-Amraoui noted that RNA interference (RNAi) approaches have also been investigated to counteract the common connexin 26 defects that leads to deafness DFN3. This is an autosomal dominant progressive deafness. RNAi can be used to selectively inhibit expression of the mutant allele. This was achieved using liposomes containing inhibitory RNA that bound to the mutant allele.

A specific microRNA was developed to reduce expression of a function mutation in the gene TMC1 transmembrane channel like 1. It is known to be required for normal function of cochlear hair cells. Mutations in this gene have been associated with progressive postlingual hearing loss and profound prelingual deafness. Gene editing approaches were used in mouse studies to target DFNA36 mutation in TMC1.

9.22.2 Efforts to promote hair cell regeneration

These include overexpression of the MOTH1 transcription factor that is important for differentiation of hair cells. MOTH1 overexpressing genes were introduced into the cochlea.

Local delivery of growth factor such as BDNF and neurotrophin NTF into the cochlea have also been shown to be beneficial in animal models.

Animal models of deafness, particularly mouse models, have been utilized to investigate possibilities for gene replacement, gene editing, and RNA-based therapeutics.

9.23 Therapy of cystic fibrosis including genetic approaches

In an editorial published in 2020, Pedemonte noted that prior to 2012 therapy for cystic fibrosis (CF) focused on cooccurring infections, deficits in mucus clearance in lungs, and intestinal malabsorption (Pedemonte, 2020).

Subsequently, treatment with CFTR (cystic fibrosis transmembrane conductance regulator) gene product modulators or potentiators followed. These included ivacaftor, or combinations including tezacaftor, ele:acaftor, and ivacaftor. They noted that in vitro demonstration of modulator efficacy was not always reflected in in vivo efficacy.

Pedemonte concluded that there is still an urgent need for novel drugs. Some of the new approaches included potentiators of expression other gene products, for example, TMEM16a also known as anomactin ANO1, a calcium-activated chloride channel.

Pedemonte noted that airway surface homeostasis is important. Mitash et al. (2020) considered molecular methods including the use of microRNA-based approaches to increase production of components necessary to promote airway surface homeostasis. Specific microRNAs were shown to impact viability of mRNA of certain genes involved in epithelial surface homeostasis including CFTR, ANO1, EnaC (SLNN1A sodium channel) SLC26A9 (Solute carrier 26A9). TGFB1 (transforming growth factor B1) was also shown to impact surface homeostasis through recruitment of specific microRNAs to CFTR. Mitash et al. (2020) noted that elucidation of effects of specific microRNAs on mRNA of gene involved in surface homeostasis could provide insights into new therapeutic approaches.

Maule et al. reviewed gene therapy approaches to cystic fibrosis. A critically important discovery in cystic fibrosis was that it arose due to damaging mutation in a specific gene that encodes a cyclic AMP-regulated chloride channel CFTR. Because of the importance of this channel, particularly in lung epithelial cells, manifestation of CFTR malfunction occurs primarily in the lung and also in other body tissues.

Maule et al. (2020) noted that 352 different CFTR mutations have been found to be disease-causing. Partial progress in treatment of cystic fibrosis resulted from use of small molecules that improved trafficking or modification of mutant CFTR protein. However, 10% of cases were noted to not respond to small molecule therapy.

They noted that there are severe challenges to gene therapy including production of large quantities of mucus and the fact that epithelial cells are constantly being renewed. Maule noted that gene editing is now being considered. In this approach, site-specific DNA cleavage will be carried out followed by target integration of corrected sequence. Correction of the delta F508 deletion is being investigated in vitro. Splicing mutations were reported to occur in 10% of cases and successful in vitro editing of these in CF patient cells was reported. Investigation of base editing approaches in in vitro studies was noted to be under investigation.

However, genetic approaches to therapy have also been proposed (Maule et al., 2020). Other delivery systems being considered are AAV vectors, nonviral delivery systems, engineered guide RNAs, nonviral vesicles including liposomes, peptides, peptide nucleic acid combination.

9.23.1 Ongoing problems needing to be addressed in gene therapies

One translational problem currently being investigated is that the therapeutic payload may not enter the nucleus of the cell.

An ongoing concern is that the patients may develop immunity to the delivery vehicle or the delivery product.

Development of safe and effective delivery systems is a key goal. Many of the currently available gene delivery systems were not considered optimal.

Another concern involves off-target effects. The report noted that there are currently limitations to the number and types of tissues to which agents for editing are delivered.

References

Anzalone AV, Randolph PB, Davis JR, et al: Search-and-replace genome editing without double-strand breaks or donor DNA, *Nature* 576(7785):149–157, 2019. Available from: https://doi.org/10.1038/s41586-019-1711-4. PMID: 31634902.

Appler JM, Goodrich LV: Connecting the ear to the brain: molecular mechanisms of auditory circuit assembly, *Prog Neurobiol* 93(4):488–508, 2011. Available from: https://doi.org/10.1016/j.pneurobio.2011.01.004. PMID: 21232575.

Ates I, Rathbone T, Stuart C, Bridges PH, Cottle RN: Delivery approaches for therapeutic genome editing and challenges, *Genes (Basel)* 11(10):1113, 2020. Available from: https://doi.org/10.3390/genes11101113. PMID: 32977396.

Bennett CF: Therapeutic antisense oligonucleotides are coming of age, *Annu Rev Med* 70:307–321, 2019. Available from: https://doi.org/10.1146/annurev-med-041217-010829. PMID: 30691367.

Bennett CF, Krainer AR, Cleveland DW: Antisense oligonucleotide therapies for neurodegenerative diseases, *Annu Rev Neurosci* 42:385–406, 2019. Available from: https://doi.org/10.1146/annurev-neuro-070918-050501. PMID: 31283897.

Broeders M, Herrero-Hernandez P, Ernst MPT, et al: Sharpening the molecular scissors: advances in gene-editing technology, *iScience.* 23(1):100789, 2020. Available from: https://doi.org/10.1016/j.isci.2019.100789. PMID: 31901636.

Chandrasegaran S, Carroll D: Origins of programmable nucleases for genome engineering, *J Mol Biol* 428(5 Pt B):963–989, 2016. Available from: https://doi.org/10.1016/j.jmb.2015.10.014. PMID: 26506267.

Choo Y, Klug A: Toward a code for the interactions of zinc fingers with DNA: selection of randomized fingers displayed on phage, *Proc Natl Acad Sci U S A* 91(23):11163–11167, 1994. Available from: https://doi.org/10.1073/pnas.91.23.11163. PMID: 7972027.

Crooke ST, Witztum JL, Bennett CF, Baker BF: RNA-targeted therapeutics, *Cell Metab* 27(4):714–739, 2018. Available from: https://doi.org/10.1016/j.cmet.2018.03.004. PMID: 29617640.

Delmaghani S, El-Amraoui A: Inner ear gene therapies take off: current promises and future challenges, *J Clin Med* 9(7):2309, 2020. Available from: https://doi.org/10.3390/jcm9072309. PMID: 32708116.

Ernst MPT, Broeders M, Herrero-Hernandez P, et al: Ready for repair? Gene editing enters the clinic for the treatment of human disease, *Mol Ther Methods Clin Dev.* 18:532–557, 2020. Available from: https://doi.org/10.1016/j.omtm.2020.06.022. eCollection 2020 Sep 11. PMID: 32775490.

Ferrari S, Jacob A, Beretta S, et al: Efficient gene editing of human long-term hematopoietic stem cells validated by clonal tracking, *Nat Biotechnol.* 38(11):1298–1308, 2020. Available from: https://doi.org/10.1038/s41587-020-0551-y. Epub 2020 Jun 29. PMID: 32601433.

Fry LE, McClements ME, MacLaren RE: Analysis of pathogenic variants correctable with CRISPR base editing among patients with recessive inherited retinal degeneration, *JAMA Ophthalmol.* 139(3):319–328, 2021. Available from: https://doi.org/10.1001/jamaophthalmol.2020.6418. PMID: 33507217.

Fry LE, Peddle CF, Barnard AR, McClements ME, MacLaren RE: RNA editing as a therapeutic approach for retinal gene therapy requiring long coding sequences, *Int J Mol Sci* 21(3):777, 2020. Available from: https://doi.org/10.3390/ijms21030777. PMID: 31991730.

Gaudelli NM, Komor AC, Rees HA, et al: Programmable base editing of A•T to G•C in genomic DNA without DNA cleavage, *Nature.* 551(7681):464–471, 2017. Available from: https://doi.org/10.1038/nature24644. PMID: 29160308.

Gomez IG, MacKenna DA, Johnson BG, et al: Anti-microRNA-21 oligonucleotides prevent Alport nephropathy progression by stimulating metabolic pathways, *J Clin Invest* 125(1):141–156, 2015. Available from: https://doi.org/10.1172/JCI75852. PMID: 25415439.

Guilford P, Ben Arab S, Blanchard S, et al: A non-syndrome form of neurosensory, recessive deafness maps to the pericentromeric region of chromosome 13q, *Nat Genet* 6(1):24–28, 1994. Available from: https://doi.org/10.1038/ng0194-24. PMID: 8136828.

Gupta SC, Awasthee N, Rai V, et al: Long non-coding RNAs and nuclear factor-κB crosstalk in cancer and other human diseases, *Biochim Biophys Acta Rev Cancer* 1873(1):188316, 2020. Available from: https://doi.org/10.1016/j.bbcan.2019.188316. PMID: 31639408.

Hammond SM, Aartsma-Rus A, Alves S, et al: Delivery of oligonucleotide-based therapeutics: challenges and opportunities, *EMBO Mol Med.* 13(4):e13243, 2021. Available from: https://doi.org/10.15252/emmm.202013243. Epub 2021 Apr 6. PMID: 33821570.

High KA, Roncarolo MG: Gene therapy, *N Engl J Med* 381(5):455–464, 2019. Available from: https://doi.org/10.1056/NEJMra1706910. PMID: 31365802.

Howard KA, Rahbek UL, Liu X, Damgaard CK, et al: RNA interference in vitro and in vivo using a novel chitosan/siRNA nanoparticle system, *Mol Ther* 14(4):476–484, 2006. Available from: https://doi.org/10.1016/j.ymthe.2006.04.010. PMID: 16829204.

Kaczmarek JC, Kowalski PS, Anderson DG: Advances in the delivery of RNA therapeutics: from concept to clinical reality, *Genome Med* 9(1):60, 2017. Available from: https://doi.org/10.1186/s13073-017-0450-0. PMID: 28655327.

Kantor A, McClements ME, MacLaren RE: CRISPR-Cas9 DNA base-editing and prime-editing, *Int J Mol Sci* 21(17):6240, 2020. Available from: https://doi.org/10.3390/ijms21176240. PMID: 32872311.

Lewis CA Jr, Crayle J, Zhou S, et al: Cytosine deamination and the precipitous decline of spontaneous mutation during Earth's history, *Proc Natl Acad Sci USA* 113(29):8194–8199, 2016. Available from: https://doi.org/10.1073/pnas.1607580113. PMID: 27382162.

Li D, Sun L: MicroRNAs and polycystic kidney disease, *Kidney Med.* 2(6):762–770, 2020. Available from: https://doi.org/10.1016/j.xkme.2020.06.013. eCollection 2020 Nov-Dec. PMID: 33319200.

Liu JJ, Orlova N, Oakes BL, et al: CasX enzymes comprise a distinct family of RNA-guided genome editors, *Nature.* 566(7743):218–223, 2019. Available from: https://doi.org/10.1038/s41586-019-0908-x. PMID: 30718774.

Luciani A, Freedman BS: Induced pluripotent stem cells provide mega insights into kidney disease, *Kidney Int* 98(1):54–57, 2020. Available from: https://doi.org/10.1016/j.kint.2020.04.033. PMID: 32571490.

Maule G, Arosio D, Cereseto A: Gene therapy for cystic fibrosis: progress and challenges of genome editing, *Int J Mol Sci* 21(11):3903, 2020. Available from: https://doi.org/10.3390/ijms21113903. PMID: 32486152.

Meng L, Ward AJ, Chun S, et al: Towards a therapy for Angelman syndrome by targeting a long non-coding RNA, *Nature.* 518(7539):409–412, 2015. Available from: https://doi.org/10.1038/nature13975. PMID: 25470045.

Mitash N, E Donovan J, Swiatecka-Urban A: The role of microRNA in the airway surface liquid homeostasis, *Int J Mol Sci* 21(11):3848, 2020. Available from: https://doi.org/10.3390/ijms21113848. PMID: 32481719.

Montes M, Sanford BL, Comiskey DF, Chandler DS: RNA splicing and disease: animal models to therapies, *Trends Genet* 35(1):68–87, 2019. Available from: https://doi.org/10.1016/j.tig.2018.10.002. PMID: 30466729.

Nidetz NF, McGee MC, Tse LV, et al: Adeno-associated viral vector-mediated immune responses: understanding barriers to gene delivery, *Pharmacol Ther.* 207:107453, 2020. Available from: https://doi.org/10.1016/j.pharmthera.2019.107453. Epub 2019 Dec 11. PMID: 31836454.

Pausch P, Al-Shayeb B, Bisom-Rapp E, et al: CRISPR-CasΦ from huge phages is a hypercompact genome editor, *Science.* 369(6501):333–337, 2020. Available from: https://doi.org/10.1126/science.abb1400. PMID: 32675376.

Pedemonte N: Editorial: Special issue on "Therapeutic Approaches for Cystic Fibrosis", *Int J Mol Sci* 21(18):6657, 2020. Available from: https://doi.org/10.3390/ijms21186657. PMID: 32932926.

Pellenz S, Phelps M, Tang W, et al: New human chromosomal sites with "Safe Harbor" potential for targeted transgene insertion, *Hum Gene Ther.* 30(7):814–828, 2019. Available from: https://doi.org/10.1089/hum.2018. PMID: 30793977.

Porto EM, Komor AC, Slaymaker IM, Yeo GW: Base editing: advances and therapeutic opportunities, *Nat Rev Drug Discov* 19(12):839–859, 2020. Available from: https://doi.org/10.1038/s41573-020-0084-6. PMID: 33077937.

Roberts TC, Langer R, Wood MJA: Advances in oligonucleotide drug delivery, *Nat Rev Drug Discov* 19(10):673–694, 2020. Available from: https://doi.org/10.1038/s41573-020-0075-7. PMID: 32782413.

Russell S, Bennett J, Wellman JA, et al: Efficacy and safety of voretigene neparvovec (AAV2-hRPE65v2) in patients with RPE65-mediated inherited retinal dystrophy: a randomised, controlled, open-label, phase 3 trial, *Lancet.* 390(10097):849–860, 2017. Available from: https://doi.org/10.1016/S0140-6736(17)31868-8. PMID: 28712537.

Saha K, Sontheimer EJ, Brooks PJ, et al: The NIH somatic cell genome editing program, *Nature.* 592(7853):195–204, 2021. Available from: https://doi.org/10.1038/s41586-021-03191-1. PMID: 33828315.

Sardh E, Harper P, Balwani M, et al: Phase 1 trial of an RNA interference therapy for acute intermittent porphyria, *N Engl J Med* 380(6):549–558, 2019. Available from: https://doi.org/10.1056/NEJMoa1807838. PMID: 30726693.

Shirley JL, de Jong YP, Terhorst C, Herzog RW: Immune responses to viral gene therapy vectors, *Mol Ther* 28(3):709–722, 2020. Available from: https://doi.org/10.1016/j.ymthe.2020.01.001. PMID: 31968213.

Stenson PD, Mort M, Ball EV, et al: The Human Gene Mutation Database (HGMD®): optimizing its use in a clinical diagnostic or research setting, *Hum Genet* 139(10):1197–1207, 2020. Available from: https://doi.org/10.1007/s00439-020-02199-3. PMID: 32596782.

Yang J: Patisiran for the treatment of hereditary transthyretin-mediated amyloidosis, *Expert Rev Clin Pharmacol* 12(2):95–99, 2019. Available from: https://doi.org/10.1080/17512433.2019.1567326. PMID: 30644768.

Zittersteijn HA, Gonçalves MAFV, Hoeben RC: A primer to gene therapy: progress, prospects, and problems, *J Inherit Metab Dis* 44(1):54–71, 2021. Available from: https://doi.org/10.1002/jimd.12270. PMID: 32510617.

Further reading

Pickar-Oliver A, Gersbach CA: The next generation of CRISPR-Cas technologies and applications, *Nat Rev Mol Cell Biol* 20(8):490–507, 2019. Available from: https://doi.org/10.1038/s41580-019-0131-5. PMID: 31147612.

CHAPTER 10

Public health applications of genetics including newborn screening and documentation of gene environment interactions

10.1 Recessive disorders carrier screening in specific populations

Antonarakis (2019) reviewed aspects of carrier screening for genetic disorders in specific populations. He noted the remarkable achievements of Stamatoyannopoulos in providing education and services to families in a small village in Greece, Orchomenos, where 23% of individuals were heterozygous for the sickle hemoglobin mutation and 1 in 100 neonates were reported to be born with sickle cell hemoglobin disorder. Antonarakis also noted the follow-up establishment of large-scale voluntary screening programs to reduce disease burden.

Antonarakis also noted regions with a high frequency of beta thalassemia. In Cyprus, the carrier frequency was reported to be 15%, in Greece and Italy 10%, in Sardinia it was 8.5%, in South-east Asia 8.5%, in Punjab 7.6%, and 4.6% in Bengal.

Harteveld and Higgs (2010) reported that alpha thalassemia was especially frequent in Mediterranean countries, South-East Asia, the Middle-East, and India. In recent decades, the frequency of alpha thalassemia was reported to have increased in North European countries and in North America.

In 2013 Hoppe reported that isoelectric focusing and high-pressure liquid chromatography were frequently been used in newborn screening for hemoglobinopathies.

10.2 Newborn screening and hemoglobinopathies

Kato et al. (2018) reviewed sickle cell disease and noted that globally between 300,000 and 400,000 infants with sickle cell disease were born each year. They emphasized that early diagnosis was critical. Early penicillin treatment is often advocated. Kato et al. noted that hydroxycarbamide (sometimes referred to as hydroxyurea) treatment, blood transfusion, and hematopoietic stem cell transplantation can reduce disease severity.

In 2018 Kato et al. noted that noninvasive prenatal diagnosis was still considered investigational. Some couples were reported to choose preimplantation genetic testing.

In areas where newborn screening is not available, initial diagnosis was noted to usually occur around 21 months of age.

In 2018 Lobitz et al. reported recommendations from a Pan-European conference regarding newborn screening for sickle cell disease The report noted that sickle cell disease was an increasing global health problem and emphasized that newborn screening enables early commencement of care.

Newborn screening for sickle cell disease is available in the United States, Europe, Brazil, India, and in some countries in Africa.

In 2019 Nkya et al. reported results of a pilot program in Tanzania to screen newborns for sickle cell disease. They noted that in Tanzania 11,000 annual births of infants with sickle cell disease occurred. In the pilot program they screened 3981 newborns, 31 were found to have sickle cell disease, 505 had sickle cell trait, and 26 infants had other hemoglobinopathies.

Their study led to recommendation for newborn screening for sickle cell disease, with enrollments in comprehensive care and for guidelines for outpatient clinics. They also noted the importance of healthcare information for mothers.

Bender et al. (2020) reported that many newborn screening programs in the United States did not directly screen for alpha thalassemia. The level of hemoglobin Barts was used to reflect alpha thalassemia The percentage of Hb Barts detected correlates with alpha thalassemia severity.

In a 2017 review, Sabath reported that hemoglobinopathy diagnoses were most frequently made using protein-based techniques including electrophoresis and chromatography. However, Sabath noted that protein-based methods were not always useful in the diagnosis of complex hemoglobinopathies, for example, combinations of alpha and beta thalassemia. In addition some forms of beta thalassemia were not detected using standard protein-based methods. Sabath emphasized that molecular genetic resting was important in diagnosing thalassemia.

Testing procedures for hemoglobinopathy determination were presented by Wiesinger et al. (2020). Their methods include high resolution mass spectrometry.

Coppinger and O'Loughlin (2019) reviewed information on newborn screening for sickle cell disease and thalassemias in the National Health Service in Britain. Data analyses revealed that improved oversight was needed to ensure appropriate handover from the screening program to the treatment service.

They emphasized that it was important to review the status of patients along the care pathway.

A prototype system was developed to factor in elements of the flow of information, the system included:

Newborn screening lab: Creation of result record.

Nursing care: Conformation of result receipt; notify patients.

Medical care: Confirm result receipt; notify patients; record status of treatment.

Comprehensive notification also involved passage of results from newborn screening laboratory to rare disease registry, national hemoglobinopathy registry, and communication with designated treatment centers.

Antonarakis listed other disorders for which carrier screening is recommended. These include cystic fibrosis (CF) due to CF transmembrane regulator (CFTR) mutations, Fragile X mental retardation, Tay − Sachs disease, spinal, muscular atrophy.

A notable achievement of carrier screening was the reduction of 90% of new cases of Tay − Sachs disease that followed support of initiation of screening in the Ashkenazi Jewish population (Kaback et al., 1993).

Other recessive disorders for which screening programs have been instituted in some countries include adrenoleukodystrophy, X-Linked muscular dystrophy, immunodeficiencies.

There are estimates that suggest that every individual is a carrier of more than 20 different disorders. Antonarakis noted that most protein coding genes are haploinsufficient, therefore most heterozygotes are unaffected. 5% of genes were noted to tolerate homozygous loss of function mutations.

The UK Deciphering Developmental Disabilities study (Wright et al., 2019) reported that 3.6% of cases involved autosomal recessive mutations in known disease-related genes. De novo coding mutation was identified in 49% of affected individuals.

Antonarakis (2019) noted that most proteins are haplosufficient and that every individual is estimated to be a carrier of more than 20 diseases. Rausell et al. (2020) reported 190 genes in which loss of function variants are present in more than 1% of the human population, that did not lead to disease; 41 of these genes included olfactory genes.

10.3 Cystic fibrosis

There are recommendations from the CF foundation that newborn screening positive tests should be followed by tests to include demonstration of dysfunction of the CFTR with a sweat chloride test or other current CFTR functional tests. These include nasal potential difference test and intestinal epithelial current measurement.

A specific type of sweat test for CF involves stimulation of sweat production with pilocarpine, collection of sweat, and measurement of chloride. The nasal potential difference test and intestinal current measurement tests are also defined as functional tests of the CFTR. These tests were reviewed by Bagheri-Hanson et al. in 2014. In these tests epithelial sodium channels were blocked using amiloride and cyclic AMP. CFTR chloride transport was stimulated with isoproterenol. Cholinergic chloride transport was then measured. Nasal epithelial cells were used in respiratory system testing. Intestinal testing was done using rectal biopsy samples.

The gene with mutations leading to CF was isolated in 1989 by Rommens et al. and was noted to encode the protein CFTR (CF transmembrane conductance regulator).

In 2014 De Boeck et al. reported on CFTR mutations identified in patients of European ancestry. CF prevalence was noted to be approximately 1 in 3000 in Europe, North America, and Australia, 1 in 6000 in the Middle-East, and 1 in 7000 in South Africa.

Identification of the type of CFTR mutations present was noted to potentially have therapeutic relevance (Quon and Rowe, 2016). At least one F508 deletion (F508del) was most commonly present. Mutations were separated into different classes.

CFTR is defined as a regulated anion channel with particularly high expression on the apical surface of epithelial cells, especially in airways and pancreatic ducts. Impaired CFTR function impacts chloride transport.

Quon and Rowe reported that 242 of the approximately 2000 identified CFTR mutations were reported to be pathogenic in this autosomal recessive disease. Six categories of mutations were defined: mutations leading to absence of synthesis, mutation with altered processing of the gene product, mutations that impacted gating or conductance, mutations with low synthesis.

Mutations were classified into classes as follows:

Class I mutations, unstable truncated mRNA with no synthesis of protein.

Class II mutations that impacted processing, leading to impaired folding or rapid degradation of CFTR.

Class III, reduced channel opening or regulation.

Class IV, reduced chloride conductance in the channel.

Class V, reduced CFTR synthesis, splicing mutations.

Class VI, reduced receptor stability and increased receptor turnover.

10.4 Molecular-based therapeutics

Small molecules have been designed to target CFTR defects. These include:

CFTR potentiators such as Ivacaftor, that increase flow through the channels and are particularly useful for class II mutation, for example, CFTR 1625 $G > A$. Ivacaftor was also reported to increase channel opening and chloride transport.

CFTR correctors facilitate CFTR protein folding, for example, in Class II mutations that include F508del. Lumacaftor acts as a facilitator.

Read-through agents have also been developed for treatment of mutations that lead to premature stop codons. These include modified aminoglycosides and searches continue for nontoxic alternatives to aminoglycosides that include Ataluren.

CFTR combination therapies may be particularly important, since F508del is often present along with other mutations.

CFTR modulators being investigated include phosphor di-esterase inhibitors, sildenafil, and guanylate cyclase stimulators.

Gene therapy trials with liposomal carriers of the gene administered intranasally are being investigated. Gene therapy with delivery with other nanoparticles is also being investigated.

CFTR database cftr2.org allows entry of the mutation and information can potentially be retrieved on sweat chloride measurement, lung function, and information related to pancreatic insufficiency, predisposition to Pseudomonas infection, and also information regarding possible therapies.

In CF, pan ethnic screening is recommended given evidence that the carrier frequency for deleterious mutations in the CFTR is 1 in 25 in whites, 1 in 58 for Hispanics, 1 in 24 in the Ashkenazi population, 1 in 61 in Africans, and 1 in 94 in Asians.

The exon 7 deletion mutation F508Del is the most common mutation in CFTR and in addition 38 pathogenic variants commonly occur. New mutations have sometimes been reported.

10.5 Newborn screening, United States

The American College of Medical Genetics (ACMG) provides updates and information on newborn screening (https://www.ncbi.nlm.nih.gov/books/NBK55827/).

It is important to note that different states in the United States differ to the extent to which the ACMG documented screening tests are carried out.

In the ACMG guidelines disease categories screened for are ordered alphabetically, the screening algorithm (mode of screening) and information for possible follow-up to positive screening are included (ACT sheet).

10.5.1 Aminoacidurias

Argininemia, arginosuccinic aciduria, citrullinemia, pyruvate carboxylase deficiency, decreased citrulline, homocystinuria, hypermethioninemia, adenosylhomocysteine hydrolase deficiency, maple syrup urine disease, hydroxyprolinemia, phenylketonuria, biopterin cofactor biosynthesis defect, biopterin cofactor regeneration defect, tyrosinemia types I, II, and III (Fig. 10.1).

10.5.2 Endocrine disorders

Primary congenital hypothyroidism, secondary congenital hypothyroidism, thyroxine binding globulin deficiency, congenital adrenal hyperplasia, 21 hydroxylase deficiency.

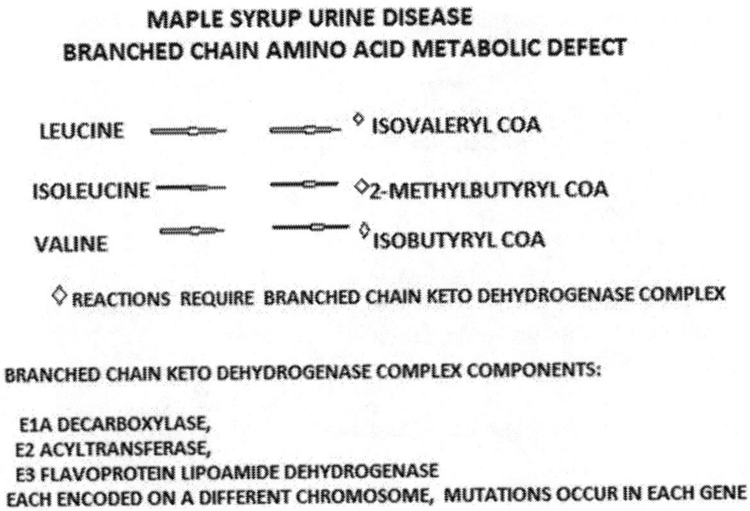

MAPLE SYRUP URINE DISEASE
BRANCHED CHAIN AMINO ACID METABOLIC DEFECT

LEUCINE ⟶ ⟶ ◊ ISOVALERYL COA

ISOLEUCINE ⟶ ⟶ ◊ 2-METHYLBUTYRYL COA

VALINE ⟶ ⟶ ◊ ISOBUTYRYL COA

◊ REACTIONS REQUIRE BRANCHED CHAIN KETO DEHYDROGENASE COMPLEX

BRANCHED CHAIN KETO DEHYDROGENASE COMPLEX COMPONENTS:

E1A DECARBOXYLASE,
E2 ACYLTRANSFERASE,
E3 FLAVOPROTEIN LIPOAMIDE DEHYDROGENASE
EACH ENCODED ON A DIFFERENT CHROMOSOME, MUTATIONS OCCUR IN EACH GENE

FIGURE 10.1

Defect in maple syrup urine disease.

10.5.3 Fatty acid oxidation defects

Carnitine uptake deficiency, carnitine palmitoyl transferase deficiency type CPT I and type CPT II, glutaric acidemia, methylmalonic acidemia, long chain fatty acid dehydrogenase deficiency, medium chain fatty acid dehydrogenase deficiency, short chain fatty acid dehydrogenase deficiency, ethylmalonic acidemia, isobutyryl Co A dehydrogenase deficiency.

10.5.4 Galactosemia

Genetic diseases: Biotinidase deficiency, critical congenital heart disease, CF, hearing loss.

Hemoglobinopathies: Sickle cell disease, sickle carrier, thalassemia beta, thalassemia alpha, thalassemia Barts, hemoglobin variant, hemoglobin E, hemoglobin CC.

Immunodeficiencies: Severe combined immunodeficiency (in development).

Lysosomal storage diseases: Fabry disease (in development), Krabbe disease (in development), Nieman-Pick disease (in development), Pompe disease.

Muscular skeletal diseases: Duchenne muscular dystrophy pathogenic variant, elevated creatine kinase muscle type.

Organic acidemias: Beta ketothiolase deficiency, biotinidase deficiency, holocarboxylase deficiency, HMG CoA lyase deficiency, 2methyl-3hydroxybutyryl CoA deficiency, 3methylglutaconic acid, 3methylcrotonylglycinuria, glutaric acidemia, isovaleric acidemia, malonic acidemia, propionic acidemia, short-branched chain acyl CoA dehydrogenase deficiency.

Urea cycle disorders: Vasquez-Loarte et al. (2020) noted that ornithine transcarbamylase deficiency and carbamoyl phosphate synthetase deficiency are screened in six states in the United States. They proposed inclusion of additional urea cycle disorders, given evidence for the importance of early treatment of these disorders (Fig. 10.2).

10.6 Methods

Adhikari et al. (2020) reported that tandem mass spectrometry was used for newborn screening for rare inborn errors of metabolism. A specific project NBseq was undertaken to investigate the possibility of using whole exome sequencing for newborn screening. The project involved use of dried blood spots from 4.5 million newborns born in California who had undergone newborn screening.

Based on analysis of results of the study the authors concluded that whole exome sequencing was insufficiently sensitive as a primary screening test. However, it had value as a secondary test for infants who tested positive on mass spectrometry testing.

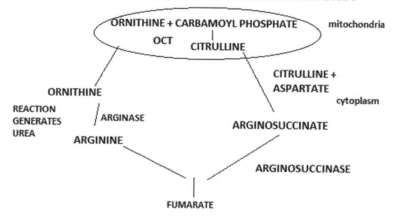

FIGURE 10.2

Urea cycle disorders.

10.6.1 Newborn screening for immunodeficiencies

In 2005 Chan and Puck reported that T-cell receptor excision circles (TRECs) were DNA byproducts of T-cell receptor recombination. They reported that patients with severe combined immunodeficiency make few or no T-cells. Therefore TRECS are present at very low levels or may be absent. They developed a PCR-based test to quantitate TRECs that were shown to detect lymphopenia.

Puck (2019) reviewed the use of the TREC assay on dried blood spots obtained for newborn screening and validated this as a test for severe combined immunodeficiency.

Individuals with severe combined immunodeficiency were reported to have low numbers of T-cells and defects in B-cells that lead to severe bacterial, fungal, and viral infections. The gene defect spectrum was noted to be broad (includes 216 genes) (http://nih.gov/CGD). Puck (2019) noted that screening does not lead to a definitive diagnosis. The TREC test identifies cases in need of monitoring and further analyses.

Treatment for severe combined immunodeficiencies includes antibiotics, immunoglobulin infusion, and stem cell transplantation.

It is important to note that some screening programs also screen for X-linked agammaglobulinemia (Gutierrez-Mateo et al., 2019).

10.6.2 Spinal muscular atrophy

Jędrzejowska (2020) reviewed advances in newborn screening and presymptomatic diagnosis in spinal muscular atrophy (SMA). This disease is sometimes referred to as SMA5Q and is known to lead to premature death in infancy or in some cases to severe motor disability.

Stimulated in part by new therapeutic possibilities, newborn screening for this disorder has been implemented in the United States, Germany, Belgium, Australia, and is being considered in other countries. Diagnosis is based on genetic testing. Treatments implemented are molecular-based and include SMN2 gene modification, splice site altering therapy, and gene therapy. Jędrzejowska noted that more precise predictors or biomarkers were needed to monitor therapeutic successes.

Other measures implemented include increased carrier screening for SMA.

Breakthrough therapies for SMA were reviewed by Chen (2020). Oligonucleotide therapies were designed to activate expression of exon 7 in the gene that encodes SMN2; production of SMN2 protein can compensate for loss of SMN1. Other forms of therapy include gene therapy to replace SMN1.

Howell (2021) noted that questions were being raised as to whether exome sequencing should be used in newborn screening and also questions arose as to what tests should be included in newborn screening.

10.6.3 Screening for X-linked disorders

Al-Zaidy et al. (2017) considered aspects of including Duchenne muscular dystrophy (DMD) in newborn screening and noted further that some programs have proposed including tests for limb-girdle muscular dystrophies.

In 2017 Al Zaidy et al. noted that the only available treatment for DMD was administration of glucocorticoids; other treatments being investigated included use of oligonucleotides to promote target exon skipping of exons with mutations and in some cases stop codon read-through applications were being applied.

DMD clinical manifestations occur in males. However, newborn screening that also includes females has a possibility of prompt detection of carrier females. It is, however, also important to note that one-third of cases of mutations leading to Duchenne muscular dystrophy are new mutations.

Other X-linked disorders that are screened for in some programs include Fragile X syndrome, adrenoleukodystrophy, Bruton type immunodeficiency, X-linked Alport syndrome, and X-linked protoporphyria.

Lee et al. (2020) carried out a study to determine the efficiency of customized polymerase chain assay to determine repeat pathogenic repeat expansion in the FMR1 gene. They reported results of analyses on dried blood spots in a study of 963 samples and detected 957 normal specimens and six specimens with premutation expansions.

Riley and Wheeler (2017) noted that very early identification of Fragile X syndrome is important in connecting individuals affected with this disorder in a timely manner with medical and support services.

Turk et al. (2020) reviewed pathology, diagnostic testing, newborn screening testing, and therapies in X-linked adrenoleukodystrophy. Pathology in this condition occurs in the adrenal cortex, in Schwann cells, and in the brain.

The disease is characterized by increased levels of very long chain fatty acids that lead to cell death and to clinical manifestations in homozygous females and in hemizygous males.

Treatments include stem cell transplant. Clinical trials are also ongoing to investigate the value of statins (e.g., Lovastatin) and PPAR gamma (peroxisome proliferator receptor gamma) antagonists. Other investigations ongoing include tests of the efficiency of hematopoietic stem cell lentiviral gene correction. Jangouk et al. (2012) reported that clinical manifestations occur in 20% $-$50% of female carriers with X-linked adrenoleukodystrophy.

Bruton tyrosine kinase (BTK) deficiency leads to an immunodeficiency in males that is characterized by reduced numbers of B-lymphocytes and immunoglobulin deficiency. Search for BTK mutations has been carried out by genomic amplifications and Sanger sequencing. Aadam et al. (2016) reported on a large series of North Africa patients with X-linked agammaglobulinemia and BTK mutations. Collins et al. (2018) reported rapid multiplexed proteomic sequencing from dried blood spots to detect this disorder. They noted that there are treatments for this disorder that lead to optimal outcomes when treatment is commenced early.

Screening for X-linked Alport syndrome in women has been developed and is based on urinary 3-hydroxyproline excretion (Bartosch et al., 1991).

Glucose 6 phosphate dehydrogenase (G6PD) deficiency is screened for in China. After screening for G6PD populational deficiency incidence, the Sports Medicine Organization has reported that in the US population G6PD deficiency incidence was between 4% and 7%. WHO has recommended G6PD screening in countries where the population incidence of deficiency is more than 5%.

10.7 Newborn screening in other parts of the world

Lloyd-Puryear et al. (2018) noted that newborn screening programs are expanding in many parts of the world.

Therrell and Padilla (2018) reviewed newborn screening in developing countries. They noted that many developing countries do not have newborn screening programs. The most effective newborn screening program was noted to be screening for congenital hypothyroidism. Screening for hemoglobinopathies and G6PD deficiency was noted to be important in Sub-Saharan Africa.

Therrell and Padilla emphasized the importance of expanded screening for inherited metabolic conditions in regions with high consanguinity.

Oster and Kochilas (2016) noted that newborn screening for critical congenital heart disease was adopted in 46 states in the United States in 2015. This program was specifically designed to detect the following core heart conditions: coarctation of the aorta, double outlet right ventricle (RV), Ebstein anomaly defined as a malformation of the tricuspid valve with myopathy of the RV that presents with variable anatomic and pathophysiologic characteristics, hypoplastic left heart syndrome, interrupted aortic arch, pulmonary atresia, transposition of the great vessels, tricuspid atresia, truncus arteriosus, anomalous pulmonary return. The initial screening was pulse oximetry.

In 2020 Martin et al. reviewed newborn screening for congenital heart disease and they specifically noted strategies for use of pulse oximetry in this screening.

10.8 Expanded carrier screening

Kraft et al. (2019) reviewed aspects of carrier screening for genetic disorders beyond newborn screening. They noted ancestry-based screening in certain populations including Mediterranean populations with hemoglobinopathies and Ashkenazi Jewish populations for Tay — Sachs disease and also subsequent initiation of CF screening.

They noted that there is consensus that newborn screening may be particularly advantageous. However, they stated that there is lack of consensus regarding which tests should be included in later screening. They concluded that key issues that should be taken into account include the prevalence of deleterious gene mutations, the aspects of the phenotype induced by mutations, the penetrance of those mutations, and the effects of resulting disorders on quality of life.

Another issue raised was whether or not a customer seeking screening should be able to decide on the screening that they wanted.

Kraft et al. also emphasized that expanded screening may increase disability-based discrimination.

Components of decision-making regarding inclusion of additional conditions in newborn screening include evidence that the clinical test has analytical and clinical validity and utility and that there is effective treatment for the conditions screened.

10.9 Genetic disorders with high frequency in certain populations

It is interesting to consider the origins of altered genetic disease frequencies in certain populations. Factors that may play roles include possible heterozygote advantages of specific mutations in specific environments or advantage of heterozygotes in coping with specific infections. Other factors include selective migrations, isolation of some population groups, and preferential mating.

Important information on genetic disorders in individuals with Jewish ancestry can be obtained in the database: https://www.jewishgeneticdiseases.org/jewish-genetic-diseases/

10.9.1 Tay – Sachs disease due to hexosaminidase mutations

Tay – Sachs disease is known to occur with higher frequency in the Ashkenazi Jewish population than in other populations. For several decades screening for Tay – Sachs loss of function was based on enzyme assays of hexosaminidase in cells (white blood cells fibroblasts). In 1988 Myerowitz and Costigan reported that the most common mutation of hexosaminidase gene (HEXA) was a 4 base-pair insertion in exon 11. In a less common juvenile onset for Tay – Sachs disease an arg178his mutation was found in HEXA protein (Fig. 10.3).

Myerowitz (1997) reported heterogeneity in Tay – Sachs disease and identified 39 disease causing mutations. An insertion mutation that included 4 bases in exon 11 was found in 80% of carriers of Tay – Sachs disease in the Ashkenazi Jewish population.

A 7.5-Kb deletion that encompassed exon 1 of Hex A was reported to be the major Tay – Sachs mutation in French Canadians.

It is important to note that Tay – Sachs disease has also been reported in non-Jewish individuals. In 2019 Cecchi et al. carried out full exome sequencing in a pan-ethnic cohort to maximize carrier testing across ethnicities.

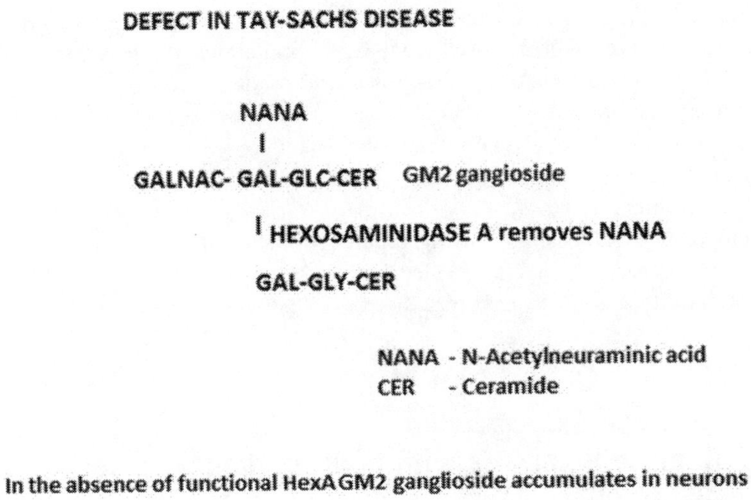

DEFECT IN TAY-SACHS DISEASE

NANA
|
GALNAC- GAL-GLC-CER   GM2 gangioside

| HEXOSAMINIDASE A removes NANA

GAL-GLY-CER

NANA - N-Acetylneuraminic acid
CER   - Ceramide

In the absence of functional HexA GM2 ganglioside accumulates in neurons

FIGURE 10.3

Defect in Tay — Sachs disease.

Akler et al. (2020) reported results of reproductive carrier screening in Ashkenazi Sephardi and Mizrahi Jewish patients. Results of their study revealed that 64.6% of individuals were carriers of one or more of 96 diseases screened for. Common diseases reported across all Jewish groups included: CF, FMF inclusion body myopathy, Fragile X mental retardation, glycogen storage type II, phenylalanine hydroxylase deficiency (PKU), retinitis pigmentosa, Smith — Lemli — Opitz disorders, SMA, Tay — Sachs disease, Wilson's disease.

Diseases common to only the Ashkenazi group included: 3 phosphoglycerate dehydrogenase deficiency, abetalipoproteinemia, Alport syndrome, arthrogryposis, Bloom syndrome, Canavan disease, CPT deficiency types I and II, chorioacanthocytosis.

Historic founder effects and genetic drift were proposed as key factors leading to increased frequency of specific diseases in the Ashkenazi population.

Shi et al. (2017) reported results of comprehensive carrier screening for 85 pathogenic mutations involved in 29 different diseases in the Ashkenazi Jewish population.

Their study included 2252 individuals with self-reported 100% Ashkenazi Jewish ancestry.

Disease and gene	Specific mutation/s	Average carrier frequency
Abetalipoproteinemia ABL	p.G865; pS738fs	1/185
Alport syndrome COL4A3	p.L14-L21 del.	1/192

(Continued)

Continued

Disease and gene	Specific mutation/s	Average carrier frequency
Arthrogryposis (AMRS) SLC35A3	p.S296G; p.Q172*	1/453
Bardet – Biedl syndrome 2	p.D104A; p.R632P	1/139
Carnitine palmitoyl transferase deficiency (CPT2)	p.S113L; p.Q143fs; p.R124*; p238fs	1/64
Congenital amegakaryocytic Thrombopenia (CAMT MPL)	c.79 + 2T > A	1/57
Congenital glycosylation disorders (PMM2)	p.P113L; p.F119L; p.R141H; p.V321M	1/79
Dyskeratosis congenita RTEL1	p.R1264H; p.R998*, p.G763V; p.M516I; p.R981W	1/165
Ehlers-Danlos syndrome VIICEDS VIIC ADAMTS2	p.Q225*p.W795*	1/187
Galactosemia GALT and 9 different mutations	C-1089 + 789 del	1/156
Multiple sulfatase deficiency SUMF1	p.S155P	1/279
Phosphoglycerate deficiency PHGDH	p.V490M	1/453
Polycystic kidney disease PHD4 other mutations at lower frequency	p.A1254fs	1/105
Retinitis pigmentosa	p.K42E	1/117
Smith-Lemli-Opitz syndrome14 other mutations lower frequency	C964-G > C	1/40
Tyrosinemia TYRSN and FAH5 other mutations at lower frequency	p.P261L	1/143
Wilson's disease ATP7B	p.E1064A (3 at lower frequency)	1/67
Peroxisomal biogenesis defect PEX2	p.R119*	1/227

10.9.2 Summary of predominant manifestations of disorders with increased incidence in Ashkenazi population and mode of inheritance (AR) autosomal recessive (AD) autosomal dominant

Abetalipoproteinemia AR, malabsorption, celiac syndrome, malabsorption of lipid soluble vitamins can lead to neuropathy, retinal degeneration.

Alport syndrome COL4A3 AD, nephritis, hypertension.

Arthrogryposis AMRS SLC35A3 AR limb malformations, hip dislocation, hypotonia, delayed development.

Bardet − Biedl 2 BBS2 AR dysmorphism. Obesity, cognitive impairment, polydactyly.

CPT deficiency CPT2 AD, AR muscle pain on prolonged exercise or on fasting, myoglobinuria may lead to kidney damage.

Congenital glycosylation disorder PMM2 AR thrombocytopenia, hypoplastic bone marrow.

Dyskeratosis congenita RTEL1 AD/AR muscle dystrophy, leukoplakia, abnormal skin pigmentation may present with immunodeficiency, microcephaly, growth retardation, enteropathy, pancytopenia, bone marrow failure.

Ehlers − Danlos VII ADAMTS2 AR dermatosparaxis type, joint hyperextensibility, velvety skin, extensible skin, easy bruising.

Galactosemia GALT AR neonatal jaundice, hepatosplenomegaly, hyperglycemia, cataracts, developmental delay.

Multiple sulfatase deficiency SUMF1 AR has features of mucopolysaccharidoses, skeletal anomalies, organomegaly, developmental delay.

Phosphoglycerate deficiency PHGDH AR microcephaly, seizure, developmental delay.

Polycystic kidney disease PKD1 AD renal cysts, liver cysts, acute and chronic nephrolithiasis, renal disease.

Retinitis pigmentosa DHDDS AR impaired night vision, impaired peripheral vision, pigmentary retinal degeneration.

Smith − Lemli − Opitz syndrome DHCR7 AR metabolic malformation syndrome, congenital anomalies, hypotonia, micrognathia, cleft palate, polydactyly.

Tyrosinemia FAH fumarylacetoacetate hydrolase AR liver disease, renal tubular dysfunction, hypophosphatemic rickets.

Wilson's disease ATP7B AR build-up of liver copper leading to cirrhosis, ring on periphery of cornea due to copper deposition, nephrocalcinosis.

Peroxisomal biogenesis defect PEX2 AR developmental delay, hypotonia, liver dysfunction, visual impairment, hearing impairment.

Tay − Sachs disease clinical manifestations in the infantile form include visual impairment, marked startle response, rigidity, developmental impairment paralysis, and death by 2−3 years.

Sometimes there is later onset of Tay − Sachs disease commencing in later childhood with gait disturbance, muscular atrophy, speech difficulties.

Adult form of Tay − Sachs disease has been described with muscle atrophy, tremor, poor coordination, and psychiatric manifestations (Kaback et al., 1993).

10.9.3 Mutation heterogeneity in Tay − Sachs HEXA mutations in the Ashkenazi population

Even within the Ashkenazi Jewish population several different HEXA mutations have been identified as being responsible for Tay − Sachs disease. They include duplications, splice site mutations, and mutations that alter amino acid code. Notable mutations include:

c.1274-C1277 dup TATC Ter
C1421 + 1G > C
C805G > A p.Gly269Ser
C533G > A p.Arg 178His
c.745C > t p.Arg 249Trp

In late onset Tay − Sachs disease one mutation fairly commonly reported led to c.805G > A, p.G269S (p.Gly269Ser).

It is important to note that Tay − Sachs disease due to HEXA mutations does occur in individuals with no known Ashkenazi ancestry.

Rivas et al. (2018) carried out an analysis of the Gnom AD database that contains information of frequency of alleles in the general population without disease. In considered different populations they noted increased frequency of certain pathogenic alleles (likely in heterozygous state) in the Jewish population.

Gaucher disease GBA Asn409Ser eightfold increase in frequency.

Tay − Sachs hexosaminidase A C.1421 + 1G > C eightfold increase in frequency.

Canavan disease Asparto-acylase ASPA P.Glu285Ala 12-fold increase in frequency.

Other pathogenic mutations that predominate in Jewish population are as follows:

Parkinson's disease LRRK2 P.Gly2019Ser.
CF CFTR P.Tyr1282ter.
Peroxisome biogenesis defect PEX p.Arg119ter.
Familial Mediterranean fever MEFV pval726Ala.
Familial dysautonomia IKBAP (ELP1) c.2204 + 6T > C.

10.10 Genetic disorders with increased frequency in other specific populations

It is important to note that certain genetic disorders differ in frequencies in different countries, and it is particularly important to be aware of altered frequency of specific genetic disorders in population isolates.

10.10.1 Aspartylglucosaminuria

This lysosomal storage disease was reported to occur with high frequency in the Finnish population (Arvio and Mononen, 2016). They reported that a specific mutation C163S in the aspartyl glucosaminidase protein in homozygous state leads to this disorder and that this mutation was present in 98% of cases of aspartylglucosaminuria in Finland. This disorder impacts physical growth, development, and behavior.

10.10.2 Familial Mediterranean fever (FMF)

This is an autosomal recessive disorder leading to inflammation in serosal tissue and to arthritic states. The disorder can lead to inflammation in synovia of joints, inflammation in pleura, or sometimes in the peritoneum. It occurs with high frequency in Middle-East and North African regions and also in Iranian population and in some Indian populations. This is an autosomal recessive condition and affected individual are homozygotes or compound heterozygotes for pathogenic MEFV mutations.

The gene, MEFV, impacted in this disease is reported to encode pyrin, an inflammation sensor, and mutations in MEFV decrease the activation threshold of pyrin.

In some cases there are deletions in the MEFV gene; frequently reported mutations include M694V. In 2020 Ait-Idir and Djerdjouri reported differences in the MEFV mutation profiles in North Africa. In Algeria p.M694I was most common and p.M680I and p.M694V were less frequent.

Park et al. (2020) presented evidence that mutations in MEFV were positively selected in ancient populations since they provided resistance to the microorganism *Yersinia pestis*.

10.11 Porphyrias

Elder et al. (2013) reviewed the incidence of porphyrias in Europe. They noted that the incidence of acute intermittent porphyria was similar across European countries except for Sweden where there is a higher incidence of that disorder.

Variegate porphyria due to decreased levels of protoporphyrinogen oxidase occurs with increased frequency in a population isolate in South Africa.

The porphyrias will be discussed in greater detail in the Chapter 11: Analysis of variants associated with abnormal drug responses, genetics, and genomics in drug design.

10.12 Factor V Leiden

This mutation leads to thrombophilia and hypercoagulability of blood. Van Cott et al. (2016) noted that a specific protein, activated protein C, (PROC), and its cofactor S (PROS1) serve to inactivate Factor Va and Factor VIIa to promote anticoagulation.

Activated protein C and its cofactor normally cleave Factor Va and Factor VIIa at a specific site and therefore inhibit coagulation. A specific mutation in Factor V ARG506GLN (R506Q) was shown to lead to resistance to this cleavage and inactivation. Other mutations in Factor V may also alter coagulability.

Van Cott et al. reported that Factor V Leiden is a founder mutation that is present in heterozygous form in 5% of Caucasian individuals. Homozygosity for this mutation was reported to occur in 1 in 5000 individuals in the Caucasian population.

This specific mutation leads to increased risk for venous thrombosis and risk was reported to be increased three to sevenfold in heterozygotes and 80-fold in homozygotes. Importantly, both genetic and environmental factors were reported to be implicated in risk. Oral contraceptive use was reported to particularly increase thrombosis risk in females with Factor V Leiden.

De Stefano et al. (1998) reported that the Factor Va Arg506 Gln mutation was present in 5% of Caucasians and was virtually absent in Africans and Asians.

10.12.1 APOL1 apolipoprotein L1 mutation in Africans and African Americans

Raghubeer et al. (2020) reviewed information on the APOL1 variants. Specific variants in APOL1 leading to generation of new isoforms were noted to likely have increased resistance to Trypanosome infections in certain African countries. However, there is now evidence that these same variants exacerbate chronic kidney disease and exacerbate manifestations of *Lupus erythrematosis* (Chokshi et al., 2019).

In 2010 Genovese et al. reported that in the United States individuals with African ancestry have a higher frequency of kidney disease than Americans with European ancestry. The kidney disease in this population was noted to more frequently include glomerulosclerosis leading to hypertension and end-stage renal disease. African Americans with this disorder were noted to have increased frequency of a specific variant in the APOL1 gene on chromosome 22. The strongest signal was obtained for a 2-locus allele, termed G1 (613743.0001), comprised of two nonsynonymous coding variants: rs73885319 (ser342 gly) and rs60910145 (ile384 met), both in the last exon of APOL1. These two alleles were noted to be perfect linkage disequilibrium.

In 2021 Friedman and Pollak emphasized that the APOL1 variant increased rates of nondiabetic nephropathy.

10.13 Population-wide screening of adults

The US Centers for Disease Control and Prevention and a number of other public health organizations recommend genome screening for familial hypercholesterolemia

(FH), hereditary breast and ovarian cancer (HBOC), and Lynch syndrome that includes increased frequency of gastrointestinal cancers.

Patel et al. (2020) carried out comprehensive analysis of DNA sequencing data derived from 49,738 participants in a UK Biobank study. They also reviewed hospital and health records back to 1957, cancer registry data back to 1971, and death registry data to 2006.

The gene sequencing data were examined to detect occurrence of pathogenic or likely pathogenic variants in genes known to play roles in FH, familial HBOC, and Lynch syndrome. Deleterious mutations in nine different genes were considered to be primarily important in the etiology of these diseases.

Pathogenic disease mechanisms in FH were defined as preventing clearance of low-density lipoproteins. Deleterious mutations in genes associated with HBOC have been determined to disrupt DNA damage repair mechanisms. Pathogenic mechanisms in Lynch syndrome include mutations that disrupt DNA mismatch repair.

Importantly, clinical management guidelines and treatment options exist for individuals found to be at risk for these four conditions.

Patel et al. noted that decreased DNA sequencing costs have led to increased application of DNA sequencing applications in a number of different healthcare systems. One remaining problem has to do with variant classification and clear definition of variants as pathogenic or likely pathogenic.

In their study whole exome DNA sequencing data generated by the UK Biobank was reanalyzed in Boston, MA. Specific genes analyzed in this project included the following:

FH: APOB Apolipoprotein B, LDLR low-density lipoprotein receptor, PCSK9 proprotein convertase subtilisin/kexin type 9.

Hereditary breast and ovarian cancer: BRCA1, BRCA2 DNA repair—associated.

Lynch syndrome: MLH1, MSH2, MSH6, PMS2 DNA mismatch repair.

Analysis of sequencing results revealed pathogenic variants that increase risk for the designated disorders in 8777 of 49,738 individuals sequenced in the Biobank UK project. Variants classified as pathogenic in Clin Var data and variants and rare missense variants with allele frequencies <0.005 were documented further. The study determined that the 232 variants met most stringent pathogenicity criteria; these included 50 for FH, 133 for HBOC syndrome, and 49 for Lynch syndrome.

Patel et al. noted that in FH, blood levels of LDL cholesterol serve as an additional marker. However, they noted a broad overlap in LDL cholesterol levels between risk gene mutation carriers and noncarriers. Analyses of clinical health records revealed that 28% of the hypercholesterolemia risk gene carriers had documented arteriosclerotic disease and 9.4% of individuals not identified as hypercholesterolemia risk gene carriers developed arteriosclerotic disease.

For HBOC 36% of female risk variant carriers and 12% of male risk variant carriers developed associated cancers.

For Lynch syndrome 36.8% of risk variant carriers and approximately 18% of nonrisk gene carriers developed Lynch syndrome-associated cancers.

The authors emphasized that clinicians may be falsely reassured by a negative DNA test and that additional efforts are required to identify nongenetic factors that contribute to genetic risk.

It is also important to note the continued importance of including family history in risk assessment. Bylstra et al. (2021) reported data that demonstrated that the collection of comprehensive family history and genomic data are complementary in risk assessment.

10.13.1 Promoting diverse population screening

Abul-Husn et al. (2021) noted that populations of non-European ancestry were particularly underrepresented in genomic medicine research. They undertook a pilot program that included genomic screening for FH, HBOC, Lynch syndrome, and added a test for hereditary transthyretin amyloidosis. The initial program included 7461 participants of diverse ancestry in the New York City population, and 692 participants indicated that they wished to receive information on pathogenic or likely pathogenic variants. These individuals included 74 participants found to have HBOC risk variants, eight with FH risk variants, and six with Lynch syndrome risk variants. Transthyretin risk variant testing revealed pathogenic variants in 34 individuals and 33 of these individuals were of African American ancestry. The pathogenic risk variant was V142I.

It is of interest to note that the transthyretin V142I TTR mutation has also been reported in Italian individuals and Spanish ancestry (Cappelli et al., 2016).

Soper et al. (2021) reported that in patients at risk for cardiac failure the TTR V142I mutation occurred in up to 4% of African American patients and in 1% of Hispanic/LatinX patients.

Damrauer et al. (2019) had reported results of a study of V122I hereditary transthyretin amyloidosis in individuals with heart failure.

Akinboboye et al. (2020) reported that in an analysis of patients with suspected hereditary transthyretin amyloidosis the Val112Ile mutation occurred in African American patients.

10.14 Hemochromatosis

This is a disorder characterized by iron overload. Grosse et al. (2018) noted that in populations of European ancestry, hemochromatosis was most commonly associated with homozygosity with a specific allele C282Y (Cys282Tyr) in HFE (homeostatic iron regulator). This disorder was reported to occur with a frequency of 1 in 300 in non-Hispanic whites in the United States. It is clear that pathogenesis in this disorder is due to the combined impact of the pathogenic variants and relatively high dietary iron intake.

The clinical manifestations of hemochromatosis include elevated iron saturation levels of transferrin and increased levels of serum ferritin. Excess iron can also be deposited in tissues including cardiomyocytes, liver, and pancreatic islet cells.

10.15 Other disorders that illustrate the impact of gene environment interactions

Quillen et al. (2019) reported a growing number of genes that have been shown to be involved in sun sensitivity and in addition there are growing numbers of methods for analysis of skin pigmentation and sun sensitivity. Most important gene products involved in skin pigmentation were noted to include melanocortin receptor MC1R and melanin transporter SLC24A5. The ASIP signaling protein that causes hair follicle melanocytes to synthesize pheomelanin, a yellow pigment, was reported to have mutations that are associated with sun sensitivity in individuals in Iceland and the Netherlands.

10.16 Human genetic variation and pathogen sensitivity

Quintana-Murci and Clark (2013) and Karlsson et al. (2014) emphasized that genetic variants in specific populations influence sensitivity to pathogens and infections. Examples included the TLR1 Toll receptor noted to impact susceptibility to leprosy infections and the FUT2 variants that influenced sensitivity to Noro virus infections.

In 1976 Miller et al. postulated that the presence of the Duffy blood group antigen promoted erythrocyte *Plasmodium vivax* malaria infections. In 2006 Langhi and Bordin reported that the Duffy blood group antigen is a receptor for proinflammatory cytokines and is also a receptor for the *Plasmodium vivax* parasite. Individuals negative for Duffy blood group were found to have a pathogenic point mutation in the receptor.

Researchers noted that there is a high frequency of the mutation leading to Duffy negativity in West Africa where the frequency of *Plasmodium vivax* malaria is low.

10.17 Genetic and environmental factors and additional aspects of population screening

10.17.1 Severe visual impairment in children

Solebo et al. (2017) reported evidence regarding the prevalence of blindness in children and noted incidence of 12–15 per 10,000 in very poor regions, 3–4 per 10,000 in affluent countries.

Most common causes of blindness in children in high-income countries included cerebral-based visual impairment and optic nerve defects. In low-income countries most common causes of childhood blindness included retinal ophthalmoplegia of prematurity and cataract. Nutritional vitamin A supplementation programs have been reported to lead to decrease in nutritional-related childhood visual impairment and blindness. In specific regions parasitic infections (toxocariasis) that led to blindness are decreasing in frequency. Rates of blindness due to vertical transmission of Rubella infections from mother to fetus are also noted to be decreasing.

10.17.2 Iodine deficiency, hypothyroidism

Giordano et al. (2019) reported that despite improvements in iodine intake through iodization of salt legislated in Italy in 2005, mild iodine deficiency still existed in certain regions of Italy.

Iodine deficiency was also noted to exist in other countries across the globe including Finland, Korea, Vietnam, Mozambique, Madagascar.

Goiter in children was noted to be an important indicator of iodine deficiency.

References

Aadam et al., 2016   Aadam Z, Kechout N, Barakat A, et al: X-linked agammagobulinemia in a large series of North African patients: frequency, clinical features and novel BTK mutations, *J Clin Immunol* 36(3):187–194, 2016. Available from: https://doi.org/10.1007/s10875-016-0251-z. PMID: 26931785.

Abul-Husn et al., 2021   Abul-Husn NS, Soper ER, Braganza, et al: Implementing genomic screening in diverse populations, *Genome Med* 13(1):17, 2021. Available from: https://doi.org/10.1186/s13073-021-00832-y. PMID: 33546753.

Adhikari et al., 2020   Adhikari AN, Gallagher RC, Wang Y, et al: The role of exome sequencing in newborn screening for inborn errors of metabolism, *Nat Med* 26(9):1392–1397, 2020. Available from: https://doi.org/10.1038/s41591-020-0966-5. Epub 2020 Aug 10. PMID: 32778825.

Ait-Idir and Djerdjouri, 2020   Ait-Idir D, Djerdjouri B: Differential mutational profiles of familial Mediterranean fever in North Africa, *Ann Hum Genet* 84(6):423–430, 2020. Available from: https://doi.org/10.1111/ahg.12404. Epub 2020 Aug 20. PMID: 32818295.

Akinboboye et al., 2020   Akinboboye O, Shah K, Warner AL, et al: DISCOVERY: prevalence of transthyretin (TTR) mutations in a United States-centric patient population suspected of having cardiac amyloidosis, *Amyloid* 27(4):223–230, 2020. Available from: https://doi.org/10.1080/13506129.2020.1764928. PMID: 3245653.

Akler et al., 2020   Akler G, Birch AH, Schreiber-Agus N, et al: Lessons learned from expanded reproductive carrier screening in self-reported Ashkenazi, Sephardi, and Mizrahi Jewish patients, *Mol Genet Genomic Med* 8(2):e1053, 2020. Available from: https://doi.org/10.1002/mgg3.1053. Epub 2019 Dec 27. PMID: 31880409.

Al-Zaidy et al., 2017 Al-Zaidy SA, Lloyd-Puryear M, Kennedy A, et al: A roadmap to newborn screening for Duchenne muscular dystrophy, *Int J Neonatal Screen* 3(2):8, 2017. Available from: https://doi.org/10.3390/ijns3020008. PMID: 31588416.

Antonarakis, 2019 Antonarakis SE: Carrier screening for recessive disorders, *Nat Rev Genet* 20(9):549–561, 2019. Available from: https://doi.org/10.1038/s41576-019-0134-2. PMID: 31142809.

Arvio and Mononen, 2016 Arvio M, Mononen I: Aspartylglycosaminuria: a review, *Orphanet J Rare Dis* 11(1):162, 2016. Available from: https://doi.org/10.1186/s13023-016-0544-6. PMID: 27906067.

Bagheri-Hanson et al., 2014 Bagheri-Hanson A, Nedwed S, Rueckes-Nilges C, Naehrlich L: Intestinal current measurement vs nasal potential difference measurements for diagnosis of cystic fibrosis: a case-control study, *BMC Pulm Med* 14:156, 2014. Available from: https://doi.org/10.1186/1471-2466-14-156. PMID: 25280757.

Bartosch et al., 1991 Bartosch B, Vycudilik W, Popow C, Lubec G: Urinary 3-hydroxyproline excretion in Alport's syndrome: a non-invasive screening test? *Arch Dis Child* 66(2):248–251, 1991. Available from: https://doi.org/10.1136/adc.66.2.248. PMID: 2001113.

Bender et al., 2020 Bender MA, Yusuf C, Davis, et al: Newborn screening practices and alpha-thalassemia detection—United States, 2016, *MMWR Morb Mortal Wkly Rep* 69(36):1269–1272, 2020. Available from: https://doi.org/10.15585/mmwr.mm6936a7. PMID: 32915167.

Bylstra et al., 2021 Bylstra Y, Lim WK, Kam S, et al: Family history assessment significantly enhances delivery of precision medicine in the genomics era, *Genome Med* 13(1):3, 2021. Available from: https://doi.org/10.1186/s13073-020-00819-1. PMID: 33413596.

Cappelli et al., 2016 Cappelli F, Frusconi S, Bergesio F, et al: The Val142Ile transthyretin cardiac amyloidosis: not only an Afro-American pathogenic variant? A single-centre Italian experience, *J Cardiovasc Med (Hagerstown)* 17(2):122–125, 2016. Available from: https://doi.org/10.2459/JCM.0000000000000290. PMID: 26428663.

Cecchi et al., 2019 Cecchi AC, Vengoechea ES, Kaseniit KE, et al: Screening for Tay−Sachs disease carriers by full-exon sequencing with novel variant interpretation outperforms enzyme testing in a pan-ethnic cohort, *Mol Genet Genomic Med* 7(8):e836, 2019. Available from: https://doi.org/10.1002/mgg3.836. PMID: 31293106.

Chan and Puck, 2005 Chan K, Puck JM: Development of population-based newborn screening for severe combined immunodeficiency, *J Allergy Clin Immunol* 115(2):391–398, 2005. Available from: https://doi.org/10.1016/j.jaci.2004.10.012. PMID: 15696101.

Chen, 2020 Chen TH: New and developing therapies in spinal muscular atrophy: from genotype to phenotype to treatment and where do we stand? *Int J Mol Sci* 21(9):3297, 2020. Available from: https://doi.org/10.3390/ijms21093297. PMID: 32392694.

Chokshi et al., 2019 Chokshi B, D'Agati V, Bizzocchi L, et al: Haemophagocytic lymphohistiocytosis with collapsing lupus podocytopathy as an unusual manifestation of systemic lupus erythematosus with APOL1 double-risk alleles, *BMJ Case Rep* 12(1):bcr-2018–227860, 2019. Available from: https://doi.org/10.1136/bcr-2018-227860. PMID: 30642866.

Collins et al., 2018 Collins CJ, Chang IJ, Jung S, et al: Rapid multiplexed proteomic screening for primary immunodeficiency disorders from dried blood spots, *Front*

Immunol 9:2756, 2018. Available from: https://doi.org/10.3389/fimmu.2018.02756. eCollection 2018. PMID: 30564228.

Coppinger and O'Loughlin, 2019 Coppinger C, O'Loughlin R: Newborn sickle cell and thalassaemia screening programme: automating and enhancing the system to evaluate the screening programme, *Int J Neonatal Screen* 5(3):30, 2019. Available from: https://doi.org/10.3390/ijns5030030. eCollection 2019 Sep. PMID: 33072989.

Damrauer et al., 2019 Damrauer SM, Chaudhary K, Cho JH, et al: Association of the V122I hereditary transthyretin amyloidosis genetic variant with heart failure among individuals of African or Hispanic/Latino ancestry, *JAMA* 322(22):2191–2202, 2019. Available from: https://doi.org/10.1001/jama.2019.17935. PMID: 31821430.

De Boeck et al., 2014 De Boeck K, Zolin A, Cuppens H, et al: The relative frequency of CFTR mutation classes in European patients with cystic fibrosis, *J Cyst Fibros* 13 (4):403–409, 2014. Available from: https://doi.org/10.1016/j.jcf.2013.12.003. PMID: 24440181.

De Stefano et al., 1998 De Stefano V, Chiusolo P, Paciaroni K, Leone G: Epidemiology of factor V Leiden: clinical implications, *Semin Thromb Hemost* 24(4):367–379, 1998. Available from: https://doi.org/10.1055/s-2007-996025. PMID: 9763354.

Elder et al., 2013 Elder G, Harper P, Badminton M, et al: The incidence of inherited porphyrias in Europe, *J Inherit Metab Dis* 36(5):849–857, 2013. Available from: https://doi.org/10.1007/s10545-012-9544-4. PMID: 23114748.

Friedman and Pollak, 2021 Friedman DJ, Pollak MR: APOL1 nephropathy: from genetics to clinical applications, *Clin J Am Soc Nephrol* 16(2):294–303, 2021. Available from: https://doi.org/10.2215/CJN.15161219. PMID: 32616495.

Genovese et al., 2010 Genovese G, Friedman DJ, Ross MD, et al: Association of trypanolytic ApoL1 variants with kidney disease in African Americans, *Science* 329(5993):841–845, 2010. Available from: https://doi.org/10.1126/science.1193032. PMID: 20647424.

Giordano et al., 2019 Giordano C, Barone I, Marsico S, et al: Endemic goiter and iodine prophylaxis in Calabria, a region of Southern Italy: past and present, *Nutrients* 11 (10):2428, 2019. Available from: https://doi.org/10.3390/nu11102428. PMID: 31614658.

Grosse et al., 2018 Grosse SD, Gurrin LC, Bertalli NA, Allen KJ: Clinical penetrance in hereditary hemochromatosis: estimates of the cumulative incidence of severe liver disease among HFE C282Y homozygotes, *Genet Med* 20(4):383–389, 2018. Available from: https://doi.org/10.1038/gim.2017.121. PMID: 28771247.

Gutierrez-Mateo et al., 2019 Gutierrez-Mateo C, Timonen A, Vaahtera K, et al: Development of a multiplex real-time PCR assay for the newborn screening of SCID, SMA, and XLA, *Int J Neonatal Screen* 5(4):39, 2019. Available from: https://doi.org/10.3390/ijns5040039. eCollection 2019 Dec. PMID: 33072998.

Harteveld and Higgs, 2010 Harteveld CL, Higgs DR: Alpha-thalassaemia, *Orphanet J Rare Dis* 5:13, 2010. Available from: https://doi.org/10.1186/1750-1172-5-13. PMID: 2050764.

Hoppe, 2013 Hoppe CC: Prenatal and newborn screening for hemoglobinopathies, *Int J Lab Hematol* 35(3):297–305, 2013. Available from: https://doi.org/10.1111/ijlh.12076. PMID: 23590658.

Howell, 2021 Howell RR: Ethical issues surrounding newborn screening, *Int J Neonatal Screen* 7(1):3, 2021. Available from: https://doi.org/10.3390/ijns7010003. PMID: 33435435.

Jangouk et al., 2012 Jangouk P, Zackowski KM, Naidu S, Raymond GV: Adrenoleukodystrophy in female heterozygotes: underrecognized and undertreated, *Mol Genet Metab* 105(2):180–185, 2012. Available from: https://doi.org/10.1016/j.ymgme.2011.11.001. PMID: 22112817.

Jędrzejowska, 2020 Jędrzejowska M: Advances in newborn screening and presymptomatic diagnosis of spinal muscular atrophy, *Degener Neurol Neuromuscul Dis* 10:39–47, 2020. Available from: https://doi.org/10.2147/DNND.S246907. eCollection 2020. PMID: 3336487.

Kaback et al., 1993 Kaback M, Lim-Steele J, Dabholkar D, et al: Tay–Sachs disease–carrier screening, prenatal diagnosis, and the molecular era. An international perspective, 1970 to 1993. The International TSD Data Collection Network, *JAMA* 270 (19):2307–2315, 1993. PMID: 8230592.

Karlsson et al., 2014 Karlsson EK, Kwiatkowski DP, Sabeti PC: Natural selection and infectious disease in human populations, *Nat Rev Genet* 15(6):379–393, 2014. Available from: https://doi.org/10.1038/nrg3734. PMID: 24776769.

Kato et al., 2018 Kato GJ, Piel FB, Reid CD, et al: Sickle cell disease, *Nat Rev Dis Primers* 4:18010, 2018. Available from: https://doi.org/10.1038/nrdp.2018.10. PMID: 29542687.

Kraft et al., 2019 Kraft SA, Duenas D, Wilfond BS, Goddard KAB: The evolving landscape of expanded carrier screening: challenges and opportunities, *Genet Med* 21 (4):790–797, 2019. Available from: https://doi.org/10.1038/s41436-018-0273-4. PMID: 30245516.

Langhi and Bordin, 2006 Langhi DM Jr, Bordin JO: Duffy blood group and malaria, *Hematology* 11(5):389–398, 2006. Available from: https://doi.org/10.1080/10245330500469841. PMID: 17607593.

Lee et al., 2020 Lee S, Taylor JL, Redmond C, et al: Validation of Fragile X screening in the newborn population using a fit-for-purpose FMR1 PCR assay system, *J Mol Diagn* 22(3):346–354, 2020. Available from: https://doi.org/10.1016/j.jmoldx.2019.11.002. PMID: 31866572.

Lloyd-Puryear et al., 2018 Lloyd-Puryear MA, Crawford TO, Brower A, et al: Duchenne muscular dystrophy newborn screening, a case study for examining ethical and legal issues for pilots for emerging disorders: considerations and recommendations, *Int J Neonatal Screen* 4(1):6, 2018. Available from: https://doi.org/10.3390/ijns4010006. eCollection 2018 Mar. PMID: 33072932.

Lobitz et al., 2018 Lobitz S, Telfer P, Cela E, et al: Newborn screening for sickle cell disease in Europe: recommendations from a Pan-European Consensus Conference, *Br J Haematol* 183(4):648–660, 2018. Available from: https://doi.org/10.1111/bjh.15600. PMID: 30334577.

Martin et al., 2020 Martin GR, Ewer AK, Gaviglio A, et al: Updated strategies for pulse oximetry screening for critical congenital heart disease, *Pediatrics* 146(1):e20191650, 2020. Available from: https://doi.org/10.1542/peds.2019-1650. PMID: 32499387.

Miller et al., 1976 Miller LH, Mason SJ, Clyde DF, McGinniss MH: The resistance factor to *Plasmodium vivax* in blacks. The Duffy-blood-group genotype, FyFy, *N Engl J Med* 295 (6):302–304, 1976. Available from: https://doi.org/10.1056/NEJM197608052950602. PMID: 778616.

Myerowitz, 1997 Myerowitz R: Tay–Sachs disease-causing mutations and neutral polymorphismsin the Hex A gene, *Hum Mutat* 9(3):195–208, 1997. 10.1002/(SICI)1098-1004(1997)9:3 < 195::AID-HUMU1 > 3.0.CO;2–7. PMID: 9090523.

Myerowitz and Costigan, 1988 Myerowitz R, Costigan FC: The major defect in Ashkenazi Jews with Tay – Sachs disease is an insertion in the gene for the alpha-chain of beta-hexosaminidase, *J Biol Chem* 263(35):18587–18589, 1988. PMID: 2848800.

Nkya et al., 2019 Nkya S, Mtei L, Soka D, et al: Newborn screening for sickle cell disease: an innovative pilot program to improve child survival in Dar es Salaam, Tanzania, *Int Health* 11(6):589–595, 2019. Available from: https://doi.org/10.1093/inthealth/ihz028. PMID: 31145786.

Oster and Kochilas, 2016 Oster ME, Kochilas L: Screening for critical congenital heart disease, *Clin Perinatol* 43(1):73–80, 2016. Available from: https://doi.org/10.1016/j.clp.2015.11.005. PMID: 26876122.

Park et al., 2020 Park YH, Remmers EF, Lee W, et al: Ancient familial Mediterranean fever mutations in human pyrin and resistance to *Yersinia pestis*, *Nat Immunol* 21(8):857–867, 2020. Available from: https://doi.org/10.1038/s41590-020-0705-6. PMID: 32601469.

Patel et al., 2020 Patel AP, Wang M, Fahed AC, et al: Association of rare pathogenic DNA variants for familial hypercholesterolemia, hereditary breast and ovarian cancer syndrome, and Lynch syndrome with disease risk in adults according to family history, *JAMA Netw Open* 3(4):e203959, 2020. Available from: https://doi.org/10.1001/jamanetworkopen.2020.3959. PMID: 32347951.

Puck, 2019 Puck JM: Newborn screening for severe combined immunodeficiency and T-cell lymphopenia, *Immunol Rev* 287(1):241–252, 2019. Available from: https://doi.org/10.1111/imr.12729. PMID: 30565242.

Quillen et al., 2019 Quillen EE, Norton HL, Parra EJ, et al: Shades of complexity: new perspectives on the evolution and genetic architecture of human skin, *Am J Phys Anthropol* 168(67):4–26, 2019. Available from: https://doi.org/10.1002/ajpa.23737. PMID: 30408154.

Quintana-Murci and Clark, 2013 Quintana-Murci L, Clark AG: Population genetic tools for dissecting innate immunity in humans, *Nat Rev Immunol* 13(4):280–293, 2013. Available from: https://doi.org/10.1038/nri3421. PMID: 23470320.

Quon and Rowe, 2016 Quon BS, Rowe SM: New and emerging targeted therapies for cystic fibrosis, *BMJ* 352:i859, 2016. Available from: https://doi.org/10.1136/bmj.i859. PMID: 27030675.

Raghubeer et al., 2020 Raghubeer S, Pillay TS, Matsha TE: Gene of the month: APOL1, *J Clin Pathol* 73(8):441–443, 2020. Available from: https://doi.org/10.1136/jclinpath-2020-206517. PMID: 32404472.

Rausell et al., 2020 Rausell A, Luo Y, Lopez M, et al: Common homozygosity for predicted loss-of-function variants reveals both redundant and advantageous effects of dispensable human genes, *Proc Natl Acad Sci U S A* 117(24):13626–13636, 2020. Available from: https://doi.org/10.1073/pnas.1917993117. PMID: 32487729.

Riley and Wheeler, 2017 Riley C, Wheeler A: Assessing the Fragile X syndrome newborn screening landscape, *Pediatrics* 139(3):S207–S215, 2017. Available from: https://doi.org/10.1542/peds.2016-1159G. PMID: 28814541.

Rivas et al., 2018 Rivas MA, Avila BE, Koskela J, et al: Insights into the genetic epidemiology of Crohn's and rare diseases in the Ashkenazi Jewish population, *PLoS Genet* 14(5):e1007329, 2018. Available from: https://doi.org/10.1371/journal.pgen.1007329. eCollection 2018 May. PMID: 29795570.

Rommens et al., 1989   Rommens JM, Iannuzzi MC, Kerem B, et al: Identification of the cystic fibrosis gene: chromosome walking and jumping, *Science* 245 (4922):1059−1065, 1989. Available from: https://doi.org/10.1126/science.2772657. PMID: 2772657.

Sabath, 2017   Sabath DE: Molecular diagnosis of thalassemias and hemoglobinopathies: an ACLPS critical review, *Am J Clin Pathol* 148(1):6−15, 2017. Available from: https://doi.org/10.1093/ajcp/aqx047. PMID: 28605432.

Shi et al., 2017   Shi L, Webb BD, Birch AH, et al: Comprehensive population screening in the Ashkenazi Jewish population for recurrent disease-causing variants, *Clin Genet* 91(4):599−604, 2017. Available from: https://doi.org/10.1111/cge.12834. PMID: 27415407.

Solebo et al., 2017   Solebo AL, Teoh L, Rahi J: Epidemiology of blindness in children, *Arch Dis Child* 102(9):853−857, 2017. Available from: https://doi.org/10.1136/archdischild-2016-310532. PMID: 2846530.

Soper et al., 2021   Soper ER, Suckiel SA, Braganza GT, et al: Genomic screening identifies individuals at high risk for hereditary transthyretin amyloidosis, *J Pers Med* 11 (1):49, 2021. Available from: https://doi.org/10.3390/jpm11010049. PMID: 33467513.

Therrell and Padilla, 2018   Therrell BL Jr, Padilla CD: Newborn screening in the developing countries, *Curr Opin Pediatr* 30(6):734−739, 2018. Available from: https://doi.org/10.1097/MOP.0000000000000683. PMID: 30124582.

Turk et al., 2020   Turk BR, Theda C, Fatemi A, Moser AB: X-linked adrenoleukodystrophy: pathology, pathophysiology, diagnostic testing, newborn screening and therapies, *Int J Dev Neurosci* 80(1):52−72, 2020. Available from: https://doi.org/10.1002/jdn.10003. PMID: 31909500.

Van Cott et al., 2016   Van Cott EM, Khor B, Zehnder JL: Factor V Leiden, *Am J Hematol* 91(1):46−49, 2016. Available from: https://doi.org/10.1002/ajh.24222. Epub 2015 Nov 17. PMID: 26492443.

Vasquez-Loarte et al., 2020   Vasquez-Loarte T, Thompson JD, Merritt JL 2nd: Considering proximal urea cycle disorders in expanded newborn screening, *Int J Neonatal Screen* 6(4):77, 2020. Available from: https://doi.org/10.3390/ijns6040077. PMID: 33124615.

Wiesinger et al., 2020   Wiesinger T, Mechtler T, Schwarz M, et al: Investigating the suitability of high-resolution mass spectrometry for newborn screening: identification of hemoglobinopathies and β-thalassemias in dried blood spots, *Clin Chem Lab Med* 58 (5):810−816, 2020. Available from: https://doi.org/10.1515/cclm-2019-0832. PMID: 32031968.

Wright et al., 2019   Wright CF, West B, Tuke M, et al: Assessing the pathogenicity, penetrance, and expressivity of putative disease-causing variants in a population setting, *Am J Hum Genet* 104(2):275−286, 2019. Available from: https://doi.org/10.1016/j.ajhg.2018.12.015. PMID: 30665703.

Further reading

Kaplanis et al., 2019   Kaplanis J, Akawi N, Gallone G, et al: Exome-wide assessment of the functional impact and pathogenicity of multinucleotide mutations, *Genome Res* 29

(7):1047–1056, 2019. Available from: https://doi.org/10.1101/gr.239756.118. PMID: 31227601.

Weil et al., 2020 Weil LG, Charlton MR, Coppinger C, et al: Sickle cell disease and thalassaemia antenatal screening programme in England over 10 years: a review from 2007/2008 to 2016/2017, *J Clin Pathol* 73(4):183–190, 2020. Available from: https://doi.org/10.1136/jclinpath-2019-206317. PMID: 31771971.

CHAPTER 11

Analysis of variants associated with abnormal drug responses, genetics, and genomics in drug design

11.1 Pharmacokinetics and pharmacodynamics

Pharmacokinetics include processes involved in drug absorption distribution and elimination. Biotransformation may influence pharmacokinetics or subsequent drug actions. Pharmacodynamics relate to mechanisms of drug action (Benet et al., 1996). Drug action is dependent on interaction of drugs with components of the organism. The latter may include specific receptors and biomolecules. Regarding binding to receptors, important considerations include the strength of binding to a receptor, and the downstream effects of drug binding to the receptor, for example, does receptor binding impact lead to downstream receptor activities like activation of G-proteins or are there other effects like receptor blocking.

It is also important to consider drugs that do not bind to specific cell receptors but may bind to ions or other small molecules and then be taken up by cells or organelles. In considering movement of drugs into cells it is also important to take into account that these can include ATP dependent transporters, solute carriers, and ion channels.

11.1.1 Important examples of drugs and their binding to receptors

The DRD2 dopamine receptor is a G-protein coupled receptor that is the target of a number of different antipsychotic drugs. Wang et al. (2018) identified specific sites in the receptor that are important in drug binding; these include amino acids at specific positions. In the DRD2 receptor a specific variant was identified, rs181028, that led to a Ser to Cys change and altered binding to risperidine. A variant in the dopamine receptor, DRD3 rs 6280 a serine to glycine missense mutation, impacts response to first generation antipsychotics.

The epidermal growth factor receptors (EGFRs), also known as ERBB and HER1, when activated are known to play important roles in cell proliferation and growth and is also implicated in cancer. Downstream signaling from EGFRs involves the Map kinase signaling pathway. EGFRs were noted to occur primarily on cell membranes at the cell surface where they are anchored to the cell membrane. There is also some evidence that these receptors may occur on the membranes of certain cell organelles, for example, endosomes.

A key early discovery was that growth factor receptors have tyrosine kinase activity. There was evidence that ligand binding led to clustering of receptor subunits and ligand binding stimulated intracellular tyrosine kinase activity. Tyrosine kinase activity stimulated intracellular phosphorylation and activated cellular signaling pathways (Schlessinger, 2014).

Freed et al. (2017) reviewed ligands that bind to EGFRs and induce signaling. They included growth factors and epiregulin. EGFRs and the ERB family of receptor tyrosine kinase play important roles in cancer (Normanno et al., 2006). Activating mutation in EGFRs were reported in certain cancers including nonsmall lung cancer.

Therapeutic developments include the development of EGFRs tyrosine kinase inhibitors. Both first- and second-generation inhibitors of EGFRs have been developed. These drugs act specifically by impacting the receptor intracellular tyrosine kinase activity (Li et al., 2021). Approximately three dozen EGFR protein kinase inhibitors have FDA approval for clinical use.

11.2 Biotransformation of medicinal compounds

Biotransformation is an important process that is required for medicinal compounds to be therapeutically active. Initial phase processes were reported to be necessary to expose important active sites, in these reactions inactive products can be converted to active compounds.

The cytochrome P450 monooxygenase system was reported to play important roles in biotransformations. Benet et al. (1996) described CYP450 enzymes as heme-containing membrane proteins located in the endoplasmic reticulum. Oxidative reactions are carried out through activity of the CYP450 heme protein and P450 reductase activity in the presence of NADPH and molecular oxygen.

In humans, 12 classes of CYP450 enzymes were described. The majority of biotransformation reactions were reported to be carried out by enzymes in the CYP1, CYP2, and CYP3 families. Benet et al. reported that the largest number of drugs required transformation by CYP3A, CYP2D6, or CYP2C; a small number of drugs requires transformation by CYP1A2 and CYP2E1.

Genetic polymorphisms in the CYP450 enzymes were noted to contribute to individual differences in biotransformation and individual differences in effective drug dosages.

11.3 World-wide distribution of genetic polymorphisms in the CYP450 system

Zhou et al. (2017) analyzed whole genome and exome sequencing data to derive 176 CYP haplotypes in five different human populations including 56,945 individuals. To carry out their study Zhou et al. reviewed sequence data in the ExAC database and data from the 1000 Genomes Project. They documented the specific alleles present at the major CYP loci in populations from Europe, Africa, East Asia, South Asia, and in admixed Americans.

Specific polymorphisms were analyzed in 12 different CYP genes that account for 75% of phase I drug biotransformation. The population frequencies of polymorphisms were particularly different in the following CYP450 genes. Derived data were made available in an accessible data base.

Gene	Chromosome location
CYP2A6	19q13.2
CYP2B6	19q13.1
CYP2C8	10q23.33
CYP2C9	10q23.33 (highest number of alleles with increased activity)
CYP2C19	10q23.33
CYP2D6	22q13.2 (highest degree variation) (noted to sometimes be duplicated)
CYP3A4	7q22.1
CYP3A5	7q22.1 (lowest degree of variation in most populations)

The greatest population differences in allele frequencies occurred in *CYP2C19, CYP2D6, CYP2B6*.

CYP2D6 enzyme activity was noted to be responsible for metabolism of 25% of drugs in clinical use. Variants that reduced function of CYP2D6 enzyme were noted to reach highest frequency in East Asians (70%) and also occurred in 52% of Africans.

It is important to note that certain drugs are classified as inhibitors of CYP2D6; strong inhibitors of CYP2D6 would eliminate activity of the enzyme. Cicali et al. (2020) noted that 22 different medications are listed by the FDA as inhibitors of CYP2D6, weak or strong inhibitors.

11.4 Other factors and processes involved in biotransformation of drugs

Important other biotransformation processes include acetylation and glucuronyl transferase activities.

Acetylation reactions include N-acetyl transferases NAT1 and NAT2. Hein et al. (2018) reviewed molecular genetics of NAT1, NAT2, and their polymorphisms These enzymes were noted to catalyze N-acetylation and O-acetylation of heterocyclic amines.

N-acetyl transferases are also commonly referred to as arylamine-N-acetyltransferase and they play in metabolism of acetylated arylamine carcinogens (Sim et al., 2014).

Various polymorphisms result in the NAT enzymes that are either slow metabolizers or fast metabolizers. The slow metabolizers increased toxicity of hydralazine (apresoline, a drug to reduce blood pressure) and isoniazid, used in treatment of tuberculosis. Sim et al. reported that NAT2 is involved in acetylation of hydralazine. Para-amino salicylate used in treatment of tuberculosis is acetylated by NAT1. A specific catabolite of folic acid para-aminobenzoylglutamate was reported to be acetylated by NAT1.

NAT1 and NAT2 differ in their tissue distribution. NAT1 has widespread distribution while NAT2 is mainly expressed in liver and gut. Sim reported that polymorphisms at the active site of an acetylation enzyme influenced its activity. Specific polymorphisms were noted to impact the folding of NAT1 or NAT2. Studies by Gupta et al. (2013) revealed that the tuberculosis medication isoniazid could induce hepatotoxicity in individuals in India with the slow acetylator genotype.

Early studies reported that in an oriental population 15% of individuals were slow acetylators while in the Caucasian population 50% were slow acetylators.

Studies carried out by Cartwright et al. (1982) revealed that the slow acetylator phenotype was associated with bladder cancer in individuals exposed to aniline dyes.

Cloning of NAT1 and NAT2 encoding genes revealed polymorphisms as point mutations. NAT1 and NAT2 encoding genes were reported to be 80% identical and both mapped within a 200-kb region on chromosome 8. Sim et al. (2014) documented nine different single nucleotide variants in NAT2 and five single nucleotide variants 1in NAT1.

11.4.1 UDP glucuronyl transferase enzymes and medication biotransformation

Jarrar and Lee (2021) reviewed glucuronyl transferase enzymes. They noted that there are 22 different enzymes in this group. UGT1, UGT2, and UGT 3 are the principal enzymes and are involved in the metabolism of many prescription drugs. Addition of glucuronic acid to the molecule was noted to lead to production of beta glucuronide conjugates.

Families of genes that encode UGT occur on different chromosomes. Principle enzymes include UGT1 family on chromosome 2Q37.1, UGT2 family on chromosome 4q13.2, and UGT3 family on 5p13.3.

Specific UDP glucuronyl transferase has been reported to play roles in human diseases. Gilbert syndrome (hereditary hyperbilirubinemia), a severe liver disease, was reported to be associated with defects in UGT1A.

11.5 Drug responses, variants, genetic and environmental factors

Ingelman-Sundberg et al. (2018) undertook an analysis of variants in 208 genes involved in pharmacokinetics to estimate the presence of rare and common variants that could influence drug responses.

They noted that together physiological, pathological, environmental, and genetic factors influence drug responses. Genetic factors were reported to account for 20%–30% of variability in drug response. They noted that a substantial fraction of the genetic factors were heritable and could be explained by previously identified common genetic variants in pharmacogenes.

Their study involved genes analyses of 208 pharmacogenes in 60,706 individuals. They also noted the importance of analyses of nonexonic variants in specific genes.

Their study of pharmacogenes included 35 CYP, 17 UGTs, 17 GSTs, 8 SULTS (sulfotransferases).

Nonexon variants important in pharmacogenes included the following:
CYP1A2*1 C rs2069514
CYP1A2*1 F rs 762551
CYP2C*17 rs12248560
CYP3A4*22 rs35599367
CYP3A5*3 rs776746
UGT1A1*28 rs8175347

The authors noted that rare pharmacogene variants were enriched in variants that impact function. They identified rare variants that impact warfarin (anticoagulant) response in CYP2C9 and transporter ABCB1. ATP binding cassette subfamily B member 1 is an ATP-dependent drug efflux pump. It is responsible for decreased drug accumulation in multidrug-resistant cells.

Anderson et al. (2020) carried out a study to determine utilization of pharmacogenetic testing for the US managed care population. For their analyses they utilized data from the medical aid records and from managed care clinics. The genes tested included CYP2C19, CYP2D6, CYP2C9, VKORC1, and UGT1A.

Results of their study revealed that the rate of testing was low, 0.12% of patients were tested. However, there was evidence that the rate of testing doubled between 2013 and 2016.

11.5.1 Pharmacogenes and therapeutics

Roden et al. (2019) emphasized that information on genetic variation can be used to individualize therapy. They noted that genetic variation can influence pharmacokinetics and pharmacodynamics and that improved methods to measure drug concentrations facilitated ability to analyze drug transport, pharmacokinetics, and drug metabolism.

Roden et al. emphasized the key roles of CYP450 enzymes in activation of prodrugs, for example, bioactivation of codeine to its major metabolite morphine by CYP2D6 and activation of clopidogrel, an antiplatelet drug by CYP2C19. They also noted that other genetic and environmental factors influence effects of certain CYP450 variants.

Roden et al. drew attention to another aspect of therapeutics, the therapeutic range of a drug, that is, the margin between therapeutic dose and toxic dose.

Important in this context is the antineoplasm drug 6-mercaptopurine that is bioactivated by thiopurine-s-mercaptotransferase activity (TPMT) and xanthine oxidase. Loss of function genetic variants in TPMT1 can lead to increased toxicity of 6-mercaptopurine.

The immunosuppressant drug azathiopurine is metabolized to 6-mercaptopurine and variant in either TPMT or in the transferases in the NUDT (nudix hydrolase family) are associated with the risk of toxicity.

The antineoplastic drugs 5-fluorouracil and fluorpyrimidine were shown to have greater toxicity in individuals with specific variants in DPYD, dihydropyrimidine dehydrogenase.

Drug transport and variants in specific gene products responsible for transport of drugs into or out of cells were noted to impact levels of drugs in the body. These include drug efflux transporter OAT1B1 (SLCO1B) that removes drugs from the circulation, and variants in this transporter can impair clearance of methotrexate.

Roden et al. noted evidence that factors involved in pharmacokinetics and pharmacodynamics can impact effect of warfarin.

Rare variants that have relevance in anesthesia have also been described. These include reports of malignant hyperthermia following administration of inhaled anesthetics or succinylcholine in individuals with variants in RYR1 ryanodine receptor and in individuals with variants in the calcium ion channel CACNA1S.

Roden noted that altered drug effects due to variants that impact pharmacodynamics or pharmacokinetics are referred to as Type A adverse drug effects. Type B adverse drug effects are nondose dependent and may have an immunologic basis. Particularly relevant to Type B adverse drug effects are variants in HLA genes. One allergic drug reaction known as Stevens − Johnson syndrome is associated with epidermal necrolysis.

Genetic variants that lead to immunologically based severe drug reactions include the HLA-B* 15.02 allele that occurs particularly in the South-East Asian population and can lead to severe drug reactions following exposure to carbamazepine, an antiepileptic medication. The HLA-A*21.01 allele may lead to adverse drug reaction in Europeans. Floxacillin, an antibiotic, induces adverse drug reaction including hepatotoxicity. This was reported in individuals with HLAB* 57.01. Abacavir is used to prevent and treat human immunodeficiency virus (HIV) infections; toxicity has also been documented in individuals with HLAB* 57.01.

11.5.2 HLA typing: application of DNA sequencing

Earlier methods of typing HLA antigens were based on use of antibodies against specific HLA sites. Yin et al. (2016) reported use of long-range PCR followed by DNA sequencing for HLA typing of buccal cell-extracted DNA. They emphasized the importance of unambiguous HLA typing for hematopoietic stem cell transplantation, for organ transplantation, and for HLA association studies.

11.6 Gene variants that are disease causing and are also associated with abnormal drug reactions

11.6.1 Glucose-6-phosphate (G6PD) deficiency

In 1994 Beutler reviewed G6PD deficiency. This disorder came to light particularly after introduction of the agent primaquine for prevention and treatment of malaria when it was found that in a subgroup of individuals it induced hemolysis.

Specific features of primaquine-induced hemolysis included the presence within red blood cells of Heinz bodies, denatured protein that adhered to the red cell membrane. Red cells also assumed abnormal shapes. Severe hemolysis led to the passage of very dark urine.

Other medications that can induce hemolysis in G6PD-deficient individuals include sulfacetamide, sulfapyridine, and furadantin.

G6PD deficiency was also shown to be a factor in certain cases of neonatal jaundice.

G6PD is known to catalyze the initial reaction in the hexose monophosphate shunt, namely the oxidation of glucose-6-phosphate to 6-phosphogluconate with conversion of nicotinamide adenine dinucleotide phosphate (NADP) to NADPH. In the subsequent reaction 6-phosphogluconate is converted to ribulose-5-phosphate and ribose-5-phosphate that can then participate in nucleic acid synthesis.

Beutler noted that adequate levels of NADPH are required for oxidation of sulfhydryl groups and oxidation of glutathione, reactions that are essential for defending the red cells against peroxides and oxidative stress.

G6PD-deficient cells were noted to be highly susceptible to oxidative stress.

Primaquine sensitivity was known to be particularly severe in males with G6PD deficiency consistent with the fact that gene that encodes this enzyme is located on the X chromosome.

In 1994 Beutler noted that 442 distinct G6PD variants had been identified in a WHO study. Studies revealed that 87% of hemolytic anemia-associated variants occurred in two regions of the G6PD protein, the NADP binding site and the glucose binding site.

Frequencies of deleterious G6PD mutations differed in different populations. The G6PD Mediterranean variant achieved a frequency of 0.70 in Kurdish Jewish population, 0.045 in the Greek population, and 0.10 in African Americans.

G6PD deficiency likely promoted survival in malaria-prone regions of the world. Studies revealed that malaria parasite counts were higher in G6PD positive erythrocytes than in G6PD negative cells. There is evidence that the higher oxidation capacity of G6PD positive cells benefits parasites.

Questions have been raised about the role of G6PD variants in favism, a form of hemolysis induced in some individual by fava bean intake. Beutler noted that there was evidence that favism may represent an abnormal immune response.

Specific resources are available that list medicinal drugs that should be avoided in G6PD deficient patients, https://www.cych.org.tw/pharm/MIMS%20Summary%20Table-G6PD.pdf.

11.6.2 Drug-induced hemolytic anemias

Renard and Rosselet (2017) reviewed drug-induced hemolytic anemias, noting that they represented potentially lethal adverse drug reactions. They noted that the hemolytic reactions could result from oxidative damage to vulnerable erythrocytes or from induced thrombotic microangiopathy and could also represent immune-mediated anemias.

Nonimmune drug-induced hemolytic anemias could result when oxidative stress-sensitive red cells come into contact with drugs causing oxidative damage or drugs that promoted the generation of oxygen radicals.

They emphasized that infections could also lead to oxidative damage. Drugs particularly involved in inducing oxidative damage include primaquine, sulfacetamide, nitrofurantoin, and rasburicase.

Thrombotic microangiopathy includes thrombocytopenic purpura and the hemolysis could lead to hemolytic uremic syndrome. Drugs associated with this complication could include quinine, quinidine, antineoplastic drugs, immunosuppressant, or antiplatelet drugs.

11.7 Porphyrias

Porphyrias are heme biosynthesis disorders. Genetic forms of porphyria are usually autosomal dominant conditions. Most common genetic forms of porphyria include acute intermittent porphyria, variegate porphyria, and hereditary coproporphyria. Specific medication sensitivities occur in these disorders. Balwani et al. (2017) and Stölzel et al. (2019) reviewed porphyrias.

Variegate porphyria is due to genetic defects in a specific enzyme protoporphyrinogen oxidase (PPO). Skin manifestations include blistering and neurological manifestations can occur and may be precipitated by intake of barbiturates, sulfonamides, rifampin. Twenty-three different pathogenic mutations are documented in PPO. Founder mutations occur in specific populations particularly those with ancestral origins in the Netherlands.

Acute intermittent porphyria was determined to be due to mutations in the enzyme porphobilinogen deaminase referred to in some publications as hydroxymethylbilane synthetase. A large number of different mutation have been reported in this disorder. Some mutations are more common in specific countries.

Symptoms of the disorder include abdominal pain, tachycardia, neuropsychiatric symptoms. Symptoms may be induced by a number of different medication also by hyponatremia and dehydration.

Coproporphyria occurs due to deficiency of coproporphyrinogen oxidase. Deficiency can lead to symptoms on exposure to certain medication and particularly by progesterone. At least 63 different pathogenic mutations have been described as leading to coproporphyria.

For individuals with porphyria-related disorders medications to avoid include sulfonamides, carbamazepine, hydantoin, progesterone, estrogens, barbiturates, nitrofurantoin, valproic acid. Individuals with these disorders should also avoid fasting, dehydration, and excessive smoking that are known to induce symptoms.

11.8 Identifying therapeutic targets and developing therapies

Work required prior to therapeutic development includes analyses of disease processes, identifying potential therapeutic targets, analyzing target biology and possibilities for target modification, assessing druggability of a target, and safety issues in modification of targets. Biomarkers are required to assess impact of therapy and unmet needs following initiation of the proposed therapy.

Emmerich et al. (2021) published GOT-IT Framework Guidelines on target assessment for innovative therapeutics.

They noted that funders for projects were most often focused on unmet medical needs. The right target, the right patient population, and possibilities to target the right tissue were considered key.

Regarding the target selection, if human disease related there must be clear identification that the target selected for treatment is related to the disease. Emmerich et al. noted that critical path questions in this context included the following:

Is target perturbation a cause or a consequence of the disease?
Is target manipulation clinically relevant?
Is detection of target-dependent processes clinically relevant, and disease relevant?
Is the target selected relevant to other diseases?
Is the tissue distribution of the target known?
Are biological models available for safety testing?
Do models translate well to human?
Can adverse effects be predicted through assessment of a safety biomarker?
At what stage of disease is therapy relevant?
Is the target expression pattern known?
Do alternative or complementary pathways exist that are not influenced by the target?

Would technology or target manipulation have advantages over other treatment approaches?

Emmerich et al. also provided links to website that provide information possibly relevant to target selection, for example, Human Protein Atlas, Genotype Tissue Expression project, Cancer Genome Atlas, Genome Wide Association Studies.

Model systems clearly linked to the disease being studied were noted to be important. These include patients' primary cells, pluripotent stem cells.

Emmerich et al. also described possible target manipulation processes. These included functional knock out of the target and mutation silencing.

11.9 Translation of biomedical observations to treatments and health improvements

In a review article in 2021, Austin defined translation as the process by which biomedical observations become interventions to improve human health. Austin defined translational science as the field that analyses the translational process and stressed that progress in medical science over the previous four decades had outpaced translational and clinical science.

Austin drew attention to transformation of genetics "from an observation endeavor to a powerful and efficiency driver of science and medicine" and noted further that data science was revolutionizing data use.

Austin reviewed information in the Drug Discovery and Deployment map generated by Wagner et al. (2018). Key features of this map include:

1. Basic science research.
2. Target identification, target pharmacology, biomarker development.
3. Lead identification.
4. Lead optimization and investigational new drug application IND.
5. Clinical research.
6. Regulator review.
7. Marketing.

Austin emphasized the importance of connections and transition between basic research, preclinical research, clinical research, clinical implementation, and public health.

Austin also focused attention on the chemical space problem that as new disease targets are identified new chemical structures would be required. The slow pace of identification of new active compounds in the chemical space was emphasized.

However, in more recent decades the range of therapeutic modalities was noted to have broadened to include small molecules, therapeutic peptides, antibodies, oligonucleotides, cell therapies, and gene therapies.

Austin mentioned the importance of advocacy groups, communities, and patient advocacy groups, both in the motivation of studies and in the adoption of approved therapies.

Factors noted to be of key importance in development of new medication were clinical trial development and analyses of biomarkers to monitor clinical response.

11.10 Fragment-based drug discovery

Concepts and core principles of fragment-based drug discovery design were reviewed by Kirsch et al. (2019). Fragment-based drug discovery refers to the detection of small molecules that bind to a specific target. These small molecules, referred to as fragments, are of low complexity. Binding regions within targets may include allosteric sites (sites that activate or inhibit activity) or small binding pockets that are involved in protein — protein interactions.

Fragments were noted to usually have molecular weights lower than 300 Daltons and low molecular complexity. Libraries of fragments are available commercially or noncommercially. Novel fragments can also be designed.

11.11 Monoclonal antibodies as therapeutic agents

Expansion of possibilities to produce monoclonal antibodies followed the introduction of hybridoma technology by Köhler and Milstein (1975) when they published a paper entitled "Continuous culture of fused cells secreting antibody of predefined specificity."

They documented injection of mice with a predefined antigen. After a period they isolated antibody producing B-cells from the immunized mice. These cells were then fused with a previously established myeloma cell line that required specific medium to survive since the cells were deficient in the enzyme hypoxanthine guanine phosphoribosyl transferase. Following fusion and growth in regular cell medium rather than the specialized medium, only the fused cells (mouse B-cells fused with myeloma cell line cells) survived. These cells continued to produce antibody and were referred to as hybridoma cells.

Later developments in monoclonal antibody production included development of phage display. This included isolation of heavy and light chain immunoglobulin-encoding genes from antibody producing cells, introduction of these genes into the gene encoding phage coat protein. Phages were modified so that they produced a single antibody.

Burton et al. (1991) reported use of phage technology to isolate monoclonal antibodies to Type 1 HIV.

Castelli et al. noted that monoclonal antibodies had been developed for a number of auto-immune diseases. Specifically, antibodies were developed to target different components of the immune system. Some therapeutic antibodies were designed to block specific T- or B-cell subtypes or proinflammatory cytokines.

Monoclonal antibodies have also been developed to treat certain cancers, for example, to block overexpressed growth factor receptors EGFR and HER2 in breast cancer. Antibodies to specific hematopoietic differentiation antigens were developed to treat specific lymphoproliferative disorders.

In a 2019 review Castelli et al. noted that monoclonal antibodies had emerged as a major class of therapeutic agents. They were used primarily in treatment of oncological diseases, in infectious diseases, and additional uses were being investigated.

Emicizumab, a recombinant monoclonal antibody, was reported to be useful in treatments of certain patients with hemophilia A. It was reported by Blair (2019) to restore function of missing activated factor VIII (FVIII). This was brought about by bridging factors FIXa and FX to facilitate hemostasis.

11.12 Approaches to target identification and therapeutic design in specific genetic disorders

11.12.1 Neurofibromatosis type 1

Walker and Upadhyaya (2018) reviewed clinical manifestations, genetic defects, pathophysiology, and therapeutic approaches in neurofibromatosis type 1 (NF1). This disorder is due to dominantly acting defects in a gene that maps to human chromosome 17q11.2 and encodes the protein neurofibromin 1. Functional studies revealed that neurofibromin acts as a negative regulator of the Ras signal transduction pathway and it is classified as an RASGTPase.

NF1 was noted to manifest a high degree of clinical variability even within a single family. Affected body systems can include skin, eyes, bony skeleton, and nervous system. Cutaneous manifestations include pigmentary changes, café-au-lait macules, axillary freckling, cutaneous neurofibromas, plexiform neurofibromas, and deeper lesions that can also affect underlying nerves.

Eye abnormalities include Lisch nodules in the iris, aggregates of pigmentary cells, and optic pathway gliomas.

Other nervous system lesions that can occur include nerve sheath tumors, astrocytic neoplasms. Cognitive impairment has also been described in some patients.

There is also evidence for increased incidence of certain malignancies, for example, tumors of the breast, gastro-intestinal duodenal tumors, carcinoid tumors, and rhabdomyosarcomas. Pheochromocytomas have also been reported in some cases. Somatic NF1 mutations have been reported in some tumors.

Walker and Upadhyaya reported that approximately half of NF1 cases appear to have arisen as new mutation. They noted further that approximately 3000 different mutations have been reported to lead to this disorder.

Neurofibromin was reported to negatively regulate five forms of RAS, HRAS, KRAS, NRAS, MRAS, RRAS 1 and 2. In this pathway, RAS GTP is involved in activation of RAF effector proteins ARAF, BRAF, CRAF, RALGDS, and the phosphatidyliosiltol-3-kinase (PI3K) family.

The pathway of action includes RAS, RAF and activation of MEKERK and PI3K AKT signaling that could then activate the MTOR signaling pathway,

In the absence of neurofibromin, increased RAS activity impacts neurotransmission since it leads to increased production of gamma-amino-butyric acid that inhibits neurotransmission.

Walker and Upadhyaya noted that neurofibromin possibly had additional functions not directly related to RAS GDP function.

11.12.2 Approaches to therapy

These include approaches to inhibit activated RASGTP signaling. These included targeting the MEK/ERK signaling pathway. This turned out to have several drawback in part due to toxicity. Activation of the PI3K AKT signaling pathway that leads downstream to MTOR (mammalian target of rapamycin) activations have been initiated. These include use of MTORC1 inhibitors, for example, rapamycin or rapalogs.

Other therapeutic avenues being investigated include other signaling pathways. Targeting of the WNT beta catenin pathways is being particularly investigated for treatment of NF1-related tumors.

11.12.3 Gene-related therapies

Walker and Upadhyaya noted that the diverse spectrum of NF1 mutations complicates gene therapy approaches; 20% of mutations were noted to be nonsense mutation leading to premature termination. Aminoglycoside antibiotics are in pilot clinical trials to promote readthrough of termination that result from specific mutations.

Splicing mutations were reported to constitute 27% of NF1 disease-associated mutation. Splice blocking antisense oligonucleotides were reported to have shown promise in counteracting these mutations in clinical tests.

Gene therapy and gene replacement approaches were noted to be complicated due to the large size of the NF1 gene.

Gene editing technologies including CRISPR Cas technologies were considered to have therapeutic potential. However, the large number of different disease-causing mutation implies that personalized therapies would be necessary.

Scala et al. (2021) reported results of genetic testing in NF1 based on cDNA sequencing and/or multiple ligation dependent probe amplification. They reported that single nucleotide variants were reported in 267 of 265 (73.2%) of cases, and genomic copy number variants were detected in 5.5% of cases.

They identified novel genotype/phenotype correlations.

11.13 Inborn errors of metabolism

During the 20th century biochemical studies and analyses of metabolic processes provided information on substrates, products and cofactors involved reactions. The cofactors turned out frequently to include specific vitamins. These analyses led the way to treatment designs for certain inborn errors of metabolism determined to be due to defects in the function of particular enzymes. Therapeutic strategies that evolved from these studies included dietary manipulation with reduction in intake of specific substrates shown to be poorly metabolized in specific disorders, for example, phenylketonuria. In certain disorders, supplementation with specific vitamin-like cofactors turned out to be beneficial, for example, vitamin B12 supplementation in forms of methylmalonic acidemia, carnitine supplementation in fatty acid oxidation disorders (Fig. 11.1).

11.13.1 Mitochondrial fatty acid oxidation defects and carnitine shuttle disorders

More than 15 different enzymes were reported to have mutations that can lead to these disorders (Knottnerus et al., 2018). Symptoms can occur or become significantly worse under catabolic conditions. It is essential in these disorders that fasting be avoided and endurance exercises be limited since caloric deprivation can lead to hypoketotic hypoglycemia, rhabdomyolysis, liver dysfunction, and cardiomyopathy.

11.13.2 Glycogen storage diseases

Dietary management is important in hepatic glycogen storage diseases. Ross et al. (2020) reported that nutritional therapy is the primary treatment for glycogen storage diseases. Cornstarch, a slow-release carbohydrate source, was found to be important in therapy. Night feeding with cornstarch were shown to be important

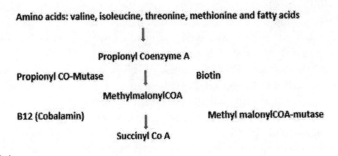

FIGURE 11.1

Methylmalonyl-coa mutase and vitamin B12.

in avoiding nocturnal hypoglycemia. Extended-release forms of cornstarch became available, for example, glycosade. Vitamin and calcium supplementation and normal protein intake for age were also reported to be important.

11.14 Lysosomal storage diseases and enzyme replacement therapy

Information of enzyme replacement therapies in lysosomal storage diseases are presented in Chapter 14, "Using insights from genomics to increase possibilities for treatment of genetic diseases." Approaches beyond enzyme replacement therapy are discussed here.

La Cognata et al. discussed therapies for lysosomal storage disorders, including hematopoietic stem cell therapies and pharmacological chaperone therapies. The goals of chaperone therapies were to provide agents that would promote appropriate folding of enzymes, improve their trafficking, and increase their catalytic activity. Success of these therapies was noted to be in part dependent on the nature of the mutations in the targeted protein.

Substrate reduction therapies have also been promoted for treatment of lysosomal storage diseases. It is interesting to note that some compounds used in chaperone therapy are closely related to small molecules proposed in substrate reduction therapy.

Chaperone therapies include miglustat and 1-deoxynojirimycin that are reported to work by correcting protein misfolding.

La Cognata et al. noted possibilities for gene therapy and projects to carry out genome editing in pluripotent stem cells. Poletto et al. (2020) reported preclinical gene editing of pluripotent stem cells in mucopolysaccharidosis and Ou et al. (2020) reported preclinical gene editing studies in Tay − Sachs and Sandhoff disease.

They noted that there were also reports of therapies in lysosomal storage diseases based on correction of splicing defects.

Dardis and Buratti (2018) reported that between 5% and 19% of lysosomal storage disease gene mutations were pre-mRNA splicing defects. Therapies that utilize antisense oligonucleotides were reported to have been used in preliminary studies on cellular models of Niemann − Pick disease and Pompe disease.

Biffi (2017) noted that brain involvement was noted to be particularly severe in Sanfilippo syndrome MPSIII and in Krabbe disease. They emphasized that particular modification of enzymes used in enzyme replacement therapy may be required to promote passage across the blood brain barrier. These modifications include coupling to the replacement enzyme of fragments of antibodies to specific blood brain barrier related proteins; the antibodies will promote binding to the blood brain barrier proteins. Other modifications include use of specific nanoparticles for delivery.

Sanfillipo syndrome can arise due to defects in the enzyme α-N-acetylglucosaminidase (NAGLU). In preclinical studies, Ribera et al. (2015) carried out gene therapy on a mouse model of this disorder. They reported delivery of AAV9 vectors encoding for NAGLU to the cerebrospinal fluid. In results of their study they reported normalization of glycosaminoglycan content and lysosomal physiology and determined that the pattern of gene expression in brain was at levels similar to that of healthy animals.

Krabbe disease, also known as globoid cell leukodystrophy, is due to deficient activity of galactosylceramidase (GALC). In a canine model of this disorder Bradbury et al. (2020a,b) investigated utility of intrathecal AAV9 encoding canine GALC administration into the cisterna magna. They reported that this restored GALC activity in the brain improved myelination and reduced neuroinflammation. When given to presymptomatic animals, this therapy prevented emergence of clinical neurological manifestations.

Biffi (2017) reviewed hematopoietic stem cell gene therapy in lysosomal storage diseases. They noted that pioneering hematopoietic stem cell gene therapy approaches in metachromatic leuko-dystrophy due to deficiency in arylsulfatase A were reported by Sessa et al. (2016). They reported that trial participants had pre-trial conditioning and then received infusion of autologous HSCs (Hematopoietic stem cells) transduced with a lentiviral vector encoding ARSA cDNA, They reported that eight patients, including seven presymptomatic patients, had prevention of disease onset or halted disease progression.

Defects in the ABCD1 gene cause adrenoleukodystrophy, an X-linked demyelinating disorder. Preliminary investigations of therapy with hematopoietic stem cells transduced with lentivirus containing the ABCD1 gene have been carried out in males with this disorder (Eichler et al., 2017). They reported that early results indicating Lenti-D gene therapy may be a safe and effective alternative to allogeneic stem-cell transplantation.

Biffi emphasized the need for optimization of current approaches to hematopoietic gene cell therapies.

11.15 Ceroid lipofuscinoses

These disorders, sometimes referred to as Batten's disease, are associated with accumulation of an autofluorescent stored protein, lipofuscin, in the brain. Patients with these disorders manifest progressive neurological deterioration. These disorders were reviewed by Kohlschütter et al. (2019). At least 14 different genes encode products that when defective lead to ceroid lipofuscinoses, The gene products function in lysosomes and include soluble and membrane-bound proteins.

The disorders lead to epilepsy and to motor abnormalities, visual defects also occur.

Genes	Gene product
CLN1	Palmitoyl protein thioesterase 1 (PPT1).
CLN2	Tripeptidyl peptidase (TPP1).
CLN3	Lysosomal/endosomal transmembrane protein, battenin.
CLN4	DNAJC5 DnaJ heat shock protein family (Hsp40) member C5.
CLN5	Intracellular trafficking protein.
CLN6	Transmembrane ER protein.
CLN7 (MFSD8)	Integral membrane protein contains a transporter domain and a facilitator domain.
CLN8	Transmembrane protein belonging to a family of proteins containing TLC domains.*
CLN9	May be a regulator of dihydroceramide synthase.
CLN10	Cathepsin D plays a role in protein turnover and in the proteolytic activation.
CLN11	Progranulin.
CLN12	ATP13A2 cation transporting.
CLN13	Cathepsin F major component of the lysosomal proteolytic system.
CLN14	KCTD7 potassium channel tetramerization domain-containing protein.

*TLC domains are postulated to function in lipid synthesis, transport, or sensing.

11.15.1 Clinical manifestations

Kohlschütter et al. noted that some forms of ceroid lipofuscinosis manifest very early in life with microcephaly and seizures. Other forms manifest within the first year of life with seizures, decreased muscle tone, and impaired social interactions.

Defective function of CLN2 was noted to cause a late infantile onset disorder with manifestation between 2 years and 4 years of life that includes motor decline, ataxia, myoclonus, then loss of motor function, loss of vision, and inability to swallow. The late infantile disorder was also noted to arise from specific mutations in CLN1, CLN5, CLN6, CLN7, CLN8, CLN14.

A juvenile onset form of the disease also occurs due to specific mutation in CLN1, 2, 5, 7, 10, and 12. Manifestation includes cognitive decline and ataxia.

Magnetic resonance imaging studies in patients with ceroid lipofuscinoses subsequently manifest with cerebral and cerebellar atrophy.

One medication that has been approved for therapy of these disorders is cerliponase. It is a recombinant proenzyme of TPP1 Tripeptidyl peptidase. It is administered by regular intracerebroventricular infusions in a surgically implanted device. Estublier et al. (2021) reported a change in the natural history of the disease in patients with CLN2 disease, when this therapy was administered in the early stages of the disease.

Kohlschütter et al. reported benefits of enzyme replacement therapy in studies on animal models of other forms of ceroid lipofuscinoses, for example, CLN3 and CLN5. They also discussed therapy in mouse models of these disorders with natural immunomodulators, for example, docosahexaenoic acid that reduced microgliosis that occurs as one of the pathological features.

11.16 Aminoacidopathies and organic acidemias

11.16.1 Methylmalonic acidemia

Chandler and Venditti (2019) reviewed aspects of the therapy of methylmalonic acidemia. This disorder arises as a result of deficiency of the enzyme methyl-malonyl-CoA-mutase (mut) that functions in mitochondria. The enzyme is known to have an obligate requirement for 5′deoxyadenosylcobalamin, a form of vitamin B12. Approaches to treatment of the disorder include supplementation with vitamin B12. Treatment utilized also includes reduced protein intake. In severe cases of the disorder, liver transplantation was sometimes undertaken as liver is the major site of activity of methyl-malonyl-CoA-mutase.

Chandler and Venditti undertook gene therapy proof of principle studies in a mouse model of the disorder. A knockout mouse model was used. Knockout of this critical enzyme led mice to excrete large quantities of methylmalonic acid in urine and they perished in early life often after 48 hours.

The gene therapy experiment involved use of an adenovirus vector into which normal 2.3 kb mouse methyl-malonyl-CoA-mutase (mut) was cloned along with a promoter sequence to ensure gene expression. Chandler and Venditti noted that this treatment prolonged to life often to 15 days or longer.

They then utilized an adeno-associated viral vector to introduce the methyl-malonyl-CoA-mutase (mut) genes. This vector was noted to remain episomal and was not integrated into genomic DNA. Other vectors investigated for gene introduction included lentiviral vectors. Use of these vectors led to greater decreases in methylmalonic acid excretion levels and improved survival.

Other therapies investigated included use of lipid nanoparticles for MRNA delivery. This therapy was reported to decrease disease manifestations; however, therapy needed to be repeated at 10–14-day intervals.

11.17 Transporter defects

In a number of different diseases, evidence emerged from early molecular pathology that specific transporters must be impaired. Examples include cystic fibrosis and Wilson's disease. However, in both of these disorders the definitive underlying defect did not emerge until the responsible genes were cloned.

11.17.1 Cystic fibrosis

Important observation relevant to this disorders emerged in 1905 when Karl Landsteiner described intestinal obstruction. Fanconi et al. (1936) described children with pancreatic and bronchial disease.

In 1953 Kessler and Andersen described that during a heat wave in New York, children with cystic fibrosis were particularly prone to development of severe dehydration associated with very low serum chloride levels. After 1953 when Di Sant Agnese et al. described methods of sweat chloride analysis, this became the standard test for cystic fibrosis.

Progress in defining the underlying defect in cystic fibrosis emerged following family studies and gene mapping and assignment of the gene responsible for cystic fibrosis to chromosome 7 by Eiberg et al. (1985). Subsequent studies by Tsui et al. (1985) led to the isolation and identification of the cystic fibrosis gene. It was found to encode for a chloride conductance regulator CFTR.

Detailed studies on the consequences of the mutations in CFTR led to definition of six classes of defects in functions of the regulator and to the development of therapies to correct the specific functional defects (Ong and Ramsey, 2016). These include lumacaftor that corrects aberrant CFTR protein folding that occurs in some forms of the disease. One mutation that leads to a protein that does not fold correctly is phe508del, the most common CFTR mutation. Another small molecule drug kalydeco improves function of CFTR protein that impacts ion conductance.

11.17.2 Wilson's disease

This disorder was first described in 1912 by Kinnear Wilson as a disease associated with liver disease and neurological manifestations. Wilson's disease was reviewed by Bandmann et al. (2015). Detailed studies of the liver and brain were reported to reveal the presence of abnormally high levels of copper. Initial treatment of the disorder included treatment with British antilewisite BAL (dimercaprol) that was used to treat heavy metal poisoning. Subsequently, other less toxic agents were used to deplete excess stored copper in Wilson's disease.

Gene mapping and cloning studies later revealed that this disease arose due to pathogenic mutations in copper transporting P-type ATPase (Bull et al., 1993; Tanzi et al., 1993).

Bandmann et al. (2015) noted that more than 500 different ATP7B mutations had been identified with this disorder (http://www.wilsondisease.med.ualberta.ca/database.asp).

Treatment includes copper chelation, with penicillamine, trientine and tetrathiomolydate, and zinc salts as zinc salts were reported to decrease copper uptake in intestinal cells. Chelation treatment was reported to be more useful in treating liver symptoms than neurological manifestations.

11.18 Complexities of mitochondrial diseases and explorations of treatments

Russell et al. (2020) reviewed mitochondrial disorders and ongoing efforts to develop therapies. Mitochondrial disorders are noted to be difficult to diagnose as they can impact a number of different organs and systems and can present at different ages.

Russell et al. noted that different disorders result from defect in functions of components of the mitochondrial oxidative phosphorylation system that is responsible for ATP generation. Further complications arise due to numbers and diversity of mitochondria in a specific cell. In any one cell, mitochondria can be homoplasmic implying that they have the same genome, or they can be heteroplasmic implying that the mitochondria in that cell have different genomes. Functional defects are also dependent on the number of mitochondria in a particular cell.

Recent reports indicate that 300 different pathogenic mutations can lead to mitochondrial diseases (Thompson et al., 2020). They documented disease-causing mutations in 37 genes in the mitochondrial genome including genes that encode OXPHOs subunits, ribosomal RNAs, and transfer RNAs.

Thompson et al. documented disease-causing variants in 295 nuclear genes, and their functions. They included OXPHOS subunits and their assembly units, proteins involved in import and processing, proteins involved in mitochondrial DNA replication and maintenance, RNA maturation and modification factors, and aminoacyl TRNA synthetases.

Pathogenic mutations in nuclear genes were noted to be particularly important in diseases with childhood onset. The mitochondrial diseases that first presented in adult life were noted to be more commonly due to mutations in the mitochondrial genome.

Regarding treatment, Russell et al. emphasized that patients with mitochondrial disease required treatments for specific clinical manifestations, for example, hypoglycemia, epilepsy, compromised cardia function.

11.18.1 Search for treatment of specific mitochondrial disorders

Russell et al. noted that advances had been made in the development of agents to increase the number of mitochondria in cells, such as enhancers of mitochondrial biogenesis. Advances were built on the discovery of master regulators of mitochondrial biogenesis, for example, PPAR peroxisomal proliferation activated receptors and their co-activator PGC1. One specific PPAR, agonist bezafibrate, has been investigated for treatment of mitochondrial myopathy. However, the PPAR system was noted to be quite complex.

Another activator under investigation is 5-aminoimidazole-4-carboxamide AICAR that can activate SIRT1 sirtuin known to regulate gene expression.

Regulators that induce mitochondrial biogenesis were noted to include NRF1 and NRF2, nuclear respiratory factors, that are known to influence gene expression.

Improvement of the NAD/NADH ratio was also investigated for treatment purposes. Reduction of the ratio of NAD to NADH was reported to be a manifestation of impaired mitochondrial function. This alteration was noted to lead downstream to increased levels of lactic acid. Treatments have been designed to increase cellular NAD levels through supplementation or through factors that enhance NAD biogenesis. One such compound is nicotinamide riboside. Provision of nicotinamide mononucleotide or a specific niacin derivative has also been investigated.

Studies on the MTOR pathway have been carried out. These is some evidence that rapamycin, an inhibitor of MTOR, has beneficial effects in enhancing mitochondrial function. Rapamycin was shown to be beneficial in treatment of Leigh syndrome.

Russell et al. noted that specific clinical trials have been initiated to study the effects of reducing levels of oxidative stress and levels of reactive oxygen species that occur in certain mitochondrial diseases. Treatments include mitochondrial antioxidants such as KH176, a form of vitamin E. Cysteamine bitartrate is another antioxidant used to treat certain mitochondrial disorders including cystinosis.

Strategies to promote electron transport include use of coenzyme CoQ10, ubiquinone. An analog of this is licensed to treat mitochondrial disorders including Leber's hereditary optic neuropathy.

Specific disorders were noted to be due to defects in mitochondrial nucleoside metabolism; these include defects in thymidine kinase. Russell et al. noted that a deoxypyrimidine analog is being investigated to treat this disorder.

Gene manipulation of the mitochondria is complicated due to multiple copies of mitochondria and of mitochondria with diverse genomes in a particular cell. Progress has been made in preimplantation genetic diagnosis to determine the possibility that some fertilized eggs may have predominantly normal mitochondria.

11.19 New approaches to cancer therapy

Following extensive genomic sequencing studies in cancer and identification of cancer-specific mutations, considerable effort was applied to discover therapeutic agents to target cancer mutations. Although several useful therapeutic agents were developed, it has over the years become increasingly clear that cancers have many different mutations, and the mutations continuously evolve, and this may result in development of resistance to a therapy that was initially successful.

However, there is evidence that in combination with other therapeutic methods, mutation targeting can be useful in treatment of certain cancers. It has become evident that particular genes tend to be mutated in more than one cancer types, for example, TP53 (p53), and efforts are ongoing to develop therapies to target those mutations. TP53 tumor suppressor protein contains domain involved in transcriptional activation, DNA binding, and it regulates gene expression, including genes involved in DNA repair and in cell cycle-related processes.

Duffy et al. (2020) noted that p53 dysfunction occurs in most malignancies and this is based on the occurrence of mutations in p53 or on downregulation of p53. Key factors responsible for p53 downregulation include MDM2 and MDM4 that can bind to p53 and reduce its activity. Agents that inhibit the binding of p53 and MDM2 or MDM4 have had some measure of success and one such agent, idasanutlin, is in clinical trials.

Zhang et al. (2018) reported that the majority of TP53 tumor mutation are missense mutations in the region of interaction of p53 with DNA. They noted further that stabilization of the conformation of p53 is essential for its activity, and for the reactivation of mutated p53.

Zhang et al. reported that a specific compound APR-246 is converted to methylene quinuclidinone that reacts with cysteine in the core p53 domain, cysteine 277, and cysteine124, and this was necessary for mutant T53 reactivation. Methyl quinuclidinone was shown to enhance thermostability of the p53 core domain.

Zhang et al. noted that various approaches involving small molecules have been investigated to reactivate mutant p53. Cys182 and Cys 277 in p53 were found to be binding sites for a sulfonylpyrimidine compound that was shown to thermostabilized p53.

11.19.1 Synthetic lethality in cancer treatment

This approach to therapy is based on the concept that a cell may be able to survive and proliferate with one defect in a specific process but multiple defects, particularly if they occur in the same or a closely related pathway, may lead to cell death.

An important application of this concept to cancer therapy was first made in the treatment of tumor with mutations in BRCA1 or BRCA2 that function in the DNA repair pathway. Tumors with BRCA1 or BRCA2 mutations that impacted DNA repair could not survive when they were treated with inhibitors of another protein involved in DNA repair PARP1 (Lord and Ashworth, 2017).

Cell cycle control is essential for the cell to proliferate. ATR protein is a serine/threonine kinase and DNA damage sensor that activates cell cycle checkpoint signaling upon DNA stress. Bradbury et al. (2020a,b) reported that ATR inhibitors may act as synthetically lethal agents in cancer cells that have mutations in the DNA repair pathways or cells that have defect in other proteins related to the cell cycle.

Other cancer therapies have been developed that exploit synthetic lethality. Cancers with mutations in the enzyme IDH1 isocitrate dehydrogenase were reported to be dependent on NRF2 (nuclear regulatory factor 2) guided synthesis of glutathione. NRF2 inhibition led to synthetic lethality in these cells.

Tumors with specific mutation in the BRAF gene, for example, V600E, were found to be particularly sensitive to inhibitors of EGFR tyrosine kinase.

11.19.2 RAS signaling pathway

The RAS signaling pathways play critical roles in cell growth and proliferation and in cancer. Specific mutations in RAS proteins are known to impact their intrinsic GTPase activity. For example, specific mutations in KRAS result in the protein remaining active and not being turned off at the appropriate time. KRAS was reported to be mutated in 95% of pancreatic cancer, in 50% of colorectal cancers, and 32% of lung cancers (Rosen, 2021). Rosen noted that pharmacologists have long sought to identify RAS inhibitors. Recently a chemical was identified that binds to a specific pocket in the RAS protein and inhibits its activity.

Skoulidis et al. (2021) reported encouraging results of therapy with a molecule that binds in this pocket and inhibits KRAS activity.

References

Anderson HD, Crooks KR, Kao DP, Aquilante CL: The landscape of pharmacogenetic testing in a United States managed care population, *Genet Med* 22(7):1247–1253, 2020. Available from: https://doi.org/10.1038/s41436-020-0788-3.

Balwani M, Wang B, Anderson KE, et al: Acute hepatic porphyrias: recommendations for evaluation and long-term management, *Hepatology*. 66(4):1314–1322, 2017. Available from: https://doi.org/10.1002/hep.29313.

Bandmann O, Weiss KH, Kaler SG: Wilson's disease and other neurological copper disorders, *Lancet Neurol* 14(1):103–113, 2015. Available from: https://doi.org/10.1016/S1474-4422(14)70190-5. PMID: 25496901.

Benet LZ, Kroetz DL, Sheiner LB: *The dynamics of drug absorption, distribution and elimination, pharmacokinetics*, Chapter 1 Goodman and Gilman's The Pharmacological Basis of Therapeutics, Ninth Edition, New York, 1996, Mc Graw Hill.

Biffi A: Hematopoietic stem cell gene therapy for storage disease: current and new indications, *Mol Ther* 25(5):1155–1162, 2017. Available from: https://doi.org/10.1016/j.ymthe.2017.03.0254.

Blair HA: Emicizumab: a review in haemophilia A, *Drugs* 79(15):1697–1707, 2019. Available from: https://doi.org/10.1007/s40265-019-01200-2. PMID: 31542880.

Bradbury A, Hall S, Curtin N, Drew Y: Targeting ATR as cancer therapy: a new era for synthetic lethality and synergistic combinations? *Pharmacol Ther* 207:107450, 2020a. Available from: https://doi.org/10.1016/j.pharmthera.2019.107450.

Bradbury AM, Bagel JH, Nguyen D, et al: Krabbe disease successfully treated via monotherapy of intrathecal gene therapy, *J Clin Invest* 130(9):4906–4920, 2020b. Available from: https://doi.org/10.1172/JCI133953. PMID: 32773406.

Bull PC, Thomas GR, Rommens JM, et al: The Wilson disease gene is a putative copper transporting P-type ATPase similar to the Menkes gene, *Nat Genet* 5(4):327–337, 1993. Available from: https://doi.org/10.1038/ng1293-327. PMID: 8298639.

Burton DR, Barbas CF 3rd, Persson MA, et al: A large array of human monoclonal antibodies to type 1 human immunodeficiency virus from combinatorial libraries of asymptomatic seropositive individuals, *Proc Natl Acad Sci U S A* 88(22):10134–10137, 1991. Available from: https://doi.org/10.1073/pnas.88.22.10134. PMID: 1719545.

Cartwright RA, Glashan RW, Rogers HJ, et al: Role of N-acetyltransferase phenotypes in bladder carcinogenesis: a pharmacogenetic epidemiological approach to bladder cancer, *Lancet* 2(8303):842–845, 1982. Available from: https://doi.org/10.1016/s0140-6736(82)90810-8. PMID: 6126711.

Chandler RJ, Venditti CP: Gene therapy for methylmalonic acidemia: past, present, and future, *Hum Gene Ther* 30(10):1236–1244, 2019. Available from: https://doi.org/10.1089/hum.2019.113.

Cicali EJ, Smith DM, Duong BQ, et al: A scoping review of the evidence behind cytochrome P450 2D6 isoenzyme inhibitor classifications, *Clin Pharmacol Ther* 108(1):116–125, 2020. Available from: https://doi.org/10.1002/cpt.1768.

Dardis A, Buratti E: Impact, characterization, and rescue of pre-mRNA splicing mutations in lysosomal storage disorders, *Genes (Basel)* 9(2):73, 2018. Available from: https://doi.org/10.3390/genes9020073. PMID: 29415500.

Duffy MJ, Synnott NC, O'Grady S, Crown J: Targeting p53 for the treatment of cancer, *Semin Cancer Biol*. Available from: https://doi.org/10.1016/j.semcancer.2020.07.005. S1044-579X(20)30160-7.

Eiberg H, Mohr J, Schmiegelow K, et al: Linkage relationships of paraoxonase (PON) with other markers: indication of PON-cystic fibrosis synteny, *Clin Genet* 28(4):265–271, 1985. Available from: https://doi.org/10.1111/j.1399-0004.1985.tb00400.x. PMID: 2998653.

Eichler F, Duncan C, Musolino PL, et al: Hematopoietic stem-cell gene therapy for cerebral adrenoleukodystrophy, *N Engl J Med* 377(17):1630–1638, 2017. Available from: https://doi.org/10.1056/NEJMoa1700554.

Emmerich CH, Gamboa LM, Hofmann MCJ, et al: Improving target assessment in biomedical research: the GOT-IT recommendations, *Nat Rev Drug Discov* 20(1):64–81, 2021. Available from: https://doi.org/10.1038/s41573-020-0087-3.

Estublier B, Cano A, Hoebeke C, et al: Cerliponase alfa changes the natural history of children with neuronal ceroid lipofuscinosis type 2: The first French cohort, *Eur J Paediatr Neurol* 30:17–21, 2021. Available from: https://doi.org/10.1016/j.ejpn.2020.12.002.

Fanconi G., Uehlinger E., Knauer C. Celiac syndrome with congenital cystic fibromatosis of the pancreas and bronchiectasis. *Wien Mes Wschnschr* 86:753–756, 1936.

Freed DM, Bessman NJ, Kiyatkin A, et al: EGFR ligands differentially stabilize receptor dimers to specify signaling kinetics, *Cell* 171(3):683–695, 2017. Available from: https://doi.org/10.1016/j.cell.2017.09.017. PMID: 28988771.

Gupta VH, Amarapurkar DN, Singh M, et al: Association of N-acetyltransferase 2 and cytochrome P450 2E1 gene polymorphisms with antituberculosis drug-induced hepatotoxicity in Western India, *J Gastroenterol Hepatol* 28(8):1368–1374, 2013. Available from: https://doi.org/10.1111/jgh.12194. PMID: 23875638.

Hein DW, Fakis G, Boukouvala S: Functional expression of human arylamine N-acetyltransferase NAT1*10 and NAT1*11 alleles: a mini review, *Pharmacogenet Genomics* 28(10):238–244, 2018. Available from: https://doi.org/10.1097/FPC.0000000000000350. PMID: 30222709.

Ingelman-Sundberg M, Mkrtchian S, Zhou Y, Lauschke VM: Integrating rare genetic variants into pharmacogenetic drug response predictions, *Hum Genomics* 12(1):26, 2018. Available from: https://doi.org/10.1186/s40246-018-0157-3. PMID: 29793534.

Jarrar Y, Lee SJ: The functionality of UDP-glucuronosyltransferase genetic variants and their association with drug responses and human diseases, *J Pers Med* 11(6):554, 2021. Available from: https://doi.org/10.3390/jpm11060554. PMID: 34198586.

Kirsch P, Hartman AM, Hirsch AKH, Empting M: Concepts and core principles of fragment-based drug design, *Molecules* 24(23):4309, 2019. Available from: https://doi.org/10.3390/molecules24234309. PMID: 31779114.

Knottnerus SJG, Bleeker JC, Wüst RCI, et al: Disorders of mitochondrial long-chain fatty acid oxidation and the carnitine shuttle, *Rev Endocr Metab Disord* 19(1):93–106, 2018. Available from: https://doi.org/10.1007/s11154-018-9448-1. PMID: 29926323.

Köhler G, Milstein C: Continuous cultures of fused cells secreting antibody of predefined specificity, *Nature* 256(5517):495–497, 1975. Available from: https://doi.org/10.1038/256495a0. PMID: 1172191.

Kohlschütter A, Schulz A, Bartsch U, Storch S: Current and emerging treatment strategies for neuronal ceroid lipofuscinoses, *CNS Drugs* 33(4):315–325, 2019. Available from: https://doi.org/10.1007/s40263-019-00620-8. PMID: 30877620.

Li T, Qian Y, Zhang C, et al: Anlotinib combined with gefitinib can significantly improve the proliferation of epidermal growth factor receptor-mutant advanced non-small cell lung cancer *in vitro* and *in vivo*, *Transl Lung Cancer Res* 10(4):1873–1888, 2021. Available from: https://doi.org/10.21037/tlcr-21-192. PMID: 34012799.

Lord CJ, Ashworth A: PARP inhibitors: synthetic lethality in the clinic, *Science* 355 (6330):1152–1158, 2017. Available from: https://doi.org/10.1126/science.aam7344. PMID: 28302823.

Normanno N, De Luca A, Bianco C, et al: Epidermal growth factor receptor (EGFR) signaling in cancer, *Gene.* 366(1):2–16, 2006. Available from: https://doi.org/10.1016/j.gene.2005.10.018. PMID: 16377102.

Ong T, Ramsey B: New therapeutic approaches to modulate and correct cystic fibrosis transmembrane conductance regulator, *Pediatr Clin North Am* 63(4):751–764, 1991. Available from: https://doi.org/10.1016/j.pcl.2016.04.006. PMID: 27469186.

Ou L, Przybilla MJ, Tăbăran AF, et al: A novel gene editing system to treat both Tay – Sachs and Sandhoff diseases, *Gene Ther* 27(5):226–236, 2020. Available from: https://doi.org/10.1038/s41434-019-0120-5. Epub 2020 January 2.

Poletto E, Baldo G, Gomez-Ospina N: Genome editing for mucopolysaccharidoses, *Int J Mol Sci* 21(2):500, 2020. Available from: https://doi.org/10.3390/ijms21020500. PMID: 31941077.

Renard D, Rosselet A: Drug-induced hemolytic anemia: pharmacological aspects, *Transfus Clin Biol* 24(3):110–114, 2017. Available from: https://doi.org/10.1016/j.tracli.2017.05.013.

Ribera A, Haurigot V, Garcia M, et al: Biochemical, histological and functional correction of mucopolysaccharidosis type IIIB by intra-cerebrospinal fluid gene therapy, *Hum Mol Genet* 24(7):2078–2095, 2015. Available from: https://doi.org/10.1093/hmg/ddu727. PMID: 25524704.

Roden DM, McLeod HL, Relling MV, et al: Pharmacogenomics, *Lancet* 394 (10197):521–532, 2019. Available from: https://doi.org/10.1016/S0140-6736(19)31276-0. PMID: 31395440.

Rosen N: Finally, effective inhibitors of mutant KRAS, *N Engl J Med* 384(25):2447–2449, 2021. Available from: https://doi.org/10.1056/NEJMe2107884. PMID: 34161711.

Ross KM, Ferrecchia IA, Dahlberg KR, et al: Dietary management of the glycogen storage diseases: evolution of treatment and ongoing controversies, *Adv Nutr* 11(2):439–446, 2020. Available from: https://doi.org/10.1093/advances/nmz092. PMID: 31665208.

Russell OM, Gorman GS, Lightowlers RN, Turnbull DM: Mitochondrial diseases: hope for the future, *Cell* 181(1):168–188, 2020. Available from: https://doi.org/10.1016/j.cell.2020.02.051. PMID: 32220313.

Scala M, Schiavetti I, Madia F, et al: Genotype-phenotype correlations in neurofibromatosis Type 1: a single-center cohort study, *Cancers (Basel)* 13(8):1879, 2021. Available from: https://doi.org/10.3390/cancers13081879.

Schlessinger J: Receptor tyrosine kinases: legacy of the first two decades, *Cold Spring Harb Perspect Biol* 6(3):a008912, 2014. Available from: https://doi.org/10.1101/cshperspect.a008912. PMID: 24591517.

Sessa M, Lorioli L, Fumagalli F, et al: Lentiviral haemopoietic stem-cell gene therapy in early-onset metachromatic leukodystrophy: an ad-hoc analysis of a non-randomised, open-label, phase 1/2 trial, *Lancet* 388(10043):476–487, 2016. Available from: https://doi.org/10.1016/S0140-6736(16)30374-9.

Sim E, Abuhammad A, Ryan A: Arylamine N-acetyltransferases: from drug metabolism and pharmacogenetics to drug discovery, *Br J Pharmacol* 171(11):2705–2725, 2014. Available from: https://doi.org/10.1111/bph.12598. PMID: 24467436.

Skoulidis F, Li BT, Dy GK, et al: Sotorasib for lung cancers with *KRAS* p.G12C mutation, *N Engl J Med* 384(25):2371–2381, 2021. Available from: https://doi.org/10.1056/NEJMoa2103695.

Stölzel U, Doss MO, Schuppan D: Clinical guide and update on porphyrias, *Gastroenterology*. 157(2):365–381, 2019. Available from: https://doi.org/10.1053/j.gastro.2019.04.050. e4.

Tanzi RE, Petrukhin K, Chernov I, et al: The Wilson disease gene is a copper transporting ATPase with homology to the Menkes disease gene, *Nat Genet* 5(4):344–350, 1993. Available from: https://doi.org/10.1038/ng1293-344. PMID: 8298641.

Thompson K, Collier JJ, Glasgow RIC, et al: Recent advances in understanding the molecular genetic basis of mitochondrial disease, *J Inherit Metab Dis* 43(1):36–50, 2020. Available from: https://doi.org/10.1002/jimd.12104.

Tsui LC, Buchwald M, Barker D, et al: Cystic fibrosis locus defined by a genetically linked polymorphic DNA marker, *Science* 230(4729):1054–1057, 1985. Available from: https://doi.org/10.1126/science.2997931. PMID: 2997931.

Wagner J, Dahlem AM, Hudson LD, et al: A dynamic map for learning, communicating, navigating and improving therapeutic development, *Nat Rev Drug Discov* 17(2):150, 2018. Available from: https://doi.org/10.1038/nrd.2017.217.

Walker JA, Upadhyaya M: Emerging therapeutic targets for neurofibromatosis type 1, *Expert Opin Ther Targets* 22(5):419–437, 2018. Available from: https://doi.org/10.1080/14728222.2018.1465931. PMID: 29667529.

Wang S, Che T, Levit A, et al: Structure of the D2 dopamine receptor bound to the atypical antipsychotic drug risperidone, *Nature* 555(7695):269–273, 2018. Available from: https://doi.org/10.1038/nature25758.

Yin Y, Lan JH, Nguyen D, et al: Application of high-throughput next-generation sequencing for HLA typing on buccal extracted DNA: results from over 10,000 donor recruitment samples, *PLoS One* 11(10):e0165810, 2016. Available from: https://doi.org/10.1371/journal.pone.0165810. eCollection 2016.PMID: 27798706.

Zhang Q, Bykov VJN, Wiman KG, Zawacka-Pankau J: APR-246 reactivates mutant p53 by targeting cysteines 124 and 277, *Cell Death Dis* 9(5):439, 2018. Available from: https://doi.org/10.1038/s41419-018-0463-7. PMID: 29670092.

Zhou Y, Ingelman-Sundberg M, Lauschke VM: Worldwide distribution of cytochrome P450 Alleles: a *meta*-analysis of population-scale sequencing projects, *Clin Pharmacol Ther* 102(4):688–700, 2017. Available from: https://doi.org/10.1002/cpt.690.

Further reading

Austin CP: Opportunities and challenges in translational science, *Clin Transl Sci.* Available from: https://doi.org/10.1111/cts.13055. Online ahead of print. PMID: 33982407.

Beis K: Structural basis for the mechanism of ABC transporters, *Biochem Soc Trans* 43(5):889–893, 2015. Available from: https://doi.org/10.1042/BST20150047. PMID: 26517899.

Beutler E: G6PD deficiency, *Blood.* 84(11):3613–3636, 1994.

Castelli MS, McGonigle P, Hornby PJ: The pharmacology and therapeutic applications of monoclonal antibodies, *Pharmacol Res Perspect* 7(6):e00535, 2019. Available from: https://doi.org/10.1002/prp2.535. PMID: 31859459.

Concolino D, Deodato F, Parini R: Enzyme replacement therapy: efficacy and limitations, *Ital J Pediatr* 44(2):120, 2018. Available from: https://doi.org/10.1186/s13052-018-0562-1. PMID: 30442189.

Di Sant'Agnese PA, Darling RC, Perera GA, Shea E: Abnormal electrolyte composition of sweat in cystic fibrosis of the pancreas; clinical significance and relationship to the disease, *Pediatrics* 12(5):549–563, 1953.

Kessler WR, Andersen DH: Heat Prostration in fibrocystic disease of the pancreas and other conditions, *Pediatrics* 8(5):648–656, 1951.

Kuznetsov IB, McDuffie M, Moslehi R: A web server for inferring the human N-acetyltransferase-2 (NAT2) enzymatic phenotype from NAT2 genotype, *Bioinformatics* 25(9):1185–1186. Available from: https://doi.org/10.1093/bioinformatics/btp121, 2009.

La Cognata V, Guarnaccia M, Polizzi A, et al: Highlights on genomics applications for lysosomal storage diseases, *Cells* 9(8):1902, 2020. Available from: https://doi.org/10.3390/cells9081902. PMID: 32824006.

Lin L, Yee SW, Kim RB, Giacomini KM: SLC transporters as therapeutic targets: emerging opportunities, *Nat Rev Drug Discov* 14(8):543–560, 2015. Available from: https://doi.org/10.1038/nrd4626. Epub 2015 Jun 26. PMID: 26111766.

Marsh SG, Albert ED, Bodmer WF, et al: Nomenclature for factors of the HLA system, 2010, *Tissue Antigens* 75(4):291–455, 2010. Available from: https://doi.org/10.1111/j.1399-0039.2010.01466.x. PMID: 20356336.

Terrón-Díaz ME, Wright SJ, Agosto MA, et al: Residues and residue pairs of evolutionary importance differentially direct signaling bias of D2 dopamine receptors, *J Biol Chem* 294(50):19279–19291, 2019. Available from: https://doi.org/10.1074/jbc.RA119.008068. PMID: 31676688.

Tomas A, Futter CE, Eden ER: EGF receptor trafficking: consequences for signaling and cancer, *Trends Cell Biol* 24(1):26–34, 2014. Available from: https://doi.org/10.1016/j.tcb.2013.11.002. Epub 2013 Nov 29. PMID: 24295852.

CHAPTER 12

Genetic and genomic medicine relevance to cancer prevention, diagnosis, and treatment

12.1 Introduction

In considering the genetics and genomics of cancer it is important to take into account germline and somatic changes and mutations. Germline mutations can lead to specific cancers and to specific syndromes that include the presence of tumors. Germline mutations can potentially also be passed on to the offspring.

12.2 Genes with germline mutations predisposing to cancer listed in order of frequency

Mandelker et al. (2019) reported on germline-focused analyses in tumor sequencing. They emphasized that a sizeable proportion of mutations in tumors are of germline origin and emphasized that germline-focused tumor analysis should be included as part of tumor analysis.

They documented well-defined genes with variants of germline origin and noted that these variants included small insertion deletion and single nucleotide variants.

Mandelker et al. reported 1494 pathogenic variants in 65 cancer susceptibility genes. This study led to recommendations for genetic testing.

12.3 Gene products with germline mutations that can lead to cancer and function of these products

1. TP53 tumor protein p53 responds to diverse cellular stresses to regulate expression of target genes (Fig. 12.1).
2. APC regulator of WNT signaling pathway, cell migration, adhesion, transcriptional activation, apoptosis.
3. KRAS Kirsten Ras homolog proto-oncogene, GTPase.

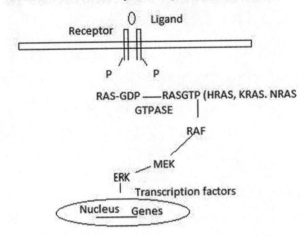

FIGURE 12.1

KRAS pathway.

4. RB1 transcriptional corepressor 1, negative regulator of the cell cycle, and tumor suppressor.
5. PTEN phosphatase and tensin homolog, phosphatidylinositol-3,4,5-trisphosphate 3-phosphatase.
6. BRCA2 DNA repair associated, involved in maintenance of genome stability.
7. NF1 neurofibromin 1, neurofibromatosis type 1, juvenile myelomonocytic leukemia, Watson syndrome*.
8. CDKN2 cyclin-dependent kinase inhibitor 2 A, important tumor suppressor gene.
9. MUTYH mutY DNA glycosylase, involved in oxidative DNA damage repair.
10. ATM serine/threonine kinase, this protein is an important cell cycle checkpoint kinase.
11. CDH1 cadherin 1, calcium-dependent cell − cell adhesion protein.
12. STK11 serine/threonine kinase 11, regulates cell polarity and functions as a tumor suppressor.
13. BRCA1 DNA repair-associated.
14. NRAS proto-oncogene, GTPase, shuttles between the Golgi apparatus and the plasma membrane.
15. BAP1 BRCA1-associated protein 1 may be involved in regulation of transcription, cycle, and growth.

16. SMAD4 may be involved in regulation of transcription, regulation of cell cycle, and growth.
17. CHEK2 checkpoint kinase 2, cell cycle checkpoint regulator, and putative tumor suppressor.
18. MEN1 Menin 1, associated multiple endocrine neoplasia (MEN) type 1, histone and epigenetic modifier.
19. VHL von Hippel − Lindau tumor suppressor component of the protein complex with ubiquitin ligase activity.
20. SMARCA4 SWI/SNF-related, matrix-associated, actin-dependent regulator of chromatin.
21. ERCC3 ERCC excision repair 3, TFIIH core complex helicase subunit, nucleotide excision repair.
22. KIT proto-oncogene, receptor tyrosine kinase phosphorylates proteins involved in proliferation.
23. NF2 neurofibromin 2 interacts with cell surface and proteins involved in regulating ion transport.
24. TSC1TSC complex subunit 1, interacts with and stabilizes the GTPase activating protein tuberin.
25. RUNX1 transcription factor 1, thought to be involved in development of normal hematopoiesis.
26. TSC2 TSC complex subunit 2, believed to be a tumor suppressor and is able to stimulate specific GTPases.
27. PALB2 partner and localizer of BRCA2, permits intranuclear localization and accumulation of BRCA2.
28. NBN nibrin involved in DNA double-strand break repair and DNA damage-induced checkpoint activation.
29. BAP2 (BAIAP2) protein disulfide isomerase (PDI) inhibitor, PDI inhibitors in glioblastoma 30759340.
30. RAD50 double-strand break repair protein, required for nonhomologous joining of DNA ends.
31. MSH6 helps in the recognition of mismatched nucleotides prior to their repair, highly conserved protein.
32. TGBR1 transforming growth factor beta receptor 1, serine/threonine protein kinase.
33. SMARCB1 SWI/SNF-related, matrix-associated, actin-dependent regulator of chromatin, subfamily b.
34. PTCH1 patched 1, encoded protein is the receptor for the secreted hedgehog ligands.
35. HRAS proto-oncogene, GTPase, functions in signal transduction pathways.
36. POLE DNA polymerase epsilon, catalytic subunit, involved in DNA repair and l DNA replication.
37. MLH1 heterodimerize with mismatch repair PMS2, to form part of the DNA mismatch repair system.
38. HOXB13 homeobox B13 transcription factor.

39. MSH2 mutS homolog 2, frequently mutated in hereditary nonpolyposis colon cancer (HNPCC).
40. BARD1 BRCA1-associated RING domain 1 may harbor oncogenic mutations in breast or ovarian cancer.
41. SDHA succinate dehydrogenase complex flavoprotein subunit A mitochondria.
42. SMAD3 protein functions in the transforming growth factor beta signaling pathway.

*Watson syndrome café au lait spots, pulmonic stenosis.

12.4 Specific syndromes that include the presence of tumors

12.4.1 Multiple endocrine neoplasia and associated gene defects

Kamilaris and Stratakis (2019) reviewed MEN type 1, defined as a rare dominant disorder characterized by predisposition to tumors of the parathyroid, enteropancreatic system, and anterior pituitary. Nonendocrine tumors can also occur, angiomas, fibromas, and lipomas. In 17% of patients, tumors were reported to occur before 21 years of age.

The germline mutations encountered in MEN1 include frameshift mutations, nonsense mutations, exon deletions, and splicing defects. Some patients manifest whole or partial gene deletions.

12.5 Retinoblastoma and RB1

The retinoblastoma gene RB1 was reviewed by Dyson in (2016) who noted that it was the first molecularly defined tumor suppressor gene. Deletions on chromosome 13q14.1–13q14.3 were found to predispose to retinoblastoma. However, a second mutation on the normal homologous chromosome was required for development of the tumor leading to the two hit tumorigenesis hypothesis of Knudson (1971) (Fig. 12.2).

The RB1 gene product was noted to regulate transcription and to be a negative regulator of cell proliferation. Dyson noted that there was little information available relating to the gene or proteins to which RB1 binds. However, one important interaction is with E2F transcription factors. There is evidence that RB1 is involved in control of cell cycle progression. Loss of RB1 function leads to defects in cell cycle.

Dyson noted that different sites are phosphorylated in RB1 protein, and it is possible that the specific site phosphorylated and the extent of phosphorylation impact RB1 protein function.

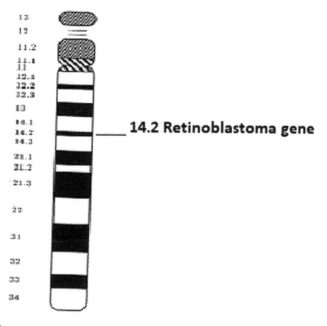

FIGURE 12.2

Retinoblastoma locus (chromosome 13).

Retinoblastoma is a tumor of early childhood. Rodriguez-Galindo et al. (2015) reported that there are two predominant forms, the hereditary form is bilateral or multifocal due to germline mutations in the RB1 gene and a more common unilateral form occurs that is nonhereditary.

Mutations in RB1 have been found in several tumors including retinoblastoma, osteosarcoma, and small cell lung cancer.

12.6 Germline succinate dehydrogenase gene mutations and cancer predisposition

Dubard Gault et al. (2018) reviewed germline mutations in succinate dehydrogenase SDH and cancers in adults and children. Succinate dehydrogenases form mitochondrial respiratory complex II and play key roles in mitochondrial respiratory and metabolic functions. Pathogenic germline mutations in SDHA, SDHB, SDHC, and SDHD have been reported to predispose to tumors including familial paraganglioma, pheochromocytomas, and gastrointestinal stromal tumors (GIST).

Dubard Gault et al. specifically reported cases of cancer with germline mutations in SDHA I, a recurrent mutation encountered in this protein was c.91 C > T, p/Arg31Ter.

Other cancer predisposing mutations were also encountered. Cancer types include colon adenocarcinoma, endometrial cancer, urothelial carcinoma prostate cancer, multifocal GIST, neuroblastoma, and pheochromocytoma.

Dubard Gault noted that if individuals are found to have germline SDH mutations, tumor screening is important. This includes imaging and biochemical testing for elevated plasma or urine concentrations of neuroendocrine-derived compounds, which include metanephrines and catecholamines.

12.7 Hereditary gastrointestinal cancers

In considering hereditary gastrointestinal cancer syndromes Syngal et al. (2015) emphasized the importance of family history evaluation in the context of genetic counseling. These guidelines were noted to be particularly relevant in familial adenomatous polyposis, MUTH-associated polyposis Lynch syndrome, Peutz−Jeghers syndrome, Cowden syndrome, hereditary gastric cancer, and hereditary pancreatic cancer. The specific syndrome and associated germline genes that may be mutated and functions of the gene products are listed below.

1. Familial adenomatous polyposis APC tumor suppressor protein that acts as an antagonist of the Wnt signaling pathway.
2. MUTYH DNA glycosylase involved in oxidative DNA damage repair.
3. Peutz−Jeghers syndrome serine/threonine kinase 11.
4. Juvenile polyposis, SMAD4 signal transduction factor, BMPR1A morphogenetic protein receptor.
5. Cowden syndrome PTEN a phosphatidylinositol-3,4,5-trisphosphate 3-phosphatase.
6. Hereditary pancreatic syndrome BRCA1/2; CDKN2A cyclin-dependent kinase inhibitor 2A, PALB2 partner and localizer of BRCA2; ATM cell cycle checkpoint kinase.
7. Hereditary gastric cancer CDH1 encodes a classical cadherin of the cadherin superfamily calcium-dependent cell − cell adhesion protein.
8. Autosomal dominant colorectal cancer microsatellite instability MLH1, MSH2, MSH6 mutL homologs. PMS mismatch repair homologs, EPCAM epithelial cell adhesion molecule.

12.7.1 Pediatric cancers

Pui et al. (2019) reviewed somatic and germline genomics in acute lymphoblastic leukemia (ALL) in the pediatric population. Key points that they emphasized included advances in transcriptional sequencing that led to expanded delineation of ALL subtypes that had relevance for therapeutic approaches. In addition, they drew attention to cooperative mutations that contribute to the heterogeneity of clinical responses even within specific tumor subtypes.

Pui et al. also reviewed specific germline mutations that influenced risk of developing ALL. They noted that identification of germline gene defects increased possibilities for counseling and early detection.

Childhood ALL was noted to have a high cure rate in developed countries. Pui et al. noted the importance of screening for both somatic and germline mutations in this cancer as these data influence patient management.

Cytogenetic abnormalities, including structural abnormalities and copy number variants, were noted to occur and their detection was noted to be useful in subtype classification. However, genomic sequencing and particularly transcriptome sequencing were noted to provide additional information with therapeutic relevance. Known chromosome abnormalities in lymphoblastic leukemias included hyperdiploidy and hypodiploidy. BCR-ABL fusion, ETV-RUNX1 fusions, TCF3-PBX1 fusions, and KMT2A rearrangements were also important to detect.

B-cell lymphoblastic leukemia with intrachromosomal amplification of chromosome 21 was reported. Other rearrangements detected by cytogenetics included MEF2D and ZNF384 rearrangements.

Specific genes that harbored germline and somatic mutations, their products, and functions are listed below:

1. PAX5, ETV6, and IKZF1 transcription factors.
2. BCR activator of RhoGEF and GTPase 22q11.23.
3. ABL1 ABL proto-oncogene 1, nonreceptor tyrosine kinase, 9q34.1, role in cell division.
4. ETV6 ETS variant transcription factor 6 12p13, required for hematopoiesis.
5. RUNX1 transcription factor involved in hematopoiesis, 21q22.1.
6. TCF3 transcription factor 3, plays a critical role in lymphopoiesis 19p13.3.
7. PBX1 transcription factor 1q23.3.
8. KMT2A lysine methyltransferase 2 A, chromatin modifications transcriptional activation 11q23.3.
9. Deleterious germline mutation also occurred in PAX5, ETV6, and IKZF2.
10. PAX5 paired box transcription factor 5, B-cell lineage specific activator protein, 9p13.2.
11. IKZF2 IKAROS family zinc finger 2 transcription factors involved in lymphocyte development, 2q34.
12. TAL1 TAL bHLH transcription factor 1, erythroid differentiation factor 1p3.3.
13. Pui et al. noted that an autosomal dominantly inherited predisposition to B-cell leukemia was reported in families with a p.G183S missense mutation in PAX5. The predisposition is transmitted in an autosomal dominant manner but biallelic inactivation of PAX5 was reported to be necessary for leukemogenesis to occur.
14. ETV6-associated leukemic thrombocytopenia was identified as an autosomal dominant condition.
15. Pui et al. also documented the frequencies of the different genetic subtypes of ALL.

16. Hyperdiploid 20%.
17. ETV-RUNX 20%.
18. BCR-ABL 12%.
19. KMT2A rearrangement 5%.
20. TAL1 5%.
21. PAX5 4%.

Pui et al. also documented genetic syndromes associated with increased frequency of development of ALL, these included:

1. Ataxia telangiectasia ATM.
2. Down syndrome chromosome 21.
3. Familial platelet disorder and myeloid malignancy RUNX1.
4. Fanconi anemia, eight different genes.
5. Mismatch repair syndromes, MLH1, MSH2, MSH6, NBN Nijmegen breakage syndrome, PMS1.
6. NF2 neurofibromatosis.
7. Noonan syndrome PTPN11.

12.8 Germline mutations and developmental origins of cancer

12.8.1 Medulloblastoma

Waszak et al. (2018) reported an international study of patients with medulloblastoma. They noted that medulloblastoma is an embryonal tumor of the cerebellum. It may occur sporadically but is also seen in rare disorders including Gorlin syndrome, Fanconi syndrome, and Li − Fraumeni syndrome. Their study documented four molecular subgroups of medulloblastoma. One associated with WNT (signaling protein) mutations, another associated with sonic hedgehog mutations and two additional groups, Groups 3 and 4.

Results of their study on 1022 patients led to validation of six additional medulloblastoma predisposition genes; rare variants were noted to occur in:

1. APC regulator of WNT signaling pathway.
2. BRCA2 DNA repair associated.
3. PALB2 partner and localizer of BRCA2.
4. PTCH1 Patched 1 component of the hedgehog signaling pathway.
5. SUFU negative regulator of hedgehog signaling.
6. TP53 tumor protein contains transcriptional activation, DNA binding, and oligomerization domains.
7. In medulloblastoma groups MB3 and MB4 predisposition variants were not identified.

To identify germline and tumor mutations, whole exome sequencing was carried out. Germline mutations included frameshift, stop codon, start codon lost, canonical splice site mutation, and exon or gene deletions. Mutations with a general population incidence higher than 0.1% were removed from consideration. Somatic mutations including single nucleotide variants, copy number alterations, and structural variants were also analyzed.

Waszak et al. reported results from an international study of medulloblastoma. Finding are summarized below.

Damaging germline mutations were identified in 11% of patients and occurred across 32 genes. Six genes were reported to show an excess of germline mutations. They included APC, BRCA2, PALB2 PTCH1, SUFU, and TP53 (functions listed above). A single patient manifested germline mutations in MSH6.

These discoveries led to development of guidelines for genetic counseling in families of patients with medulloblastomas.

Zhang et al. (2015) carried out studies on 1120 patients with major forms of pediatric cancer. They carried out exome sequencing on noncancerous tissue to identify germline cancer predisposing genes. Their study involved analyses of 565 genes including 49 genes associated with autosomal dominant cancer syndromes. These included 11 genes associated with Rasopathies; 29 genes associated with autosomal recessive cancer predisposition disorders; 23 tyrosine kinase gene; 58 tumor suppressor genes; and 395 genes reported to be associated with cancer. Results of this study revealed the following:

1. Pathogenic variants or possibly pathogenic variants were detected in 21 genes. In order of frequency of involvement, genes were TP53 mutations in 50 patients; APC mutations in six patients; BRCA2 mutations in six patients; NF1 mutation in four patients; PMS2 mutations in four; RB1 mutation in three; RUNX1 mutations in three (PMS mismatch repair component; RB1 RB transcriptional corepressor 1; RUNX1 RUNX family transcription factor 1).
2. Cancer types associated with germline TP53 mutations included adrenocortical tumors, hypodiploid ALL, and choroid plexus cancer.
3. Zhang et al. noted that in pediatric cancers family history did not necessarily predict cancer predisposition.

12.9 Osteosarcoma

Osteosarcoma is reported to be the most common bone tumor in children and adolescents. Mirabello et al. (2020) carried out a study to identify pathogenic germline variants in cancer susceptibility genes associated with osteosarcoma and they investigated germline genetics in 1244 patients with osteosarcoma. In total 28% of patients with osteosarcoma had a rare germline mutation in a cancer predisposing gene.

Mirabello et al. selected 238 cancer susceptibility genes for sequencing. Genes with the highest number of pathogenic germline mutations are listed below.

12.9.1 Autosomal-dominant gene mutations

1. TP53 tumor protein contains transcriptional activation, DNA binding, and oligomerization domains.
2. CDK2NA2 cyclin-dependent kinase inhibitor 2A.
3. MEN1 Menin is a scaffold protein that functions in histone modification and epigenetic gene regulation.
4. VHL von Hippel – Lindau tumor suppressor ubiquitination and degradation of hypoxia-inducible factor.
5. POT1 protection of telomeres 1, binds to the TTAGGG repeats of telomeres.
6. APC regulator of WNT signaling pathway.
7. MSH2 mismatch repair gene.
8. **Autosomal recessive gene mutation.**
9. RECQL4 RecQ like helicase 4 unwind double-stranded DNA into single-stranded DNAs.
10. **X-linked gene mutations.**
11. DKC1 unwinds double-stranded DNA, active in telomerase stabilization, and maintenance.
12. GPC3 cell-surface glycoprotein, may play a role in the control of cell division and growth regulation.
13. WAS actin nucleation, transduction of signals from receptors on the cell surface to the actin cytoskeleton.

12.9.2 Syndromic genes associated with osteosarcoma

1. RB1 RB transcriptional corepressor 1, negative regulator of the cell cycle.
2. RECQL4 RecQ like helicase 4, unwinds DNA to single stranded DNA.
3. RPL32A 60S ribosomal protein L32–1R.
4. RPL5 Ribosomal protein L5. RPS19 ribosomal protein S19. RPS7 ribosomal protein S7.

12.9.3 Developmental origins of pediatric cancers

In 2019 Filbin and Monje (2019) stressed the developmental origins of pediatric cancers. They noted the lower overall mutational burden in pediatric cancers as compared with adult cancers and stressed epigenetic dysregulation in stem cell and precursor cells.

In the category of childhood cancers they included lymphomas, leukemias, retinoblastomas, central nervous system (CNS) tumors, sarcomas of bone and soft tissue such as rhabdoid tumors, and renal and liver tumors.

Sequencing studies revealed that fusion genes occurred more frequently in childhood cancers and epigenetic dysregulation was more common. Fusions were reported to activate key developmental genes, for example, the neurotrophic growth factor receptor NTRK1. This kinase is a membrane-bound receptor that, upon neurotrophin binding, phosphorylates itself and members of the MAPK pathway.

Point mutations in histone genes were noted to be more common. There was evidence for decreased H3K27 trimethylation due to dysfunction of methyltransferase activity. Abnormal patterns of H3K27 methylation were reported in brain tumors including ependymomas and medulloblastomas. Oncogenic histone alterations were also reported to occur in myeloid leukemias.

Pediatric cancers were also reported to have an increase in mutations in SETD2 histone lysine methyltransferase, KDM6A lysine demethylase 6A, and in chromatin-related components and modifiers.

B-cell lymphoid cancers in childhood were noted to sometimes arise in the prenatal stage as evidenced by the early presence of fusion gene. Transcription factor gene fusions involved ETV6-RUNX1, MLL-AFF4 (KMT2A-AFF4), AML1-ETO, and RUNX1 − RUNX1T. Chromosomal translocations and rearrangements involving RUNX1 are well-documented and have been associated with several types of leukemia.

Additional factors were noted to perhaps be necessary to further convert cells with these fusions to cancerous cells.

Transient myeloproliferative disease and acute megakaryoblastic leukemia were noted to occur in 5%−10% of neonates with Down syndrome but was reported to spontaneously resolve.

Malignant rhabdoid tumor was noted to occur as a soft tissue tumor in young children and was noted to have mutation in SNF5 (SMARCB1), a component of the chromatin remodeling complex.

Filbin and Monje noted evidence that high-grade gliomas in the midbrain regions of thalamus and pons and in the spinal cord arise from precursor cells in the oligodendroglial lineage and have mutations in histone H3.

12.10 Genetic alterations in cancers in children, adolescents, and young adults

Gröbner et al. (2018) presented results of a study of genetic alterations in 961 tumors in children, adolescents, and young adults. In 149 cancer-driver genes they identified mutations including small changes, copy number changes, and structural changes.

Importantly 7%−8% of children were reported to have cancer predisposing germ line variants. Hereditary predisposition variants were noted to be particularly common in adrenocortical tumors, in hypodiploid ALL, high grade gliomas, medulloblastomas, and retinoblastoma. Genes with cancer predisposition mutation included LZTR1 leucine zipper-like transcription regulator 1;TSC2 tuberous sclerosis 2; CHEK2 checkpoint kinase 2 cell cycle checkpoint regulator; SDHA succinate dehydrogenase complex flavoprotein subunit A.

Germline variants with cancer predisposing genes included MSH2, MSH6, PMS2, and double-stranded DNA break repair genes TP53, BRCA2, CHEK2.

Carriers of TP53 mutation also had increased risk for treatment-induced secondary tumors. Germline mutations also occurred in transcription genes VHL von Hippel − Lindau and LZTR1 leucine zipper-like transcription regulator 1.

Mutation processes and signatures in cancer were analyzed. Mutations were predominant C > T transitions. A unique mutation signal was identified and designate P1, it occurred in a CCC/CCT context.

Gröbner et al. noted that a canonical double-stranded DNA break signature was linked to chromothripsis with widespread chromosome fragmentation.

Cancer-driver genes were noted to include TERT promoter mutation and other significantly mutated genes with nonsilent mutations including TP53, H3F3A (histone 3), SMARCA4 chromatin regulator, CCND3 cyclin D3 (cell cycle regulator), MYC (protooncogene transcription factor).

In pediatric cancers they established that there were 879 significantly mutated genes. Some of these overlapped with significantly mutated genes in adult cancer. Others were predominantly mutated in pediatric cancers. These included H3FA (H3−3A), ID3 inhibitor of DNA binding 3, HLH protein, MYC, CCND3, SMARCB1 regulator of chromatin, BCOR transcription repressor, DROSHA ribonuclease 3, [catalyzes the initial processing step of microRNA (miRNA) synthesis].

Ma et al. (2018) analyzed data from a pediatric pan cancer study that included leukemias and solid tumors. Their study included paired tumor and normal tissue samples from 1699 patients in the Children's Oncology Group. Their analyses included genome-wide sequence data, whole exome sequencing, and exome sequencing.

They noted high consistency in calling of point mutations in driver genes in whole genome and exome sequencing. Whole genome sequencing enabled detection of copy number changes and structural variants, and these were noted to occur frequently in driver genes in pediatric cancer. In addition, integrative analyses of copy number abnormalities and structural events in whole genome sequencing revealed chromothripsis in some cases with widespread rearrangements. Chromothripsis was noted to occur in 11% of cases including osteosarcomas, Wilms tumors, neuroblastomas, acute lymphoid leukemia, and acute myeloid leukemia (AML).

Ma et al. identified biologic pathways frequently disrupted by driver mutations. These included particularly cell cycle and epigenetic pathways. Some driver gene mutations impacted signaling pathways: RAS, JAK-STAT, PI3K. Two novel KRAS isoforms were identified in 70% of leukemias.

In solid tumors, disrupted pathways included the following ALK receptor tyrosine kinase, NF1 neurofibromin 1, and PTEN phosphatase and tensin homolog.

12.11 Adult cancers, driver gene mutations, and passenger gene mutations

In 2013 Vogelstein et al. (2013) reported that 140 genes constituted cancer-driver genes and that a typical tumor harbored 2 − 8 driver gene mutations. They classified 10−12 signaling pathways disrupted in cancer and noted that these pathways impaired three core processes: genome maintenance, cell fate, and cell survival.

Vogelstein et al. documented the mutation types in five different cancer types, colorectal cancer, breast cancer, pancreatic cancer, glioblastoma, and medulloblastoma. In each of these tumor types the majority of significant changes were reported to involve single base substitutions. Other changes that occurred included indels, amplification, deletions, and translocations.

In distinguishing between driver mutations and passenger mutation, Vogelstein et al. noted that a driver mutation conferred a specific growth advantage. Passenger mutations did not necessarily contribute a growth advantage. Both driver and passenger mutation could occur within one specific gene. Vogelstein et al. emphasized that the majority of mutations in a neoplasm were immaterial to tumor initiation; they arose secondarily and constituted passenger mutations.

In documenting key cellular processes disrupted in cancer they referred to pathways that disrupted cell fate. These included NOTCH and hedgehog signaling, chromatin modifications and transcription regulators. Other key processes included those involved with genome maintenance and DNA damage control and key proteins included TP53, ATM, MLH1, and MSH2. Also key were proteins that impacted cell survival: TGFbeta, MAPK, STAT (signal transducers and activators of transcription), PI3K, RAS, and proteins that impacted apoptosis.

In considering future clinical approaches to therapy, they noted that each tumor has 30−70 protein altering mutations and that these could constitute a pathway for immunotherapy. Particularly important would be abnormal proteins presented in the HLA context.

They noted that driver mutations provided a pathway for early diagnosis through analysis of genes or their products. They documented 140 driver genes.

Vogelstein et al. also proposed continued research in analyses of environmental influences and genetic alterations associated with cancer. They proposed that cancer is caused by 2−8 sequential mutations that develop during a 20−30-year period.

In 2014 Lawrence et al. (2014) analyzed a total of 5000 tumors within 21 different tumor types and reported well-established cancer-driver genes. They included:

1. TP53 tumor protein 53.
2. PIK3CA phosphatidylinositol-4,5-bisphosphate 3-kinase catalytic subunit alpha.
3. PTEN The protein encoded by this gene is a phosphatidylinositol-3,4,5-trisphosphate 3-phosphatase.
4. RB1 RB transcriptional corepressor 1.
5. KRAS encodes a protein that is a member of the small GTPase superfamily.
6. NRAS proto-oncogene, GTPase.
7. BRAF B-Raf proto-oncogene, serine/threonine kinase.

8. CDKN2A cyclin dependent kinase inhibitor 2A.
9. FBXW7 constitutes one of the four subunits of ubiquitin protein ligase complex.
10. ARID1A thought to regulate transcription of certain genes by altering the chromatin structure.
11. MLL2 (KMT2D) histone methyltransferase that methylates the Lys-4 position of histone H3.

They added additional 10 genes impacted in at least three tumor types.
1. ATM important cell cycle checkpoint kinase.
2. CASP8 caspase plays a central role in the execution phase of cell apoptosis.
3. CTCF CCCTC-binding factor, can impact transcription.
4. ERBB3 member of the epidermal growth factor receptor (EGFR) family of receptor tyrosine kinases.
5. HLA histocompatibility antigen.
6. HRAS proto-oncogene, GTPase.
7. IDH1 isocitrate dehydrogenases catalyze the oxidative decarboxylation of isocitrate to 2-oxoglutarate.
8. NF1 neurofibromin 1 negative regulator of the ras signal transduction pathway.
9. NFE22L (NRF2) transcription factor regulates genes which contain antioxidant response elements.
10. PIK3R1 phosphoinositide-3-kinase regulatory subunit 1.

In listing functions of impacted genes and their products, Lawrence et al. included cell proliferation, products that impact genome stability, chromatin regulators, RNA processing, products involved in apoptosis, protein homeostasis, and immune evasion.

Lawrence et al. noted that cancer precision medicine would require a comprehensive catalog of cancer genes to help determine the pathway disrupted in a particular tumor.

Uprety and Adjei (2020) reported that RAS oncogenes, including KRAS, HRAS, NRAS, were the frequently mutated oncogenes in cancer. Mutated RAS was reported to be activated with GTPase activity that stimulated to downstream signaling.

12.12 DNA damage and repair

Chatterjee and Walker (2017) reviewed mechanisms of DNA damage and repair emphasizing the relevance to cancer. Five major DNA damage repair pathways are known. They include base excision repair, nucleotide excisions repair, mismatch repair, repair of single-strand DNA breaks, repair of double-stranded DNA breaks. Repair of DNA strand breaks can occur via homologous recombination and nonhomologous recombination. Trans-lesion DNA synthesis was noted to require activity of particular polymerases.

DNA repair pathway defects are noted to be disrupted in several forms of cancer.

DNA damage can be induced by endogenous reactive oxygen species and also by exogenous chemical and physical agents including ultraviolet light that leads to formation of pyrimidine dimers. One form of DNA damage involves base deamination. Deamination of cytosine in the CG context is one of the most common mutations and leads to conversion of CG to TA.

Reactive oxygen species can damage DNA bases and can also lead to single-strand DNA breaks. A specific repair pathway, the base excision repair pathway, repairs oxidized bases; single-strand DNA breaks can be repaired by the single-strand break repair pathway. A specific repair pathways exists to repair double-stranded DNA breaks.

Another form of DNA damage includes the developments of interstrand cross-links.

In 2010 Ciccia and Elledge (2010) reported on DNA damage repair that utilizes specific polymerases and kinases including the ATM kinase that is mutated in ataxia telangiectasia. Significant contributions to our understanding of DNA damage and repair were made by Lindahl, Modrich, and Sancar, who were awarded the Nobel Prize in 2015 for their work (see Kunkel, 2015).

Krokan and Bjørås (2013) reviewed base excision repair.

Another form of DNA damage involves collapsed replication forks and these can be repaired by AT and RECQ helicases. Other forms of DNA damage are also noted to be repaired by RECQ helicases.

Kim and D'Andrea (2012) reported that DNA cross-links require repair by proteins in the Fanconi BRCA pathway.

The mismatch repair system is responsible for the repair of base mismatches or small deletions and insertions and was reported to detect such lesions, to excise them and then to repair them (Fishel and Kolodner, 1995). In 1996 Leach et al. (1996) reported expression of the human mismatch repair gene hMSH2 in normal and neoplastic tissues. MSH2, MSH3, and MSH6 were reported to recognize mismatches and MLH1, MLH3, PMS1, and PMS3 were reported to be important in their repair (Li and Modrich, 1995).

Alexandrov and Stratton (2014) reported that sequence analyses of mutational changes in cancer provided insight into the mutational inducing processes leading to cancer and pathogenic mechanisms in cancer. They presented evidence that GG to AT transversion predominated in smoking-induced cancers, CC GG to TT AA double mutations were reported to be common in UV light-associated cancers.

12.13 Lymphomas and leukemia

Young et al. (2019) reported that signals from B-cell receptors play roles in the proliferation of cells in several forms of B-cell tumors and that therapies designed to target the B-cell receptor pathway are useful in treating many, but not all, lymphomas.

Young et al. noted that there are more than 70 different forms of lymphomas. The B-cell receptor was noted to be involved in the etiology of many of the B-cell lymphomas. This origin led to the development of therapeutic agents that target signaling downstream of the B-cell receptor. Young et al. noted that more recent studies have been directed at signal transduction factors that promote lymphoma growth. These include TLR9 (Toll like receptor 9) and MYD88 (MYD88 innate immune signal transduction adaptor). One common form of lymphoma is diffuse large B-cell lymphoma (DLBCL). There is evidence that different types of DLBCL exist, and the gene expression patterns differ in the different subgroups. Different subgroups are also characterized by different genetic changes that include mutations, deletions, translocations, and fusions.

Young et al. noted that studies are being carried out to identify specific oncogenic targets in B-cell lymphomas.

Roschewski et al. (2020) reported on molecular classification of the DLBCL and noted striking genetic heterogeneity of these aggressive lymphomas.

They noted that distinct subtypes of DLBCL were found to include different genetic subtypes. The different DLBCL subtypes were noted to have different primary anatomic origins including primary CNS lymphomas PCNSL, primary cutaneous leg type, intra-vascular large B-cell lymphomas. Included in this category are primary mediastinal B-cell lymphomas.

Studies revealed that in 10%–15% of patients with DLBCL, an oncogenic rearrangement occurred in the MYC oncogene. In 8% of patients, rearrangements of MYC with BCL2 and BCL6 apoptosis regulators occurred.

Other abnormalities found included mutations in EZH2 transcription regulator, and amplifications in REL protooncogene transcription factor.

A number of these changes were reported to be therapeutically targetable.

Roschewski et al. noted that in refractory cases or in relapsed cases of DLBCL immunotherapy approaches were used in therapy.

Wang et al. (2019) noted that Hodgkin included different entities and that characterization of the molecular alterations in the neoplastic cell and in their environment have led to therapeutic advances. Hodgkin Reed − Sternberg cells have been considered to be typical of Hodgkin lymphoma for many years.

They noted that a number of inflammatory lymphoproliferative conditions mimic Hodgkin lymphoma.

Hodgkin lymphoma is now known to include different disease forms, nodular lymphocyte predominant Hodgkin's lymphoma NLPHL, nodular sclerosis, classical Hodgkin lymphoma.

The incidence of NLPHL was reported to peak in the fourth decade of life but it does sometime occur in children. Typically patients present with lymphadenopathy without systemic symptoms. NLPHL was reported to not impact mediastinal or axial lymph nodes. Histologically it was noted to be a histocyte rich B-cell lymphoma. Familial risk for this disease was reported in some studies.

The characteristic cells were reported to show multiple chromosomal abnormalities including rearrangements of BCL6 in half of the cases.

Wang reported that different subtypes of Hodgkin lymphoma occur including NSCHL, nodular sclerosis classic Hodgkin lymphoma, noted to primarily affect young adults 15–34 years of age with female predominance.

The mixed cellularity type, MCCHL, was noted to have one incidence peak in children and a second peak after 60 years.

Lymphocyte depleted Hodgkin disease, LDCHL, was noted to be more common in developing countries and to frequently be associated with Epstein–Barr Virus (EBV) infection and EBV antigens were detectable in Reed–Sternberg cells. Patients presented with cervical, mediastinal, supraclavicular, and axillary lymphadenopathy. Hematogenous spread and extra nodular involvement was noted to sometimes occur and to involve the lung and brain.

Wang et al. noted that molecular studies in Hodgkin lymphoma revealed rearrangements and damaging mutations within immunoglobulin gene. Cytogenetic studies revealed aneuploidy that impacted chromosome 2p, 2q, 17q, 19q, and 20q and loss of 6q, 13q. Recurrent mutations were present in JAK STAT pathway genes.

Wang noted that significant efforts continue to be made to therapeutically target that pathway.

EBV B-cell proliferation disorders were noted to sometimes mimic classical Hodgkin disease. EBV positive mucocutaneous ulcer was noted to sometimes occur along with EBV cell proliferation.

Peripheral T-cell lymphomas were noted to sometimes have clinical features similar to Hodgkin lymphoma.

Shannon-Lowe et al. (2017) reviewed EBV-associated lymphomas. They noted that EBV was initially reported to be associated with Burkitt lymphoma but that it became recognized as being linked to a number of lymphoproliferative conditions and B, T, and NK lymphomas. They noted evidence for complex interplay of viral gene expression and cellular genetic changes, in pathogenesis of EBV-associated conditions.

EBV virus was defined as a gamma-1 herpes virus with widespread population distribution. Initial infections involved virus replication in epithelial cells and B-cells in the nasopharynx. The virus could lead to transformation of B-cell and to latent infection in B-cells. Some EBV infections were noted to lead to infectious mononucleosis and to T-cell responses.

Shannon-Lowe et al. noted that three types of B-cell malignancies were associated with EBV: Burkitt lymphoma, Hodgkin lymphoma, and DLBCL. They noted evidence of latent EBV protein expression, including expression of the antigen EBNA1 in Burkitt lymphoma, and also in the lymphocyte depleted form of Hodgkin lymphoma.

12.14 Myeloid leukemia

Haferlach and Schmidts (2020) noted that the diagnosis of myeloid leukemia relies on cytomorphology, immunophenotyping, cytogenetics, and molecular genetics. They noted that cytogenetics is useful for classification and is also used to monitor therapeutic response and to monitor clonal evolution. Molecular genetics was noted to optimize classification and monitor residual disease. It was also reported to aid in the development of targeted therapies.

The application of whole genome sequencing or whole exome sequencing was noted to facilitate analyses of sequence changes and structural and numerical changes. Transcriptome analyses facilitates detection of fusion transcripts and identification of specific gene expression changes.

Epigenetic changes were noted to play important roles in leukemogenesis. Recurrent abnormalities in specific epigenetic regulators were identified. These included DNMT3A DNA methyltransferase 3 alpha; ASXL1 ASXL transcriptional regulator 1; TET2 tet methylcytosine dioxygenase 2; KMT2 lysine methyltransferases; IDH1, IDH2 isocitrate dehydrogenases.

Haferlach and Schmidts documented the WHO classification of AML and associated genetic abnormalities and fusions.

In AML these included chromosomal abnormalities: t(8:21) q22-q22.1 RUNX1-RUNX1T1; inv 16 p13 1q22, AML t16;16 p13.1q22.

Acute promyelocytic leukemia abnormalities included fusions:

1. PML-RARA PML nuclear body scaffold, RARA retinoic acid receptor alpha.
2. DEK-NUP DEK protooncogene nucleoporin 214 retinoic acid receptor t(6,9) translocation.
3. GATA2 MECOM GATA zinc finger transcription factor, MECOM transcriptional regulator and oncoprotein.
4. CBFB-MYH11 core binding transcription factor Myosin-like heavy chain associated with AML of the M4Eo subtype.
5. BCR-ABL1 BCR activator of RhoGEF and GTPase BCR-ABL fusion protein ABL1 protooncogene. The unregulated tyrosine kinase activity of BCR-ABL1 contributes to the immortality of leukemic cells.

Mutations and abnormalities also occur in the following:

1. FLT3 Fms-related receptor tyrosine kinase 3. The activated receptor kinase subsequently phosphorylates and activates multiple cytoplasmic effector molecules in pathways involved in apoptosis, proliferation, and differentiation of hematopoietic cells in bone marrow.
2. KMT2A lysine methyltransferase 2A undergoes rearrangement in leukemia. It is a transcriptional coactivator that plays an essential role in regulating gene expression during early development and hematopoiesis.
3. RBM15 RNA binding motif protein 15 has repressor function in several signaling pathways and has been implicated in megakaryocytic AML.

Specific germline mutations that predispose to AML were reported in the following genes:

1. CEBPA activity of this protein can modulate the expression of genes involved in cell cycle regulation.
2. DDX4 putative RNA helicases.

3. RUNX1 transcription factor that binds to the core element of many enhancers and promoters.
4. ANKRD26 ankyrin repeat domain 26 functions in protein − protein interactions.
5. ETV6 ETS variant transcription factor 6 required for hematopoiesis and maintenance of the developing vascular network. This gene is known to be involved in a large number of chromosomal rearrangements associated with leukemia.
6. GATA2 GATA binding protein 2, plays an essential role in regulating transcription of genes involved in the development and proliferation of hematopoietic and endocrine cell lineages.

12.15 Providing insight into cancer-inducing mechanisms

Zhan et al. (2019) noted that CRISPR-Cas applications in cancer research can potentially provide methods to promote understanding of cancer mechanisms and identify therapeutic targets. They specifically addressed the use of investigation of specific genomic changes on functions. Zhan et al. emphasized that increased analyses of the genetic landscape in cancer has identified altered genes. However, it is frequently not clear how these changes contribute to cancer generation and progress.

Crispr-Cas studies were noted to primarily be used to promote removal of specific nucleotides and sequences from DNA. More recently CRISPR-CAS applications have been used to investigate transcription changes through targeting of transcription start sites. In addition CRISP-CAS has been used to altered methylation.

12.16 Genome sequencing in cancer

One of the first reports of cancer genome sequencing was published by Ley et al. (2008). They sequenced cytogenetically normal myeloid leukemia cells and noted the importance of comparing sequence of tumor cells with the patient's normal cells. They discovered 10 genes with acquired mutations. Two genes identified with mutations had been previously reported to be mutated in other patients with myeloid leukemia. These genes encoded FLT3 and NPM1. FLT3 fms-related receptor tyrosine kinase 3, regulates hematopoiesis. NPM1 nucleophosmin 1 is involved in centrosome duplication, protein chaperoning, and cell proliferation.

Fletcher (2020) in an article on the cancer genome noted that there is evidence that even in an individual cancer type there is evidence for substantial genetic heterogeneity.

Cancer genome sequencing and mutation analyses have drawn attention to the diverse cellular processes and epigenetic regulation alterations in cancer.

Nakagawa and Fujita (2018) emphasized that whole genome sequencing in cancer could promote understanding of driver mutations. They emphasized that further information was needed on nonprotein coding genome regions and on regulatory genomes in cancer. They also emphasized integration of RNA sequencing, epigenomic analyses, and immunogenomic information, and correlation of this information with clinical and pathological information.

They noted that much evidence had been placed on specific driver mutations since the work of Vogelstein et al. However, they noted that there remained "a long tail of rare driver genes."

12.17 Cell-free DNA analyses in testing for tumors

Bettegowda et al. (2014) reported detection of cell-free DNA (cfDNA) in blood samples of more than 75% of 640 patients with different advanced cancer types. They utilized digital polymerase chain reactions to search for cancer-specific mutations. They reported that use of cfDNA was more reliable than detection of circulating cancer cells.

Stewart et al. (2018) reviewed the value of cfDNA analyses for diagnoses of primary tumors and tumor metastases. They also noted that cfDNA can readily be measured to determine response to therapy. Stewart et al. discussed analyses in different body fluids, blood plasma urine, cerebrospinal fluid (CSF).

Stewart et al. noted both successes and limitations of tissue-based clinical testing. Limitations included the fact that biopsy of tissue was needed.

They noted that the majority of cfDNA in plasma is derived from leukocytes. There is evidence that cfDNA is highly fragmented; it was noted to be largely derived from nucleosomes and transcription factors.

The small proportion of circulating DNA derived from tumors was referred to as ctDNA. The quantities of ctDNA were noted to be a function of the size of the tumor and the stage of cancer with higher amounts in metastatic tumors. The levels of ctDNA were noted to be indicative of prognosis.

Stewart et al. noted that methodologies are being applied to enrich for ctDNAs; one method is based on size selection of DNAs prior to sequencing.

Different tests were being applied to ctDNA; they included PCR, for example, for determination of tumor-specific changes. Tests of multiple genes on gene panels are also carried out. Reliability of assays are directly related to the mutant allele fraction present in the sample.

Stewart et al. noted that in addition to nucleic acid analyses, methylation analyses were being carried out on cfDNA to identify cancer-specific methylation changes. They noted that cell-free RNA (cfRNA) analyses were also being carried out and were useful in analysis of fusions between genes.

Davis et al. (2020) reported the value of analysis of circulating tumor DNA analyses early after treatment initiation to validate molecular response and to minimize side effects and costs of ineffective treatment.

12.18 Cell-free studies including transcriptome analyses

Larson et al. (2021) reported results of cell-free transcriptome analyses in plasma from 46 cases of breast cancer, 30 cases of lung cancer, and in 89 cancer-free individuals. They compared results obtained on circulating cfDNA with RNA expression in matched tumor tissue.

Larson et al. documented optimal methods for isolation and preservation of (cf)RNA from patient plasma. They also characterized the parameters of (cf)RNA they isolated. Their analyses revealed that (cf)RNA consisted primarily of ribosomal and mitochondrial DNA, and that 2% of (cf)RNA was messenger RNA. They generated CDNA from the messenger. CDNA was depleted of sequences corresponding to hemoglobin, beta 2 microglobulin, and RN7SL (RNA component of signal recognition particle 7SL1) that were abundant.

To focus on cancer-specific RNA they focused analyses on cfRNA-derived CDNAs that were absent from control samples. They then analyzed those derived from breast and lung cancer patient plasma.

Further analyses revealed that the tumor content of specific genes predicted level of expression of those genes in cfRNA.

Mattox et al. (2019) noted that studies on CSF presented important minimally invasive methods for diagnosis and monitoring of treatment efficacy of malignancies in the CNS.

They noted that diagnosis of CNS tumors often involves tissue acquisition that includes open biopsy or stereotactic biopsy. Both of those procedures were noted to carry morbidity risk.

Mattox et al. emphasized the potential for detecting cfDNA and cfRNA and specific proteins in CSF to improve diagnosis and guide clinical management of CNS tumors. They noted that it was important to select specific biomarkers for analysis. In cfDNA they noted the importance of considering driver tumor mutations. Studies included polymerase chain reaction-based analyses. In some studies DNA sequencing was carried out. The following genes were frequently analyzed:

1. TP53 tumor protein coding; IDH1 isocitrate dehydrogenase (NADP(+)) 1; TERT telomerase reverse transcriptase; NF2 neurofibromin 2; PIK3R1 phosphoinositide-3-kinase regulatory subunit 1; PTCH1 patched 1; PTEN phosphatase and tensin homolog; AKT1 AKT serine/threonine kinase 1; BRAF B-Raf proto-oncogene, serine/threonine kinase; NRAS proto-oncogene, GTPase; KRAS proto-oncogene, GTPase; EGFR epidermal growth factor receptor.
2. Detection of tumor specific cfDNA in CSF was known to be in part dependent on the location of the tumor. Brain stem gliomas were reported to often lead to CSF cfDNA changes. Meningiomas and craniopharyngiomas were also noted to be detectable through presence in CSF of nucleic acids indicative of the BRAF V600E mutation. Detection of this mutation was noted to be especially important given known response of these tumors to BRAF inhibitors.

CSF nucleic acid studies were also noted to be applied to metastatic brain tumors.

Mattox et al. noted that many studies focused on detection of point mutation and small deletion and insertions. However, amplification of genes, for example, EGFR, within tumors was important to determine. They noted that analysis of cfRNA was more challenging. Studies of microRNAs were also noted to be important.

Studies have also been carried out to investigate potential protein biomarkers in CSF. Particular proteins found to be amplified in certain cases included B2M beta 2 microglobulin; VEGFB vascular endothelial growth factor B; SPARCL1 SPARC like 1 that has broad expression in brain.

12.19 Somatic mutations in cancer

In 2019 Tate et al. (2019) reviewed COSMIC, a Catalog of Somatic Mutations in Cancer: http://cancer.sanger.ac.uk.cosmic. They noted that the catalog included data on 6 million coding mutations identified in 1.4 million tumor samples. Data included not only coding mutations but also information on nonprotein coding DNAs, information on copy number variants, on gene fusions, and also information on drug-resistant mutations.

COSMIC was noted to also include information on 719 cancer-driver genes and their functions. In the case of TP53, the spectrum across the 340 mutant TP53 genes was documented. Amino acids at three positions, 175, 200, and 220 were most commonly mutated. However, tumor-related TP53 mutations were documented from position 70 to 350. Insertions and deletions were also documented across the gene, with deletions particularly prominent in one position.

Specific cancer-related functions of specific genes were documented. In COSMIC, information on functions of cancer-related genes expands as new information becomes available.

Drug-resistant mutations are documented, and on the latest version of COSMIC information on 2134 drug-resistant mutations were documented.

Where available COSMIC was noted to include information not only on tumor mutations but also on clinical findings and treatments. Prediction of small molecule binding sites on specific mutated genes was included.

12.20 Breast cancer risk genes

In 2021 the Breast Cancer Association Consortium et al. (2021) reported a comprehensive study of sequence analyses on samples obtained from 60,466 women with breast cancer and on 53,461 controls. Sequence data were obtained on 34 genes reported to harbor mutations that increase cancer risk. Genes associated with the highest level of risk, listed in alphabetical order, included: ATM, BARD1, BRCA1, BRCA2 CHEK2, PALB2, PTEN, RAD51C, RAD51D. It is interesting to note that seven of these nine genes are involved in DNA damage detection and repair.

12.21 Whole genome sequencing of metastatic solid tumors

Priestley et al. (2019) presented results of sequencing from normal tissue and metastatic solid tumors from 2520 individuals. Genomic changes recorded included single nucleotide variants, multiple nucleotide variants, insertion, deletion, and structural genomic abnormalities.

They reported that the mutational burden in metastatic solid tumors varied across different cancer types. Mutational load was particularly increased in tumors deficient in DNA repair processes.

Copy number alteration analyses revealed amplification of established oncogenes EGFR epidermal growth factor receptor, CCNE1 cyclin E1, CCD1 cyclin D1, and MDM2 protooncogene that targets tumor suppressor genes.

Specific chromosome arms with amplification included 1p, 5p, 8q, and 20q. The 5p amplification involved TERT telomerase reverse transcriptase and 8q harbored the MYC oncogene transcription factor.

Analyses were also carried out to determine loss of heterozygosity (LOH). LOH of TP53 varied with tumor type and was noted to be particularly prominent in ovarian cancer. LOH on chromosome 3p was found in 90% of kidney tumors and LOH on chromosome 10 was common in the CNS glioblastoma multiforme. Homozygous LOH on autosomes was reported to be rare.

Whole genome duplication was reported to occur in 56% of metastatic tumors and were particularly high in esophageal tumors.

Priestley et al. set out to identify significantly mutated genes in metastatic tumors. These were found to include MAP3K21 (MLK4) in colorectal tumors. In breast cancer, FPM1 (FOG1) that encodes a transcription factor was significantly mutated. ZFP36L1 is significantly mutated and underexpressed in bladder cancer and reduction of ZFP36L1 expression was associated with worse survival in patients with breast cancer.

In lymphoid cancer HLAB was significantly mutated and TP53 was significantly mutated across cancers.

Priestley et al. noted that 10 driver genes were significantly more impacted in metastatic cancers than in primary cancers. Different cancers varied with respect to the number of amplified drivers.

The top five drivers included CHEK2, BRCA2, MUTYH, BRCA1, ATM. Pathogenic hotspots were also identified in KRAS, PIK3CA, BRAF, NRAS,

TERT (telomerase reverses transcriptase), and TSR genes that encode ribosome maturation factor.

Priestley et al. reported that in 31% of cases of metastatic cancer, an actionable candidate was predicted with sensitivity to a specific drug.

They emphasized that genomic sequencing of tumors is challenged in meeting clinical needs. However, the sequence analyses of metastatic tumors revealed that in 62% of patients, the genomic variants identified could be used to stratify patients toward therapies.

A specific mutation KRAS G12C was reported to occur in most pancreatic cancers, in half of colorectal cancers, and in one-third of cases of lung cancer (Christensen et al., 2020). A specific drug that targets this mutation was approved for cancer therapy by the US Food and Drug Agency (FDA) in May 2021 http://www.fda.gov 'drugs' drug-approvals-and-databases, May 28, 2021.

Ghimessy et al. (2020) reported that in lung cancer KRAS G12C was common in current and former smokers while KRAS G12D was more common in nonsmokers.

Cobain et al. (2021) reported on an assessment of benefits of tumor exome analyses. Their study involved whole exome analyses or target exome analyses of paired blood and tumor samples from 1138 patients. The tumor biopsies were from metastases.

In 817 patients (approximately 80%), potentially therapeutically actionable mutations were detected. However, only 132 patients received sequence-directed therapy; 49 of these patients benefited clinically and 26 were noted to have exceptionally favorable responses.

Importantly, 160 patients (15.8%) were reported to have pathogenic germline mutations. Another important finding was that next generation sequencing derived data enabled identification of the likely primary site of the tumor.

Based on this study, the authors emphasized the importance of identification of therapeutically relevant somatic findings and also identification of germline cancer predisposing mutations.

Cobain et al. noted that different studies report identification of clinically actionable genomic changes in 40% − 94% of cases. However, only 10% − 25% of patients received therapy based on sequencing information.

12.22 Therapy-related genetic and genomic information: molecular profiling and cancer therapies

Malone et al. (2020) noted the value of molecular pathology analyses in cancer to inform diagnosis, prognosis, and sometimes therapy. They expressed optimism regarding implementation of cancer precision medicine, while considering challenges and potential solutions. They noted limitations of genotype-directed therapies.

Malone et al. emphasized expansion of molecular characterization beyond genomics and noted inclusion of transcriptomics, epigenetics, and immunophenotyping.

Their review listed specific genes noted to have cancer-associated molecular changes that are matched to specific approved therapies by FDA and European Molecular Agency. The table below lists specific genes and proteins with known cancer-related mutation and the cancer types involved. It does not, however, list the anticancer agents used in therapy.

Gene/protein	Biomarker	Cancer type
ALK (receptor tyrosine kinase)	ALK translocations	Nonsmall cell lung cancer (NSCLC)
Androgen receptor (AR)	AR expression level	Prostate cancer
BCL2 (apoptosis regulator)	Expression level	Myeloid leukemia
BCR/ABL	Fusion	Myeloid leukemia
BRAF	Mutation V600E	Melanoma, NSCLC
BRCA	Mutation germline	Breast/ovarian cancer
BRCA	Somatic mutation	Breast/ovarian cancer
CKIT	Mutations	Leukemia, gastric cancer
PDGFRA/PDGFRB	Mutations	Hematological and GI cancers
Estrogen receptor	Expression level	Breast cancer
ERBB2/HER2	Overexpression	Breast, ovarian cancer
EGFR	Exon deletion, mutation	Numerous cancers and NSCLC
FGFR2/FGFR3	Mutations	Bladder cancer
FLT3	Mutation	Myeloid cancer
IDH1/2	Mutations	Myeloid cancer
MET	Amplification	Nonsmall cell lung cancer
MSI (MLH1, MSH2, MSH16 PMS)	Mutations	Selective colorectal cancers
NTRK	Fusions	Pediatric and adult tumors

Malone et al. noted that as the numbers of identified druggable mutations increase, single gene testing is being substituted with use of multigene mutation platforms. Investigations of exome sequencing and whole genome sequencing are also being undertaken.

Driver mutations were noted to occur in 40% of tumors. However, only 10%–15% of patients were reported to be treated with a genotype matched drug.

Increasingly, circulating biomarkers including circulating tumor cells and circulating cfDNA, RNA, protein, and metabolites were being assayed.

Malone et al. noted that continued efforts are undertaken to improve variant analyses and annotation and to improve clinical predication based on variant analyses. They noted that the error rate with long-read genome sequencing was still too high. RNA sequencing did provide further information on presence of novel fusions.

Analyses revealed that protein truncating variants in ATM, BRCA1, BRCA2, CHEK2, PALB2, and breast cancer were significantly associated $P = .0001$.

Pathogenic missense variants in BRCA1, BRCA2 had risk similar to those of protein truncating variants.

Missense variants for ATM, CHECK2, TP53 were associated with breast cancer risk $P = .001$.

Protein truncating variants in BARD1, RAD51C, RAD51D, and Breast Cancer Association yielded a significance value of .05.

Studies revealed that 25 other genes considered as susceptibility genes had no significant evidence.

The specific gene mutation and estrogen and progesterone sensitivity of tumors were analyzed.

A more recently recognized gene with germline mutations predisposing to breast cancer is REQ, a member of the RecQ DNA helicase family. DNA helicases are reported to be enzymes involved in various types of DNA repair, including mismatch repair, nucleotide excision repair, and direct repair.

ATM and CHEK2 pathogenic variants were correlated with estrogen receptor positive cancer.

It is important to note that recently therapy was approved for a specific KRAS mutation by the FDA.

12.23 Synthetic lethality

This concept has been applied to cancer treatment. Kaelin (2005) noted that mutations in different genes could work together to impair cellular fitness. It was also known that rapidly dividing cells were most susceptible to DNA damage. Farmer et al. (2005) demonstrated that tumors with BRCA1 and BRCA2 mutations are associated with DNA damage. Furthermore, the PARP1 protein repairs DNA breaks. They therefore proposed the use of PARP1 inhibitors to impair repair of DNA breaks.

12.24 Cancer immunotherapy

Waldman et al. (2020) drew attention to the earlier work of Paul Ehrlich who, in 1908, proposed that neoplastic cells could be eradicated by the immune system. Waldman noted that MacFarlane Burnett in 1957 proposed cancer immunosurveillance and Lewis Thomas (1982) referred to activity of the immune system in combatting cancer.

The T-cell was noted to be a major focus in cancer immunotherapy based on its capacity to induce antigen-directed cytotoxicity.

Pardoll (2012a,b) noted that in cancerous lesions many genetic and epigenetic alterations occurred and that led to the generation of novel antigens that could be recognized by T-cells. He also noted that substances that inhibit T-cells were also produced by cancers and that these substances needed to be inhibited so that T-cells could function in destruction of tumors.

T-cell also harbor an immune checkpoint receptor that downregulates their activation. One such receptor is CTLA4. CTLA4 can outcompete the stimulatory T-cell receptors (Sharma and Allison, 2015). Another immune checkpoint protein is PD1. Pardoll promoted blockade of the immunoinhibitory receptor PD-1 in cancer immunotherapy (2012).

In considering advances in cancer immunotherapy Kruger et al. (2019) noted that the two main drivers included checkpoint inhibitors (CPI) and chimeric antigen receptor T-cells CAR-T cells. Nobel Prizes were awarded in 2018 for cancer immunotherapy.

In a review in 2020, Hegde and Chen (2020) reviewed key challenges facing immunotherapy and emphasized that those required combined efforts of basic researchers and clinicians. They noted that tumors could evade the immune systems.

Lee and Kulkarni (2019) noted that the clinical response to immunotherapy differs in different patients and different cancer types That noted that presence of specific biomarkers in tumors could predict tumor responses to immunotherapies and focused their review primarily on melanoma and small cell lung cancer.

Lee and Kulkarni noted the importance of molecular profiling of tumor cells to identify predictive biomarkers. They emphasized that determination of circulating biomarkers are valuable in assessment of the likely response to checkpoint blockade. Tumor mutational burden and transcriptome signature were important to take into account.

Kruger et al. noted that there is now evidence for the combined use in cancer therapy of chemotherapeutic agents together with blockade of the immune inhibitory receptor. They noted that checkpoint blockade therapy was necessary as some cancers promote their growth by inhibiting checkpoint mechanisms.

PD1, PDL1, and CTLA4 were noted to constitute the key checkpoint components.

PD1 also known as PDCD1 programmed cell death 1 immune-inhibitory receptor expressed in activated T−cells.

CD274 (PDCD1L1), immune inhibitory receptor ligand that is expressed by hematopoietic and nonhematopoietic cells, such as T-cells and B-cells and by various types of tumor cells.

CTLA4 cytotoxic T-lymphocyte associated protein 4 protein, which transmits an inhibitory signal to T-cells.

Initial therapeutic CPI therapies included antibodies Ipilumab. In 2019, 25 different CPIs were available.

Studies have been carried out to determine the characteristics of tumors likely to be suppressed by CPI therapy. Kruger et al. noted that there is evidence that tumors with microsatellite instability and deficient mismatch repair capacity are likely to be more susceptible to PD1 and PDL1 blockade. There is also evidence that the inflammatory signature in tumors can also provide information.

Other approaches to cancer immunotherapy include isolation and propagation of tumor infiltrating and tumor-reactive immune cells including T-cells and subsequent administration of these to patients (Fig. 12.3).

12.25 CAR-T cells

One step in response to cancer involves receptors on T-cells attaching to new antigens produced by the tumor. This attachment in combination with other activities can lead to tumor cell destruction.

Specific genetic alterations in the interior of T-cells can lead to production of receptors with improved attachment to tumor antigens. Specific domains in the modified receptor include a variable extracellular domain, including an antibody-like region and an internal costimulatory domain (Graham et al., 2018). The modified receptor CAR-T cells (chimeric artificial receptor) T-cells that have been genetically modified to target specific tumor products, for example, a specific antigen produced by the tumor but not expressed by healthy cells.

Production of the CAR-T cells involves isolation of leukocytes from blood, their culture, and stimulation to divide. They are then transduced by a viral vector to recognize a specific tumor antigen encoded by the tumor.

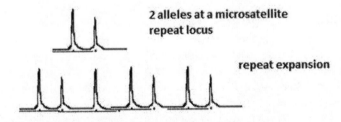

Microsatellite instability cancer predipostion

2 alleles at a microsatellite repeat locus

repeat expansion

Microsatellite instability can arise due to mutations in genes that are homologs of muts:
MLH1, MLH2
MSH2, MSH6
PMS mismatch repair system component

FIGURE 12.3

Microsatellite instability.

Kosti et al. (2018) noted that CAR-T cell immunotherapy was approved for cancer therapy by the FDA in 2017. This form of therapy was noted to have very good outcomes in the treatment of lymphoblastic leukemia and lymphoma. Successes were based in part on the development of CAR-T cells that targeted the B-cell maturation antigen.

Kosti et al. noted that the response to CAR-T cell treatment of solid tumors was variable. They noted that in solid tumors, especially tumors of epithelial origin. It was difficult to identify tumor-specific markers and antigens to target, as many were not tumor-specific and also occurred in normal tissue.

They also noted that physical barriers often limited the contact that CAR-T cells needed with tumor cells and the latter did not migrate in sufficient number to tumors cells.

In addition, specific cytokines acted to inhibit CAR-T cell migration. Within tumors there were limited interactions with cells brought about by the presence of check point molecules. However, in combined therapies with CPIs, CAR-T cells could be therapeutically more effective.

12.25.1 CAR-T cell therapy and cytokine release syndrome (cytokine storm)

Hay (2018) reviewed cytokine release syndrome and neurotoxicity that can follow and documented specific therapies that can be used to manage this syndrome. Hay noted that a deeper understanding of the pathophysiology of cytokine release syndrome and associated neurotoxicity was required.

The tumor microenvironment, including macrophage content and inflammatory response, was noted to likely be playing a role in these adverse reactions.

References

Alexandrov LB, Stratton MR: Mutational signatures: the patterns of somatic mutations hidden in cancer genomes, *Curr Opin Genet Dev* 24(100):52–60, 2014. Available from: https://doi.org/10.1016/j.gde.2013.11.014. PMID: 24657537.

Bettegowda C, Sausen M, Leary RJ, et al: Detection of circulating tumor DNA in early- and late-stage human malignancies, *Sci Transl Med* 6(224):224ra24, 2014. Available from: https://doi.org/10.1126/scitranslmed.3007094. PMID: 24553385.

Breast Cancer Association Consortium, Dorling L, Carvalho S, et al: Breast cancer risk aenes—association analysis in more than 113,000 women, *N Engl J Med* 384(5):428–439, 2021. Available from: https://doi.org/10.1056/NEJMoa1913948. PMID: 33471991.

Chatterjee N, Walker GC: Mechanisms of DNA damage, repair, and mutagenesis, *Environ Mol Mutagen* 58(5):235–263, 2017. Available from: https://doi.org/10.1002/em.22087. PMID: 28485537.

Christensen JG, Olson P, Briere T, et al: Targeting Kras(g12c)—mutant cancer with a mutation-specific inhibitor, *J Intern Med* 288(2):183–191, 2020. Available from: https://doi.org/10.1111/joim.13057. PMID: 3217637.

Ciccia A, Elledge SJ: The DNA damage response: making it safe to play with knives, *Mol Cell* 40(2):179–204, 2010. Available from: https://doi.org/10.1016/j.molcel.2010.09.019. PMID: 20965415.

Cobain EF, Wu YM, Vats P, et al: Assessment of clinical benefit of integrative genomic profiling in advanced solid tumors, *JAMA Oncol* 7(4):525–533, 2021. Available from: https://doi.org/10.1001/jamaoncol.2020.7987. PMID: 33630025.

Davis AA, Iambs WT, Chan D, et al: Early assessment of molecular progression and response by whole-genome circulating tumor DNA in advanced solid tumors, *Mol Cancer Ther* 19(7):1486–1496, 2020. Available from: https://doi.org/10.1158/1535-7163.MCT-19-1060. Epub May 5, 2020. PMID: 32371589.

Dubard Gault M, Mandelker D, DeLair D, et al: Germline SDHA mutations in children and adults with cancer, *Cold Spring Harb Mol Case Stud.* 4(4):a002584, 2018. Available from: https://doi.org/10.1101/mcs.a002584. Print 2018 Aug.PMID: 30068732.

Dyson NJ: RB1: a prototype tumor suppressor and an enigma, *Genes Dev* 30 (13):1492–1502, 2016. Available from: https://doi.org/10.1101/gad.282145.116. PMID: 27401552.

Farmer H, McCabe N, Lord CJ, et al: Targeting the DNA repair defect in BRCA mutant cells as a therapeutic strategy, *Nature.* 434(7035):917–921, 2005. Available from: https://doi.org/10.1038/nature03445. PMID: 15829967.

Filbin M, Monje M: Developmental origins and emerging therapeutic opportunities for childhood cancer, *Nat Med* 25(3):367–376, 2019. Available from: https://doi.org/10.1038/s41591-019-0383-9. PMID: 30842674.

Fishel R, Kolodner RD: Identification of mismatch repair genes and their role in the development of cancer, *Curr Opin Genet Dev* 5(3):382–395, 1995. Available from: https://doi.org/10.1016/0959-437x(95)80055-7. PMID: 7549435.

Fletcher, M: Sequencing the secrets of the cancer genome. S10 Nature Milestones December 2020, <http://www.nature.com/collections/cancer-milestones>.

Ghimessy A, Radeczky P, Laszlo V, et al: Current therapy of KRAS-mutant lung cancer, *Cancer Metastasis Rev* 39(4):1159–1177, 2020. Available from: https://doi.org/10.1007/s10555-020-09903-9. PMID: 32548736.

Graham C, Hewitson R, Pagliuca A, Benjamin R: Cancer immunotherapy with CAR-T cells—behold the future, *Clin Med (Lond)* 18(4):324–328, 2018. Available from: https://doi.org/10.7861/clinmedicine.18-4-324. PMID: 30072559.

Gröbner SN, Worst BC, Weischenfeldt J, et al: The landscape of genomic alterations across childhood cancers, *Nature.* 555(7696):321–327, 2018. Available from: https://doi.org/10.1038/nature25480. PMID: 29489754.

Haferlach T, Schmidts I: The power and potential of integrated diagnostics in acute myeloid leukaemia, *Br J Haematol.* 188(1):36–48, 2020. Available from: https://doi.org/10.1111/bjh.16360. Epub December 6, 2019. PMID: 31808952.

Hay KA: Cytokine release syndrome and neurotoxicity after CD19 chimeric antigen receptor-modified (CAR-) T cell therapy, *Br J Haematol* 183(3):364–374, 2018. Available from: https://doi.org/10.1111/bjh.15644. PMID: 30407609.

Hegde PS, Chen DS: Top 10 challenges in cancer immunotherapy, *Immunity* 52(1):17–35, 2020. Available from: https://doi.org/10.1016/j.immuni.2019.12.011. PMID: 31940268.

Kaelin WG: The concept of synthetic lethality in the context of anticancer therapy, *Nat Rev Cancer* 5(9):689–698, 2005. Available from: https://doi.org/10.1038/nrc1691. PMID: 16110319.

Kamilaris CDC, Stratakis CA: Multiple Endocrine Neoplasia Type 1 (MEN1): an update and the significance of early genetic and clinical diagnosis, *Front Endocrinol (Lausanne)* 10:339, 2019. Available from: https://doi.org/10.3389/fendo.2019.00339. eCollection 2019.PMID: 31263451.

Kim H, D'Andrea AD: Regulation of DNA cross-link repair by the Fanconi anemia/BRCA pathway, *Genes Dev* 26(13):1393–1408, 2012. Available from: https://doi.org/10.1101/gad.195248.112. PMID: 22751496.

Knudson AG: Mutation and cancer: statistical study of retinoblastoma, *Proc Natl Acad Sci U S A* 68(4):820–823, 1971. Available from: https://doi.org/10.1073/pnas.68.4.820. PMID: 5279523.

Kosti P, Maher J, Arnold JN: Perspectives on chimeric antigen receptor T-cell immunotherapy for solid tumors, *Front Immunol.* 9:1104, 2018. Available from: https://doi.org/10.3389/fimmu.2018.01104. eCollection 2018.PMID: 29872437.

Krokan HE, Bjørås M: Base excision repair, *Cold Spring Harb Perspect Biol* 5(4):a012583, 2013. Available from: https://doi.org/10.1101/cshperspect.a012583. PMID: 23545420.

Kruger S, Ilmer M, Kobold S, et al: Advances in cancer immunotherapy 2019—latest trends, *J Exp Clin Cancer Res* 38(1):268, 2019. Available from: https://doi.org/10.1186/s13046-019-1266-0. PMID: 31217020.

Kunkel TA: Celebrating DNA's repair crew, *Cell.* 163(6):1301–1303, 2015. Available from: https://doi.org/10.1016/j.cell.2015.11.028. PMID: 26638062.

Larson MH, Pan W, Kim HJ, et al: A comprehensive characterization of the cell-free transcriptome reveals tissue- and subtype-specific biomarkers for cancer detection, *Nat Commun* 12(1):2357, 2021. Available from: https://doi.org/10.1038/s41467-021-22444-1. PMID: 33883548.

Lawrence MS, Stojanov P, Mermel CH, et al: Discovery and saturation analysis of cancer genes across 21 tumour types, *Nature* 505(7484):495–501, 2014. Available from: https://doi.org/10.1038/nature12912. Epub January 5, 2014. PMID: 24390350.

Leach FS, Polyak K, Burrell M, et al: Expression of the human mismatch repair gene hMSH2 in normal and neoplastic tissues, *Cancer Res* 56(2):235–240, 1996. PMID: 8542572.

Lee EY, Kulkarni RP: Circulating biomarkers predictive of tumor response to cancer immunotherapy, *Expert Rev Mol Diagn.* 19(10):895–904, 2019. Available from: https://doi.org/10.1080/14737159.2019.1659728. Epub September 10, 2019. PMID: 31469965.

Ley TJ, Mardis ER, Ding L, et al: DNA sequencing of a cytogenetically normal acute myeloid leukaemia genome, *Nature.* 456(7218):66–72, 2008. Available from: https://doi.org/10.1038/nature07485. PMID: 18987736.

Li GM, Modrich P: Restoration of mismatch repair to nuclear extracts of H6 colorectal tumor cells by a heterodimer of human MutL homologs, *Proc Natl Acad Sci U S A.* 92(6):1950–1954, 1995. Available from: https://doi.org/10.1073/pnas.92.6.1950. PMID: 7892206.

Malone ER, Oliva M, Sabatini PJB, et al: Molecular profiling for precision cancer therapies, *Genome Med* 12(1):8, 2020. Available from: https://doi.org/10.1186/s13073-019-0703-1. PMID: 31937368.

Mandelker D, Donoghue M, Talukdar S, et al: Germline-focussed analysis of tumour-only sequencing: recommendations from the ESMO Precision Medicine Working Group, *Ann Oncol* 30(8):1221–1231, 2019. Available from: https://doi.org/10.1093/annonc/mdz136. PMID: 31050713.

Mattox AK, Yan H, Bettegowda C: The potential of cerebrospinal fluid-based liquid biopsy approaches in CNS tumors, *Neuro Oncol.* 21(12):1509–1518, 2019. Available from: https://doi.org/10.1093/neuonc/noz156. PMID: 31595305.

Ma X, Liu Y, Liu Y, Alexandrov LB, et al: Pan-cancer genome and transcriptome analyses of 1,699 paediatric leukaemias and solid tumours, *Nature.* 555(7696):371–376, 2018. Available from: https://doi.org/10.1038/nature25795. PMID: 29489755.

Mirabello L, Zhu B, Koster R, et al: Frequency of pathogenic germline variants in cancer-susceptibility genes in patients with osteosarcoma, *JAMA Oncol* 6(5):724–734, 2020. Available from: https://doi.org/10.1001/jamaoncol.2020.0197. PMID: 32191290.

Nakagawa H, Fujita M: Whole genome sequencing analysis for cancer genomics and precision medicine, *Cancer Sci* 109(3):513–522, 2018. Available from: https://doi.org/10.1111/cas.13505. PMID: 29345757.

Pardoll DM: Immunology beats cancer: a blueprint for successful translation, *Nat Immunol.* 13(12):1129–1132, 2012a. Available from: https://doi.org/10.1038/ni.2392. PMID: 23160205.

Pardoll DM: The blockade of immune checkpoints in cancer immunotherapy, *Nat Rev Cancer* 12(4):252–264, 2012b. Available from: https://doi.org/10.1038/nrc3239. PMID: 22437870.

Priestley P, Baber J, Lolkema MP, et al: Pan-cancer whole-genome analyses of metastatic solid tumours, *Nature* 575(7781):210–216, 2019. Available from: https://doi.org/10.1038/s41586-019-1689-y. Epub October 23, 2019. PMID: 31645765.

Pui CH, Nichols KE, Yang JJ: Somatic and germline genomics in paediatric acute lymphoblastic leukaemia, *Nat Rev Clin Oncol* 16(4):227–240, 2019. Available from: https://doi.org/10.1038/s41571-018-0136-6. PMID: 30546053.

Rodriguez-Galindo C, Orbach DB, VanderVeen D: Retinoblastoma, *Pediatr Clin North Am.* 62(1):201–223, 2015. Available from: https://doi.org/10.1016/j.pcl.2014.09.014. PMID: 25435120.

Roschewski M, Phelan JD, Wilson WH: Molecular classification and treatment of diffuse large B-cell lymphoma and primary mediastinal B-cell lymphoma, *Cancer J* 26(3):195–205, 2020. Available from: https://doi.org/10.1097/PPO.0000000000000450. PMID: 32496453.

Shannon-Lowe C, Rickinson AB, Bell AI: Epstein–Barr virus-associated lymphomas, *Philos Trans R Soc Lond B Biol Sci* 372(1732):20160271, 2017. Available from: https://doi.org/10.1098/rstb.2016.0271. PMID: 28893938.

Sharma P, Allison JP: The future of immune checkpoint therapy, *Science* 348(6230):56–61, 2015. Available from: https://doi.org/10.1126/science.aaa8172. PMID: 25838373.

Stewart CM, Kothari PD, Mouliere F, et al: The value of cell-free DNA for molecular pathology, *J Pathol.* 244(5):616–627, 2018. Available from: https://doi.org/10.1002/path.5048. Epub March 12, 2015. PMID: 29380875.

Syngal S, Brand RE, Church JM, et al: ACG clinical guideline: genetic testing and management of hereditary gastrointestinal cancer syndromes, *Am J Gastroenterol* 110(2):223–262, 2015. Available from: https://doi.org/10.1038/ajg.2014.435. PMID: 25645574 quiz 263.

Tate JG, Bamford S, Jubb HC, et al: COSMIC: the catalogue of somatic mutations in cancer, *Nucleic Acids Res* 47(D1):D941–D947, 2019. Available from: https://doi.org/10.1093/nar/gky1015. PMID: 30371878.

Uprety D, Adjei AA: KRAS: from undruggable to a druggable cancer target, *Cancer Treat Rev* 89:102070, 2020. Available from: https://doi.org/10.1016/j.ctrv.2020.102070. PMID: 32711246.

Vogelstein B, Papadopoulos N, Velculescu VE, et al: Cancer genome landscapes, *Science* 339(6127):1546–1558, 2013. Available from: https://doi.org/10.1126/science.1235122. PMID: 23539594.

Waldman AD, Fritz JM, Lenardo MJ: A guide to cancer immunotherapy: from T cell basic science to clinical practice, *Nat Rev Immunol* 20(11):651–668, 2020. Available from: https://doi.org/10.1038/s41577-020-0306-5. PMID: 32433532.

Wang HW, Balakrishna JP, Pittaluga S, Jaffe ES: Diagnosis of Hodgkin lymphoma in the modern era, *Br J Haematol* 184(1):45–59, 2019. Available from: https://doi.org/10.1111/bjh.15614. PMID: 30407610.

Waszak SM, Northcott PA, Buchhalter I, et al: Spectrum and prevalence of genetic predisposition in medulloblastoma: a retrospective genetic study and prospective validation in a clinical trial cohort, *Lancet Oncol* 19(6):785–798, 2018. Available from: https://doi.org/10.1016/S1470-2045(18)30242-0. PMID: 29753700.

Young RM, Phelan JD, Wilson WH, Staudt LM: Pathogenic B-cell receptor signaling in lymphoid malignancies: new insights to improve treatment, *Immunol Rev* 291(1):190–213, 2019. Available from: https://doi.org/10.1111/imr.12792. PMID: 31402495.

Zhang J, Walsh MF, Wu G, et al: Germline mutations in predisposition genes in pediatric cancer, *N Engl J Med* 373(24):2336–2346, 2015. Available from: https://doi.org/10.1056/NEJMoa1508054. PMID: 26580448.

Zhan T, Rindtorff N, Betge J, et al: CRISPR/Cas9 for cancer research and therapy, *Semin Cancer Biol* 55:106–119, 2019. Available from: https://doi.org/10.1016/j.semcancer.2018.04.001. PMID: 29673923.

CHAPTER 13

Benefits of the incorporation of genomic medicine in clinical practice

Green et al. (2020) reiterated one of the major goals of the Human Genome Project, which was to demonstrate how genomic information could be effectively used in clinical care. They emphasized that a major focus of the Genome Project was to understand biology to enhance disease knowledge.

Green et al. emphasized that through technological improvements the cost of genome sequencing has been dramatically decreased. Advances have also been made in functional genomics that include studies designed to identify genomic segments involved in the regulation of gene expression.

Green et al. emphasized that continued efforts are required to implement use of genomic medicine in clinical care.

13.1 Genetic and genomic studies in congenital anomalies and/or neurodevelopmental anomalies

13.1.1 Microarray analyses

Waggoner et al. (2018) reviewed specific genetic tests and noted that chromosomal microarray was recommended as a first-tier test in evaluation of individuals with congenital anomalies and/or neurodevelopmental disabilities. The goal of their study was the benefit of additional tests particularly when microarray testing yielded normal results.

They noted that microarray platforms include arrays with chromosome, regionally assigned oligonucleotide probes, or positionally assigned single nucleotide polymorphic markers (SNPs). These microarray platforms are particularly useful for identifying chromosome deletions and duplications. SNP microarrays can detect regions of homozygosity that may be valuable in assessing possibility of some recessive disorders. SNP microarrays may be useful in detecting some forms of uniparental disomy.

Waggoner et al. emphasized that it is important to be aware of the cytogenetic anomalies that are not detected in microarray studies. Chromosome anomalies not detected include balanced translocations, inversions, mosaicism, and deletion involving a small number of nucleotides in sections of the genome between the oligonucleotide or SNP markers. Microarray studies can also not readily detect mosaicism, the presence of one or more chromosomally different cell populations. Impaired capability of detecting balanced chromosome rearrangements is a shortcoming as these rearrangements can interrupt transcription or regulation of transcription.

Savatt and Myers (2021) reviewed genetic testing in neurodevelopmental disorders. They emphasized that determination of an underlying genetic etiology can impact clinical management and treatment, direct patients to treatment-specific resources, and clarify recurrence risk.

They noted that neurodevelopmental disabilities constitute the most prevalent chronic medical condition in the pediatric population. Neurodevelopmental-based disabilities were noted to be characterized by developmental defects in cognition, language, behavior, and/or motor skills.

Savatt and Myers referred to useful information resources regarding clinical features, genetic and genomic alteration in specific disorders. Links to these resources are listed below:

https://search.clinicalgenome.org/kb/gene-validity/
https://dosage.clinicalgenome.org/
https://www.deciphergenomics.org

They reviewed utility of chromosome microarray testing, exome sequencing, and specific testing for Fragile-X syndrome testing through analysis of the FMR1 CGG repeat.

Savatt and Myers documented important tests in neurodevelopmental disorders. These include detection of segmental genomic copy number variants on microarrays. Region assessed for copy number changes on microarrays were reported to range between 250 and 400 kb. The diagnostic yield on microarrays testing was reported to be 15%.

13.1.2 Exome sequencing

This can detect nucleotide sequence changes in protein coding regions of the genome that together constitute 1.5% − 2% of the content of the genome. Microarray capacity to detect segmental deletions or duplications in the genome is limited. The diagnostic yield of exome sequencing in the neurodevelopmental disorders was reported by Waggoner et al. to range between 25% and 38%.

Single gene testing and testing of disease-relevant panel of genes are also utilized. However, 11% of panel negative cases were reported to subsequently have been reported to have abnormalities on exome sequencing.

In both clinical microarray testing and exome sequencing variants of uncertain significance were often detected. Importantly in the course of time, as data on additional patients accrue, it may be possible to later determine whether such variants are benign or pathologically significant.

In testing of CGG repeat expansions in the FMR1 gene, the normal repeat size was listed as 44 repeats, in premutation carriers repeat size was 55–200 repeats, and full mutation repeat size was greater than 200 repeats.

Genome-wide sequencing in acutely ill infants has been demonstrated to achieve rapid diagnosis and to improve managements (Dimrock et al., 2020).

13.2 Rare disease medicine

Alkuraya (2021) traced genomic developments in identifying genes responsible for Mendelian disorders noting that this commences with positional mapping of a disease to a specific chromosome region through analysis of specific markers with known chromosome position assignments.

The availability of polymorphic markers assigned to specific chromosome regions and ability to follow segregation of those markers in individuals within families where some members had inherited diseases, opened the way to positional cloning of genes.

Availability of the complete genome sequence and search for genome sequence changes in individuals with specific genetic disease further expanded disease gene identification. Of particular value is emergence of methods to identify carriers of severe diseases that are recessively inherited in order to expand the reproductive options in carrier parents.

Alkuraya noted that the identification of carrier parents is of particular value in populations where cousin marriages are common. Carrier testing is also of value in populations where specific deleterious mutations have reached high frequency, for example, thalassemia in Cyprus.

13.3 Carrier screening

Antonarakis (2019) noted that increases in knowledge and information on recessive disease-associated genetic variants have led to increased possibilities for screening for carriers of specific genetic disorders.

He reviewed work of Stamatoyannopoulos (1972) in counseling families with sickle cell disease in a particular region of Greece where 1 in 100 neonates were born with sickle cell anemia. Antonarakis noted that other early screening programs involved families at risk in the Mediterranean area who were at risk for beta thalassemia. Another early screening program involved carrier screening for Tay – Sachs disease in Ashkenazi Jewish families and screening for cystic fibrosis in European families.

The main early objective in carrier screening were confined to counseling of at-risk individuals. In recent years options for at-risk parents have expanded due to possibilities for preimplantation genetic testing.

Early screening for specific genetic diseases involved studies on proteins such as hemoglobin altered charge and altered electrophoresis (Pauling et al., 1949); analysis of altered properties of enzymes, for example, altered heat sensitivity of hexosaminidase in Tay−Sachs disease, (Kaback et al., 1993); screening for altered physiological parameters, which increased sweat chloride levels in cystic fibrosis (Gibson and Cooke, 1959).

As genetic technologies evolved with increased elucidation of specific nucleotide changes leading to disease and increased capabilities of DNA sequence analysis, molecular screening for carrier detection has increased.

Antonarakis emphasized that novel genetic associations and nucleotide changes for specific diseases have continually been discovered as DNA sequencing is increasingly utilized in clinical medicine. It is of course important to establish if disease-causing mutations in a patient are de novo or inherited and whether inherited mutations are dominant or recessive. Many of the newly characterized recessive disease-causing mutations occur in populations where marriage between cousins or other close family relations are common.

In a number of countries a standard set of disease-causing variants are screened when screening is offered. Antonoarakis noted that the Expanded Carrier Screening program uses DNA screening to examine 417 pathogenic variants in 94 genes. One aspect for selection of variants for screening was that the disease resulting from the observed genetic changes should be severe.

Lord et al. (2019) described results of a sequencing study carried out in fetal DNA when fetuses were found on ultrasound to have physical structural abnormalities Their study involved analysis of data on 620 fetuses with structural body abnormalities and no chromosomal aneuploidy or copy number variants.

DNA analyses were carried out on fetal and parental DNA. Diagnostic genetic variants were identified in 52 fetuses. The diagnostic variants were reported to be useful in distinguishing between syndrome disorders that involved more than one organ and/or body structure and nonsyndromic disorders. Diagnostic sequence variants associated with multisystem abnormalities were found in 22 fetuses. Diagnostic variants were reported to be least common in fetuses with increased nuchal translucency as the only abnormality.

13.4 Diagnoses and management in disorders with phenotypic abnormalities

13.4.1 Human Phenotype Ontology database

Köhler et al. (2021) noted that the Human Phenotype Ontology (HPO) database was launched in 2018 to define standard terms to be used to document human disease phenotype-associated abnormalities. Input into generation of this database comes from clinicians and researchers. Ontologies for specific categories of disease are included and subcategories have also been developed.

In the 2021 review HPO was reported to cover 8000 rare disease and has 15,247 terms. There are categories relevant to 18 different body systems and additional categories include prenatal relevant categories, metabolism, growth, and neoplasms. Subcategories were developed under neurology to include epilepsy seizure types and electroencephalogram types. Subcategories were also developed for immune system disorders and for newborn abnormalities.

Search on a specific disease will yield information on the body systems and types of tissue showing abnormalities and the frequency of specific abnormalities.

13.4.2 Other resources

It is interesting to note that on-line templates exist for capture of phenotype information, family history, and medical information suitable for automated analysis; these include Phenotips (Girdea et al., 2013) and PhenoDB (Hamosh et al., 2013).

PhenoDB is useful for storage of standardized phenotype and pedigree information and for storage and interaction with DNA sequence files.

13.5 Databases of importance in searching for gene, genotype phenotype correlations

OMIM, Online Mendelian Inheritance in Man.

HPO, standardized vocabulary of phenotypic abnormalities encountered in human disease hpo.jax.org/.

dbGAP NCBI, the database of Genotypes and Phenotypes (dbGaP) was developed to archive and distribute the data and results from studies that have investigated the interaction of genotype and phenotype in humans.

Phenome Central, PhenomeCentral is a repository for clinicians and scientists working in the rare disorder community (http://www.phenomecentral.org).

Database with templates useful for collecting standardized phenotype information in a manner suitable for automated analyses include Phenotip (Girdea et al., 2013).

PhenoCB is a database useful for storage of standardized phenotype and pedigrees information for storage and correlation with sequence information (Hamosh et al., 2013).

Clinvar NCBI, ClinVar is a freely accessible, public archive of reports of the relationships among human variations and phenotypes, with supporting evidence.

DECIPHER is a web-based resource and database of genomic variation data from analysis of patient DNA. It documents submicroscopic chromosome abnormalities (microdeletions and duplications) and pathogenic sequence variants decipher.sanger.ac.uk

Orphanet, The ORPHANET is a database dedicated to providing information on rare diseases and orphan drugs (http://www.orpha.net).

Exomiser Welcome Sanger Institute, the Exomiser is a Java program that finds potential disease-causing variants from whole-exome or whole-genome sequencing data.

13.6 Genomic studies to guide diagnosis and therapy in epilepsy

13.6.1 Metabolic pathways and epilepsy

Sharma and Prasad (2017) considered pathophysiology in metabolic pathways leading to epilepsy in children. They noted evidence that GABA (Gamma aminobutyric acid) functions as an excitatory neurotransmitter from preterm period on. They noted that many of the inborn errors of metabolism impaired brain metabolism and included metabolite transport and utilization. In addition, metabolic coupling between astrocytes and neurons can be impaired. Also neurotransmitter signaling, cerebral blood flow, and passage across the blood brain barrier can be impaired. In some inborn errors of metabolism neurotoxic compounds are produced.

Epileptic encephalopathies were noted to frequently have distinct encephalographic features. Inborn errors of metabolism-associated epilepsies were noted to sometimes present at different ages. Those that present in early infancy were noted to include the following.

PNPO deficiency, epilepsy deficiency of peridoxamine-5-phosphate oxidase. These epilepsies are often Vitamin B6 responsive. ALDH7A1 (aldehyde dehydrogenase 7 family member A1) mutations may also lead to vitamin B6 responsive epilepsies.

Folate-responsive seizures are also important to recognize. These may result from deficiency of the proton-coupled folate transporter, deficiency of the folate receptor alpha, leading to an isolated cerebral folate deficiency and can lead to intractable seizures, as reported by Pope et al. (2019).

GLUT1 (SLC2A1) defects lead to impaired transport of glucose across the blood brain barrier and can lead to early severe seizures.

GOT2 glutamate-oxaloacetic transaminase 2 defects have been shown to lead to seizures. These lead to low serine levels and there is evidence that these may be responsive to treatment with serine and Vitamin B6.

Nonketotic hyperglycinemia-related seizures may present early. These arise due to defects in the mitochondrial glycine cleavage system.

ALDH7A1 aldehyde dehydrogenase 7 family member A1 was reported to undergo mutations, pyridoxine-dependent epilepsy.

van Karnebeek et al. (2018) noted that inborn errors of metabolism are not common causes of epilepsy, but that their identification was important as patients with these disorders may require specific therapies in addition to seizure medications. They emphasized that inborn errors of metabolism only account for approximately 1.1% of cases with epileptic seizures.

Van Karnebeek et al. reviewed 268 inborn errors of metabolism in which seizures occur. Importantly they noted that 74 of these conditions were treatable.

They reviewed specific diagnostic tests to identify inborn errors of metabolism. In addition to specific laboratory tests, clues to the presence of some of these disorders could be obtained in clinical examination. Important clinical assessment included family history, evidence of parental consanguinity, history of seizures, occurrence of seizures following fasting, history of developmental regression. In addition, it is important during clinical examination to note whether or not the following are present: dysmorphic features, organomegaly, ophthalmologic abnormalities, unusual body odor, history of acidotic episodes.

Van Karnebeek et al. also documented the life stage at which epileptic seizures due to inborn errors most commonly occur. In addition, the specific inborn error of metabolism was documented along with evidence as to whether specific treatment was available.

Disorder category	Numbers of disorders	Specific treatment possibility
Amino acid disorders	43	12
Defects in sterols, bile acids	7	1
Fatty acids, ketones, carnitine	9	6
Carbohydrate disorders	8	4
Lysosomal storage	9	1
Metals, metallothionine	5	3
Mitochondrial	26	7
Neurotransmitters	4	4
Peroxisomal	7	1
Purines and pyrimidine	5	2
Urea cycle	9	9
Vitamins and cofactors	23	12
Heme	1	1
Glycans, glycolipids	63	1

13.7 Diagnostic testing for inborn errors of metabolism leading to seizures

Testing on blood included metabolic panel analysis with measurements of glucose, anion gap. Liver transaminases, alkaline phosphate.

Blood gasses and lactate and pyruvate levels.

Plasma amino acids, plasma acylcarnitine, copper ceruloplasmin homocysteine.

Urine, ketones, purines, pyrimidines, creatine and metabolites, organic acids, orotic acid, oligosaccharides, sialic acid, sulfocysteine.

Blood enzymes: glucocerebrosidase, arylsulfatase, biotinidase, very long chain fatty acids.

Cerebrospinal fluid: CSF/plasma glucose ratio, lactate, pyruvate, amino acids, biogenic amines, tetrahydrobiopterin BH4, tetrahydrofolate.

Genomes, DNA sequencing, gene panel analysis, transcriptome analysis testing for alternate transcripts.

It is important to address vitamin-responsive epilepsies, pyridoxine phosphate-dependent seizures, folinic acid-responsive seizures, biotin and biopterin-related seizures.

In 2016 the EpiPM consortium reported a road map for precision medicine in epilepsies. They noted that in addition to prioritizing specific treatments, genetic studies also provided evidence for avoidance of specific epilepsy treatments. In epilepsy associated with POLG1 mutations, treatment with valproic acid could lead to acute liver failure. In addition, individuals with the HLAB*15.02 allele carbamazepine treatment can lead to adverse reactions. This variant frequently occurs in individuals of Asian descent. The Epi consortium noted evidence for implication of pathogenic variants in the specific genes that could potentially form the basis of targets for precision medicine.

13.8 Epilepsies, genetics, mechanisms, and therapy

Perucca and Perucca (2019) noted that growing insights into underlying mechanism in epilepsy had therapeutic implication for treatment. Important examples include use of ketogenic diet in cases of GLUT1 deficiency and use of pyridoxine in treatment of specific forms of epilepsy. They noted that selection of antiepileptic medication in sodium channel defects was complicated because the specific gene mutation had important implications in choice of therapy. It is important to establish if the specific gene mutation leads to a loss of function or a gain of function in the specific sodium channel.

Specific resources have been established to guide physicians in the treatment of epilepsies. These include the EpiPM consortium and the Network for Rare Epilepsies NETRE.

von Stülpnagel et al. (2021) published data on gene defects leading to epilepsy where evidence has been gathered relating to specific therapies. These include epilepsies induced by pathogenic mutations in sodium channels SCN1A, SCN2A PCDH19 (protocadherin 19) and also epilepsies due to pathogenic defects in POLG DNA polymerase gamma, catalytic subunit of mitochondrial polymerase.

CDKL5 cyclin-dependent kinase like 5, defect associated with West syndrome in infants.

FOXG1 forkhead box G1 transcription factor.

GRIN2A glutamate ionotropic receptor NMDA type subunit 2A.

COL4A ½ components of basal membranes, defects in porencephaly.

SMARCA2 SWI/SNF-related, matrix-associated, actin-dependent regulator of chromatin, subfamily a, member 2.

Williams syndrome defects in chromosome 7q11.23.

SYNGAP1 synaptic Ras GTPase activating protein 1. Located in the postsynaptic membrane to regulate synaptic plasticity and neuronal homeostasis.

SYN1 synapsin 1. Located in the postsynaptic membrane regulates synaptic plasticity and neuronal homeostasis.

This group developed a project PATRE (patients-based phenotype and evolution of therapy for rare epilepsies). The NETRE group noted that during the 15 years prior to 2021, genetic etiologies had been identified for more than half of the epilepsy types. However, they also noted that personalized therapeutic approaches were still being sorted.

13.9 Epilepsy genetics, genomics, and relevance to therapy

Hebbar and Mefford (2020) reviewed epilepsy genomics and genetics. They defined developmental epileptic encephalopathies as early onset of severe and refractory seizures associated with developmental delay or regression. The prognosis of these disorders was noted to be poor.

They emphasized progress in identifying genes involved in these heterogeneous conditions but also noted that despite advances in genetic testing, genetic diagnosis was not determined in 50% of the patients.

Overall genetic epilepsies were reported to account for 30% of cases of epilepsy. Genetic testing procedures offered include analyses on gene panels that may involve hundreds of genes, that is, gene sequencing, most frequently exome sequencing. They also noted that in some cases analysis of genomic copy number variants need to be taken into account.

Hebbar and Mefford documented genes with pathogenic variants discovered to case developmental epileptic encephalopathies. These genes frequently encode components of ion channels, or genes that encode products involved in neuronal excitability, or genes that encode products involved with neuronal inhibition. More recently, gene abnormalities in epilepsy were found to encode chromatin remodeling factors, signaling molecules that function in cells. Other genes involved in developmental epileptic encephalopathy were found to encode metabolic enzymes, transcription factors, and genes that encode products that function in mitochondria.

13.9.1 Genes involved in developmental epileptic encephalopathies

Steward et al. (2019) reported that more than 135 genes have been reported to have pathogenic variants that lead to developmental epileptic encephalopathies.

Genes reported by several groups involved the following: calcium ion channels, potassium ion channels, sodium ion channels, neurotransmitter-related genes, chromatin modeling genes.

13.9.2 Chromatin modeling genes

SMARCC2, this gene product acts as actin-dependent regulator of chromatin subfamily c member 2.

STAG2 encodes a subunit of the cohesin complex.

ACTL6B, actin-related proteins serve as a subunit of BAF, represses chromatin mediated transcription.

13.9.3 Dravet syndrome

This is a specific early infantile type of epileptic encephalopathy associated particularly with severe and prolonged seizures that can be induced by elevated temperature. This disorder is sometimes described as severe myotonic seizures of infancy. Steward et al. (2019) reported that 89% of cases of Dravet syndrome has mutation in SCN1A. Gene defects in this disorder sometimes involve SCN2A, SCN2B, SCN8A, GABRG2, GABRA1, HCN1.

Other genes reported to be involved in Dravet syndrome include: CHD2 chromatin organization modifier chromodomain helicase DNA binding protein; PCDH19 a calcium-dependent adhesion protein expressed in brain; SLC25A42 solute carrier family 25 member 42, transported across the inner mitochondrial membrane; and ATPA1 ATPase, Na+/K+ transporting, alpha 1 polypeptide.

Balciuniene et al. (2019) published results of genetic screening carried out on 151 children with epilepsy between 2016 and 2018. The analysis included initial panel gene screening and copy number variant analysis. In 16 children panel screening yielded positive results. Children who had negative panel screening underwent exome screening and four children received positive results. Therefore panel screening and exome sequencing yielded diagnostic genetic testing results in 20 children.

The author reported that diagnostic yield was highest in probands with epilepsy onset before 12 months of age. Expansion of gene examined in the panel screening subsequently yielded genetic diagnostic results in eight children. In total pathogenic mutations associated with epilepsy occurred in 28 out of 151 children.

13.10 Common epilepsies

Koeleman (2017) reviewed heritable basis of common epilepsies and reported that the genes that have pathogenic variants that lead to monogenic epilepsies can also have variants that increase risk for common epilepsies. In addition,

Koeleman noted that the genetic basis for common epilepsies was likely polygenic and heterogeneous. Koeleman noted that monogenic mutations do not explain a significant proportion of sporadic cases of epilepsy and that associated genetic factors may be of low effect size. In addition, certain genetic variants may modify the effects of pathogenic variants.

Common epilepsies were noted to occur in 1 of 200 individuals and characterized by recurrent seizures. They were noted to include generalized epilepsy, juvenile myoclonic epilepsy, idiopathic absence epilepsy, idiopathic focal epilepsy, and temporal lobe epilepsy.

Koeleman also described febrile seizures and evidence for association with variants in SCN1A and SCN2A.

13.10.1 Genomic copy number variants and epilepsies

Helbig (2015) reported that recurrent microdeletions at 15q13.3, 16p13.11, and 15q11.2 were found to be relevant risk factors for nonfamilial generalized epilepsy.

Coppola et al. (2019) studied 1225 patients diagnosed with "epilepsy plus," which they defined as epilepsy with comorbid features, including intellectual disability, neurological, and nonneurological features. One hundred and twenty patients were found to have at least one pathogenic copy number variant.

13.11 Cerebral palsy and genetic factors

Very early description of a condition that resembles what is now known as cerebral palsy was noted to have been described by Hippocrates by Panteliadis et al. (2013). The concept of cerebral palsy was noted to have been proposed by John Little who in 1863 gave a lecture to the Obstetrical Society of London, in which he proposed that injuries to the nervous system in infants could result in spasticity. This disorder became known as Little's disease. In 1887 William Osler published information on a spastic disorder in children that he referred to as cerebral palsy.

Oskoui et al. (2013) carried out comprehensive analyses of 49 reports of cerebral palsy in children born in 1985 or later. Their analyses revealed that the highest prevalence of cerebral palsy occurred in children who had birth weights between 1000 and 1499 g, and based on gestational age the highest prevalence occurred in children reported to have been born before 28 weeks. Worldwide they determined that cerebral palsy occurred in 2 − 11 per 1000 births.

13.12 Typical cerebral palsy

Fahey et al. (2017) reviewed genetic factors in cerebral palsy. They noted that in approximately one-third of cases there was no evidence of prematurity and/or hypoxic ischemic injury.

On the basis of initial studies they noted that the genetic architecture was likely to be complex. However, they postulated that development of new insights would lead to identification of new targets for therapy. Detailed recent analyses were reported to provide evidence that acute intra-partum hypoxia led to fewer than 10% of cases of cerebral palsy.

They cited evidence that genetic factors may contribute to 30% of cases of cerebral palsy. DNA defects contributing to cerebral palsy developments were noted to include nuclear DNA defects inherited or de novo, mitochondrial abnormalities, and epigenetic abnormalities.

Fahey et al. defined cerebral palsy as permanent disorders of movement and posture attributable to defects in the developing brain. They emphasized further that various factors also affect the impact of genetic mutations. Highly damaging mutations may be sufficient to lead to cerebral palsy. In some cases, however, other gene mutations and additive factors may have impact. These could be environmental factors or additional genetic changes. In some cases, cerebral palsy may be due to the cumulative effects of less damaging mutations in several genes.

Fahey et al. considered potential molecular pathway involved in the etiology of cerebral palsy and noted evidence that they included synaptic function, cortical cytoarchitecture and circuitry, neuron glia signaling, and neuroinflammation.

However, it is very important to consider the involvement of metabolic disorders as some of these are treatable. Important examples of these were noted by Fahey et al., which include neurometabolic disorders such as dopamine responsive dystonia. Other monogenic disorders leading to cerebral palsy included defects in the following:

KCNC3 potassium voltage-gated channel subfamily C member 3, an integral membrane protein that mediates the voltage-dependent potassium ion permeability of excitable membranes.

ITPR1 inositol 1,4,5, triphosphate receptor type 1, this receptor mediates calcium release from the endoplasmic reticulum.

SPTBN2 spectrin beta, nonerythrocytic 2 regulates the glutamate signaling pathway by stabilizing the glutamate transporter EAAT4 at the surface of the plasma membrane.

It is important to note that the category of cerebral palsy sometimes includes disorders defined as spinocerebellar ataxia and spastic quadriplegia.

Matthews et al. (2019) reported that the birth prevalence of cerebral palsy was 2−3 per 1000 live births. They reported that cerebral palsy was described as a permanent disorder that impairs movement and posture and leads to limitation in activity. Cerebral palsy was attributed to disturbances in brain development in the fetus or infant. Comorbidities that may occur with cerebral palsy include intellectual disability, epilepsy, autism, impairment in vision or hearing.

The study by McMichael et al. (2015) and a study by Matthews et al. (2019) emphasized genetic factors in the etiology of cerebral palsy. They studied individuals assessed between 2011 and 2016 in hospitals in Vancouver and Edmonton in Canada, who were documented to have impairment of motor function documented before 1 year of age. These children often have one or more of the following: manifestation developmental delay, abnormal neurologic features including hemiplegia, diplegia, ataxia, hypotonia, hypertonia. They documented evidence of neuroimaging abnormalities not typical of cerebral palsy, evidence of congenital abnormalities in other body regions, abnormal biochemical anomalies in other body region, abnormal biochemical finding on assessment of metabolites or neurotransmitters.

It is important to note that the cases studies in these reports included patients in the category defined as atypical cerebral palsy.

Next generation sequencing studies were carried out in these patients. Matthews et al. (2019) reported establishing a molecular diagnosis in 65% of 50 families. In 28 families, atypical cerebral palsy was reported to be due to defects in a known disease gene. In four cases, novel genes were implicated in cerebral palsy; these include gene loci encoding the following:

DHKZ diacylglycerol kinase zeta, regulates diacylglycerol levels in intracellular signaling cascade and signal transduction.

EPHA4 EPH receptor A4, belongs to the ephrin receptor subfamily of the protein-tyrosine kinase family implicated in mediating developmental events.

PALM phosphoprotein, associated with brain synaptic plasma membranes.

PLXNA2 Plexin A2 semaphorin coreceptor. Semaphorins are secreted or membrane-bound proteins that mediate repulsive effects on axon pathfinding during nervous system development.

Matthews et al. documented gene defects found in cerebral palsy patients and the modes of inheritance associated with the specific gene defect. One gene with defects, KANK1, associated with cerebral palsy was shown to be an imprinted gene and cerebral palsy occurs if the mutant gene is located on the paternally derived gene. Matthews et al. also documented other neurological defects that sometimes occurred in addition to cerebral palsy.

13.12.1 Genes with mutations leading to autosomal dominant cerebral palsy

AKT3 AKT serine/threonine kinase 3, regulators of cell signaling in response to insulin and growth factors.

ASXL1 ASXL transcriptional regulator 1 chromatin-binding protein,

ATP1A3 ATPase Na+/K+ transporting subunit alpha 3 integral membrane protein responsible for establishing and maintaining the electrochemical gradients of Na and K ions.

CHRNA1 incomplete penetrance, cholinergic receptor nicotinic alpha 1 subunit.

EHMT1/2 histone methyl transferase that methylates the lysine-9 position of histone H3. This action marks the genomic region packaged with these methylated histones for transcriptional repression.

EPHA4 ephrin receptor 2 subfamily of the protein-tyrosine kinase family important in development.

GNAO1 G protein subunit alpha o1 of G-protein signal-transducing complex.

KANK1 KN motif and ankyrin repeat domains 1, important in actin polymerization.

KCNJ6 potassium inwardly rectifying channel subfamily J member 6 important in circuit activity in neuronal cells.

KIDINS220 kinase D interacting substrate 220, controls neuronal cell survival, differentiation into exons and dendrites.

NAA10 N-alpha-acetyltransferase 10, catalytic subunit, posttranslational protein modifications.

PLXNA2 PLXNA2 semaphorin coreceptor, axon pathfinding.

SCN3A sodium voltage-gated channel alpha subunit 3 generation and propagation of action potentials.

SPAST Spastin ATPase domain protein involved in membrane trafficking, intracellular motility, organelle biogenesis, protein folding, and proteolysis.

TCF4 Transcription factor 4 broadly expressed including in neuronal development.

TUBB4A Tubulin Involved in assemble of microtubules.

13.12.2 Autosomal recessive (homozygous)

ATP8A2 ATPase phospholipid transporting 8A2 aids in generating and maintaining asymmetry in membrane lipids.

CSTB cystatin B protein is thought to play a role in protecting against the proteases leaking from lysosomes.

KMT2C lysine methyl transferase involved in transcriptional coactivation.

13.12.3 Autosomal recessive compound heterozygous

DGKZ diacylglycerol kinase zeta, involved in regulating diacylglycerol levels in intracellular signaling cascade and signal transduction.

GCDH glutaryl-CoA dehydrogenase catalyzes the oxidative decarboxylation of glutaryl-CoA.

ITPA inosine triphosphatase.

NBAS thought to be involved in Golgi-to-ER transport.

PALM Paralemmin prenylated and palmitoylated phosphoprotein that associates with the cytoplasmic face of plasma membranes.

RANBP2 RAN binding protein 2 GTP-binding protein of the RAS superfamily that is associated with the nuclear membrane.

TMEM67 transmembrane protein 67 functions in centriole migration to the apical membrane and formation of the primary cilium.

13.12.4 X-linked dominant

MECP2 methyl-CpG binding protein 2, essential to embryonic development.

WDR45 WD repeat domain 45 involved in a variety of cellular processes, including cell cycle progression, signal transduction, apoptosis, and gene regulation.

13.12.5 X-linked recessive

PAK3 p21 (RAC1) activated kinase 3, may be necessary for dendritic development and for the rapid cytoskeletal reorganization in dendritic spine.

PLP1 proteolipid protein 1 predominant component of myelin.

Matthews et al. noted that cerebral palsy has usually been defined on the basis of clinical findings. However, there is evidence that genetic evaluations are important, particularly if patients present with cerebral palsy and additional atypical findings. They noted that assessment of chromosomal microarrays and mitochondrial DNA analyses still have a place in diagnosis and that next generation sequencing is becoming increasingly important.

Jin et al. (2020) reported that gene defects that disrupt neuritogenesis and early neuronal connectivity play roles in cerebral palsy. In this context they identified defects in the following genes:

TUBA1A tubulin alpha 1a. The alpha and beta tubulins represent the major components of microtubules.

CTNNB1 catenin beta 1, participate in connections between cells.

FBXO31. This protein may have a role in regulating the cell cycle as well as dendrite growth and neuronal migration.

RHOB ras homolog family member B.

Neuritogenesis is defined as the process in which newborn neurons form growth cones that are regions for formation of axons and dendrites.

Gordon-Weeks (2017) noted that during neuritogenesis dynamic interactions occur between neuronal cytoskeleton components, actin filaments, and microtubules. There is also evidence that protein kinases are essential to neuritogenesis.

13.13 Monoamine neurotransmitter disorders

Ng et al. (2014) reviewed the phenotypic features and therapy of childhood neurotransmitter disorders. They noted that clinical features of these disorders overlap with those of other disorders including ischemic encephalopathy and cerebral palsy. They emphasized that early accurate diagnosis of monoamine neurotransmitter disorders can enable initiation of appropriate therapy.

Monoamine neurotransmitter disorder treatment involves measures that include increasing substrate activity, boosting enzyme levels, and replacing depleted monoamines. Importantly, they noted that therapy brought about significant improvement in quality of life and complete amelioration of motor symptoms in some cases.

Ng et al. reviewed monoamine neurotransmitter synthesis and function including information on dopamine, norepinephrine, and epinephrine. As neurotransmitters, they are involved in signaling in the central and peripheral nervous systems. In addition to their involvement in movement they are involved in other functions including mood and attention.

Clinical features of monoamine neurotransmitter defects were shown to lead to epilepsy, motor disorders, autonomic disturbances, and behavior changes. Importantly, severity of some clinical manifestations fluctuate at different times. They include dystonia, ptosis, oculogyric crises, brisk reflexes, and temperature instability.

Ng et al. reviewed steps in dopamine and serotonin biosynthesis. Tetrahydrobiopterin was noted to be essential for synthesis. Monoamine neurotransmitter synthesis in presynaptic nerve terminals release into the synaptic cleft, binding to postsynaptic receptors. Termination of action was followed by degradation or reuptake of the neurotransmitter.

Specific monoamine neurotransmitter disorders were found to be due to pathological mutations in the following gene products.

GCH (GCH1) GTP cyclohydrolase 1, a member of the GTP cyclohydrolase family. The encoded protein is the first and rate-limiting enzyme in tetrahydrobiopterin (BH4) biosynthesis, catalyzing the conversion of GTP into 7,8-dihydroneopterin triphosphate.

PTS 6-pyruvoyltetrahydropterin synthase catalyzes the second and irreversible step in the biosynthesis of tetrahydrobiopterin from GTP.

SPR sepiapterin reductase catalyzes the NADPH-dependent reduction of pteridine derivatives and is important in the biosynthesis of tetrahydrobiopterin (BH4).

DHPR QDPR quinoid dihydropteridine reductase gene encodes the enzyme dihydropteridine reductase, which catalyzes the NADH-mediated reduction of quinonoid dihydrobiopterin.

TH tyrosine hydroxylase involved in the conversion of tyrosine to dopamine.

AADC (DDC) dopa decarboxylase catalyzes the decarboxylation of L-3,4-dihydroxyphenylalanine (DOPA) to dopamine, L-5-hydroxytryptophan to serotonin and L-tryptophan to tryptamine. Defects in this gene are the cause of aromatic L-amino-acid decarboxylase deficiency (AADCD).

PNPO pyridoxamine 5′-phosphate oxidase rate-limiting step in the synthesis of pyridoxal 5′-phosphate, also known as vitamin B6. Vitamin B6 is a required cofactor for enzymes involved in both homocysteine metabolism and synthesis of neurotransmitters such as catecholamine.

PITX3. Transcription factor.

SLC6A4 (SERT1). Serotonin transporter transports the neurotransmitter serotonin from synaptic spaces into neurons.

SLC6A3 (DAT). Dopamine transporter mediates the active reuptake of dopamine.

Ng et al. reviewed agents that impact dopamine and serotonin metabolism. In cases where functions of specific enzymes are inefficient, monoamine substrate and cofactor levels can be altered. Ng et al. noted that levels of dopamine and serotonin are reduced in specific neurotransmitter disorders and levels can be increased by administration of levodopa and 5-hydroytryptophan. However, D-DOPA administration was not recommended for all neurotransmitter disorders. It was not considered to be of value in patients with AADC deficiency and should also not be used in cases with deficiency of dopamine transporter.

5HTP (5 hydroxytryptophan) was in cases where it was necessary to boost serotonin levels. They noted that in some cases it is important to use agents that modify metabolism of dopamine or serotonin.

Related to monoamine neurotransmitters, for example, DHPR or AADC defects in these cases folate therapy was recommended.

Ng et al. reviewed disorders of tetrahydrobiopterin (BH4) synthesis and these include SPR and PTs defects. BH4 deficiency is treatable. Diagnostic workup for defects in BH4 synthesis involves analysis of pterins.

Brennenstuhl et al. (2019) reviewed inherited disorders of neurotransmitters and noted that they arise due to defects in the synthesis, transport, or degradation of neurotransmitters and may also be due to defects in bioavailability of cofactors such as tetrahydrobiopterin. They reviewed neuronal activity and connections and the release of neurotransmitters that bound to receptors and triggered synaptic activity.

Brennenstuhl et al. noted that inherited neurotransmitter disorders have a range of clinical presentations including early onset encephalopathies and later onset moderate movement disorders. In addition, clinical manifestation of these disorders overlap with those of other disorders including cerebral palsy. They noted that there were frequent delays in diagnosis of neurotransmitter disorders and that in those disorders brain MRI findings are usually normal.

Specific molecules included in the neurotransmitter disorders include monoamine neurotransmitters that are derived from aromatic amino acids. In this group are catecholamines, dopamine, adrenaline, and noradrenaline. Synthesis of these specific catecholamines was reported to occur in dopaminergic and noradrenergic regions and early steps in the synthesis involved hydroxylation of phenylalanine and tyrosine that are dependent on the cofactor tetrahydrobiopterin. They noted that key enzymes in this pathway include aromatic amino acid decarboxylases, dopamine hydroxylase, tyrosine hydroxylase. In addition to key enzymes, specific cofactors and chaperones are necessary. Some enzymes play roles in biosynthesis while others are involved in recycling and reuptake of synthesized products.

Brennenstuhl et al. emphasized specific enzyme deficiencies known to lead either autosomal dominant or autosomal recessive inherited neurotransmitter defects.

In the categories of enzyme defects, cochaperone defects in biopterin synthesis, or recycling defects, they emphasized importance of the following:

DNAJC12 associated with complex assembly, protein folding, and export. Defects in the gene product were identified in nonBH4 deficiency hyperphenylalaninemia with neurotransmitter deficiency.

SR (SPR) SRD Sepiapterin reductase. Mutations in this gene result in DOPA-responsive dystonia.

GTPCH (GCH1) GTP cyclohydrolase. The first and rate-limiting enzyme in tetrahydrobiopterin (BH4) biosynthesis. Mutations in this gene are associated with malignant hyperphenylalaninemia and dopa-responsive dystonia.

PTPS 6-pyruvoyltetrahydropterin synthase.

DHPR (QDPR) dihydropteridine reductase, which catalyzes the NADH-mediated reduction of quinonoid dihydrobiopterin.

PCBD1 pterin-4 alpha-carbinolamine dehydratase 1, functions as a dehydratase involved in tetrahydrobiopterin biosynthesis.

Autosomal recessive primary neurotransmitter synthesis defects can result from defects in

TH tyrosine hydroxylase, involved in the conversion of tyrosine to dopamine.

AADC (DDC) dopa decarboxylase catalyzes the decarboxylation of L-3,4-dihydroxyphenylalanine (DOPA) to dopamine, L-5-hydroxytryptophan to serotonin and L-tryptophan to tryptamine. Defects in this gene are the cause of aromatic L-amino-acid decarboxylase deficiency (AADCD).

Monoamine transport defects include SLC6A3 (VMAT2) transmembrane protein that functions as an ATP-dependent transporter of monoamines, such as dopamine, norepinephrine, serotonin, and histamine.

Monoamine catabolism defects leading neurotransmitter disorders could involve:

MAOA monoamine oxidase mitochondrial enzymes, which catalyze the oxidative deamination of amines, such as dopamine, norepinephrine, and serotonin.

MAOB monoamine oxidase B located in the mitochondrial outer membrane. It catalyzes the oxidative deamination of biogenic and xenobiotic amines and plays an important role in the metabolism of neuroactive and vasoactive amines.

DBH dopamine beta-hydroxylase catalyzes the conversion of dopamine to norepinephrine, which functions as both a hormone and as the main neurotransmitter of the sympathetic nervous system.

Brennenstuhl et al. documented diagnostic testing procedures for these disorders They noted that cerebrospinal fluid samples were the gold standard for diagnosis of these disorders. Following acquisition, samples should be stored at $-70°C$ to preserve metabolites, particularly pterin including tetrahydrobiopterin. Other components to be assayed in cerebrospinal fluid include homovanillic acid, a dopamine metabolite and norepinephrine. Importantly, they also documented treatment for these disorders.

13.14 Spastic paraplegias and ataxias

The International Parkinson and Movement Society Task Force and Marras et al. (2016) reported recommendation on nomenclature for Genetic Movement Disorders. This included the following terms: parkinsonism, dystonia, inherited ataxias, spastic paraplegias, chorea, paroxysmal movement disorders, neurodegeneration with brain iron accumulation, and familial brain calcification.

Shribman et al. (2019) noted that currently spastic paraplegias are symptomatically treated. However, they proposed that with increasing understanding of the genetic and pathophysiological mechanisms, opportunities for targeted molecular therapies could emerge.

Kara et al. (2016) reviewed the genetic and phenotypic characterization of spastic paraplegias. They noted that spastic paraplegias could be classified as predominantly lower limb spasticity or complex spasticity that could include additional neurological features. They documented studies on 97 cases of complex spastic paraplegia. Thirty of the 97 cases were found to have SPG11, five cases were in the SPG7 category, four cases were in the SPG35 category, and two were in the SPG15 category.

SPG11 spastic paraplegia 11 defect in vesicle trafficking associated Spastacin 15q21.1.

SPG7 spastic paraplegia, autosomal recessive, paraplegin, an ATP-dependent proteolytic complex of the mitochondrial inner membrane that degrades misfolded proteins and regulates ribosome assembly 16q24.3.

SPG35 spastic paraplegia 35 autosomal recessive defect in Fatty acid alpha hydrolase FAAH FA2H 16q23.1.

SPG15 spastic paraplegia 15 with retinal degeneration autosomal recessive, ZFYVE26 colocalizes partially with markers of endoplasmic reticulum and endosomes, 14q24.1.

13.14.1 General clinical features

Clinical features of SPG11 include onset of walking problems at the age of 11 years with range 4–17 years, spastic ataxia, bladder problems, Parkinson features. Some patients with SPG11 had atypical presentation with toe walking, brisk reflexes, extensor plantar reflexes noted by 12 years of age.

In SPG7 defects the age of onset differed with some individuals being diagnosed early as cases of cerebral palsy and others being diagnosed later with progressive spasticity. Other manifestation of this disorder included neuropathy and optic atrophy (22 years).

Manifestations of SPG35 occurred between 5 and 22 years. Early manifestation included toe walking and balance problems. Other manifestations include dysarthria and ophthalmoplegia.

SPG15 clinical manifestations overlapped with those of SPG11 defects.

Boutry et al. (2019) noted that heterogeneity in hereditary spastic paraplegias reflected the different cellular pathways involved. These included mitochondrial dysfunction, microtubule trafficking defects, membrane defects, abnormalities in lipid metabolism, lysosome, and autophagy defects. They also emphasized overlap in manifestation between spastic paraplegias, Parkinson disease, spinocerebellar ataxia, and chorea.

Diagnostic studies in spastic paraplegias include initial sequencing for SPG11 mutation and if these are absent next generation sequencing was carried out.

13.15 Friedreich ataxia (FRDA)

This disorders was reviewed by Delatycki and Bidichandani (2019) who noted that it is the most common inherited ataxia in individuals of European ancestry. However, the population frequency was noted to be 1 in 40,000.

The average age of onset of disease manifestations was reported to be between 10 and 15 years of age. The disorder is associated with progressive ataxia (lack of control of muscle movements and gait difficulties). Other manifestations that can develop in this disorder include diabetes and cardiomyopathy. Life span is reduced, and death was reported to occur at an average age of 30 years.

Neuropathology studies in Friedreich ataxia reveal abnormalities in dorsal root ganglia, posterior columns in the spinal cord, defects in spinocerebellar tracts, and in peripheral nerves. In addition, pathological abnormalities develop in the heart and pancreas. Campuzano et al. (1996) discovered that Friedreich ataxia arises due to homozygous or compound heterozygous mutations in a gene that maps to chromosome 9 and was designated FRDA. Detailed analyses of mutation reported by Galea et al. (2016) revealed that 96% of affected individuals were homozygous for an expanded GAA repeat in intron 1 of the gene. On 4% of affected individuals, heterozygosity at this locus was found with GAA repeat expansion in one FRDA allele and a different type of mutation in the other FRDA allele.

The GAA repeat expansion was found to be unstable and this instability was particularly marked in the heart and pancreas.

The disease-causing mutations were noted to lead to decreased levels of production of the FRDA gene product frataxin. The frataxin protein was noted to be particularly abundant in mitochondria. There the full-length protein undergoes cleavage to produce mature protein that is a component of iron − sulfur (FS) clusters in the mitochondrial matrix. These FS clusters are important to a number of different functions in mitochondria, including the mitochondrial respiratory complexes I, II, and III, and the function of the Krebs cycle.

The disease manifestations are primarily due to frataxin deficiency. Gene therapy is problematic due to difficulty in targeting mitochondria.

A finding of importance is that tissues of patients with Friedreich ataxia have high levels of iron. However, iron depletion therapy led to worsening of disease manifestations. Ast et al. (2019) reported that hypoxia counteracts the impact of frataxin loss and restores biogenesis of FS complexes. This illustrates that environmental factors may influence disease severity.

Castro et al. (2019) reported that frataxin function depends on interaction with other proteins to form supercomplexes. These include NFS1 and ISCU scaffolding protein. NFS1 cysteine desulfurase supplies inorganic sulfur to these FS clusters. ISCU is defined as FS cluster assembly enzyme.

13.16 Polyglutamine cerebellar ataxias

These disorders were reviewed in Buijsen et al. (2019). They noted that there are six known forms of autosomal dominant spinocerebellar ataxias due to expansion of CAG repeats in the gene coding regions. Buijsen et al. noted that increasing knowledge of pathological mechanisms involved in these diseases is promoting investigations into therapeutic measures to slow disease progression. Key features of these disorders include progressive ataxia. Other manifestations are dysarthria and oculomotor defects. In addition, one-third of families with a clinical diagnosis of autosomal dominant cerebellar ataxia (ADCA) did not have a genetic diagnosis. Population frequency of ADCA was reported to be approximately 5 per 100,00 in several countries including Portugal, Norway, and Japan. The prevalence of different forms of ADCA differed. Important with increasing length of the CAG repeats, onset of clinical manifestations occurred earlier.

Several disease mechanisms have been proposed including disrupted gene function and toxicity of the polyglutamine derived from CAG repeats.

Different genes that manifest the CAG repeat expansions that lead to ADCA include the following:

SCA1 ATXN1 reported to bind chromatin and acts as a transcriptional repressor.

SCA2 ATXN2 reported to control distinct steps in posttranscriptional gene expression.

SCA3 ATXN3 suggested to play role in autophagy and proteostasis.

SCA6 CACNA1A subunit of calcium channel predominantly expressed in neuronal tissues.

SCA7 ATXN7 was reported to associate with microtubules and to stabilize the cytoskeleton.

SCA17 TBP (TATA box binding protein) serves as the scaffold for assembly of transcription complex.

Therapeutic approaches being considered included promotion of proteastasis through induction of autophagy in some forms of the disorder. Gene therapies including RNA interference therapy and exon skipping approaches were being investigated in some preclinical models.

The repeat expansions and polyglutamine expansions were reported to impair autophagy and activity of the ubiquitin proteosome system. Expanded polyglutamine sequences led to accumulation of large insoluble protein aggregates. These aggregates were shown to be present particularly in neurons in the cerebellum. The aggregates likely disrupt cellular function.

13.17 Autosomal recessive ataxias

Beaudin et al. (2019) reported a classification system PRAT to organize these disorders according to clinical presentation and underlying pathology and mechanisms. They included 59 different disorders in their analysis; 15 of these disorders were found to be frequent and 44 were found to be infrequent and occurred only in certain populations or in certain families.

With respect to clinical evaluation, they emphasized the importance of family history or also history of trauma and history of possible exposure to toxins They noted that it is also important to include information on age of onset of symptoms and signs.

In listing forms of recessive ataxias, they first listed those forms that occur worldwide.

Ataxia and mutated protein	Disorder
ATX FXN (frataxin)	Friedreich ataxia.
ATX ATM (AT mutated)	Ataxia telangiectasia ATM protein is a phosphatidyl inositol kinase.
ATX APTX (Aprataxin)	Early onset ataxia with oculomotor apraxia and hypoalbuminemia.
ATX SETX (senataxin)	Ataxia with axonal neuropathy 2.
ATX/HSP SACS	Autosomal recessive ataxia with Sacsin deficiency.
ATX SANDO POLG	Sensory ataxic neuropathy, opthalmoparesis POLG deficiency.

It is important to note that a number of autosomal recessive ataxias are also included in disorders classified as spinocerebellar ataxias.

Beaudin et al. classified ataxia into six categories according to the most frequently encountered clinical manifestation, and they noted the genes that were shown to be involved in each category. The categories included:

Pure cerebellar ataxias, 3 genes.

Cerebellar syndrome with motor neuron involvement, 16 genes.

Cerebellar syndrome with polyneuropathy, 9 genes.

Cerebellar syndrome with extra pyramidal involvement and oculomotor apraxia, 6 genes.

Ataxias due to metabolic or mitochondrial syndrome, 14 genes.

Cerebellar syndrome with intellectual disability, developmental delay, or dementia, 14 genes.

Considering the pathways involved, the Beaudin report noted the importance of mitochondrial dysfunction that could result from abnormal mitochondrial DNA maintenance that then results in progressive mutations leading to altered mitochondrial dynamics, altered mitochondrial protein synthesis, increase in reactive oxygen species, altered coenzyme Q10 function, and altered mitochondrial respiratory chain activity.

Other pathways involved included defects in increased susceptibility to DNA damage and defects in DNA repair.

Other disease mechanisms involved included impaired synaptic activity.

13.18 Spinocerebellar ataxias

In a review Ashizawa et al. (2018) reported that more than 40 autosomal dominant forms of spinocerebellar ataxias have been identified. They noted that analyses of the molecular pathology in a number of forms of spinocerebellar ataxia have led to development of interesting treatment strategies. This applies not only to the polyglutamine repeat forms of these disorders but also to other spinocerebellar ataxia due to abnormalities in RNA or to generation of toxic proteins.

They noted further that magnetic resonance imaging and analysis of magnetic resonance spectroscopic biomarkers facilitate assessment of disease activity.

Ashikawa documented different forms of spinocerebellar ataxia due to nucleotide repeat expansions. It is important to note that CAG repeat expansion and CAA repeat expansions represent polyglutamine expansion if in the coding region.

Gene region	Repeat expansion	Type of spinocerebellar ataxia
5′UTR	CAG	SCA12
Exon	CAG	SCA1, SCA2, SCA3, SCA6, SCA7, SCA17 DRPLA
Intron	ATTCT	SCA10
	TGGAA	SCA31
	ATTC	SCA37
	GGCCTG	SCA36
3′UTR	CTG	SCA8

Ashizawa et al. documented pathophysiological mechanisms in these disorders noting that they included development of aggregates, impairment of protein function, perturbation of neural circuit, impairment of bioenergetics, alterations in DNA repair.

They noted that therapeutic strategies could include gene editing, transcription inhibition with targeting of disease-specific transcripts, modification of splicing, blocking of translation of mutant mRNA, clearance of aberrant protein, promotion of protein folding.

Specific treatments were also aimed at the pathophysiological effects of the mutant gene and gene products.

Ashizawa et al. noted that it was important to consider the CAG repeat expansions in coding regions, the polyglutamine expansions as a distinct category. These included the gene products listed below. It is important to note that the functions of ataxins (ATXN) are listed as unknown in NCBI genes (2021).

SCA1 ATXN1.
SCA2 ATXN2.
SCA3 ATXN3.

SCA6 CACNA1A calcium ion channel.
SCA7 ATXN17.
SCA17 TBP TATA box binding protein important in initiation of transcription.
DRPLA ATN1 atrophin this protein accumulates to abnormal concentrations in dentatorubral-pallidoluysian atrophy.

Important new information relevant to therapy of polyglutamine expansion diseases has emerged. Nakamori et al. (2020) demonstrated that a specific small molecule naphthyridine azaquinolone (NAT) binds to expanded CAG repeats. This binding was shown to prevent repeat expansions and in fact enhance repeat contraction.

Expansion of trinucleotide repeats over time is thought to play important roles in polyglutamine diseases. The binding of NAT appears to be specific for expanded CAG repeats. There is evidence that expanded repeats form unusual slipouts during replication and certain DNA repair proteins act to eliminate these slipouts and the NAT was shown to prevent repair of these slipouts (Nakamori et al., 2020).

13.18.1 Nonrepeat cerebellar ataxias

SCA31 BEAN1 brain expressed associated with NEDD4, a member of a family of ubiquitin-protein ligases.
SCA19 KCND1, potassium voltage-gated channel subfamily D member, regulation of neurotransmitter release.
SCA12 PPP2R2B protein phosphatase 2 regulatory subunit Beta.
SCA10 ATXN10 may function in neuron survival, neuron differentiation, and neuritogenesis.
SCA14 PRKCG protein kinase C gamma is expressed solely in the brain and spinal cord and its localization is restricted to neurons.
SCA15 SCA16 ITPR1 inositol 1,4,5-trisphosphate receptor type 1, mediates calcium release from the endoplasmic reticulum.

13.19 Spinocerebellar ataxias

Synofzik and Schüle (2017) reported that spinocerebellar ataxias and hereditary spastic paraplegias have been designated as separate conditions. Yet some genes, identified as causing spastic paraplegias, have now been determined to lead to ataxia. They proposed that an ataxia spastic paraplegia gene spectrum exists.

These rare disorders were reported to be associated with degenerative changes that occur in Purkinje cells, spinocerebellar tracts, and corticospinal tracts. Autosomal dominant and autosomal recessive forms of spinocerebellar ataxias have been reported.

13.20 Genomic medicine in common diseases in adults

13.20.1 Hypertension

Personalized medicine and treatment of hypertension

Melville and Byrd (2019) noted that a key concept in personalized medicine is that individual patients with the same disease are different from each other and may respond differently to the same treatment. They noted, however, that for some diseases a specific treatment may be effective in a large number of individuals.

They discussed aspects of the treatment of hypertension, addressing whether or not some forms of hypertension require individualized treatment and emphasized that identification of specific hypertension phenotypes indicate the need for personalized approaches.

Elevated blood pressure was noted to be the leading risk factor for disability and death. It is important to note that level of physical activity and lifestyle are also important factors in determining blood pressure levels.

Resistant hypertension was considered to be an important factor in determining whether additional information and further evaluation were required. Resistant hypertension was defined as hypertension levels not decreasing despite good adherence to therapy and concurrent use of three antihypertensive drugs (Carey et al., 2018, American Heart Association, 2018).

Ahn and Gupta (2018) reviewed aspects of the genetics of hypertension and emphasized the importance of gene environment interactions, monogenic and polygenic factors, and possible epigenetic factors. These authors reviewed four forms of monogenic hypertension:

Glucocorticoid remediable aldosteronism (GRA) with autosomal dominant inheritance.

Apparent mineralocorticoid excess (AME) with autosomal recessive inheritance.

Congenital adrenal hyperplasia (CAH) with autosomal recessive inheritance.

Liddle syndrome with autosomal dominant inheritance.

13.20.2 Glucocorticoid remediable aldosteronism

This condition was reported to be due to a structural genomic change leading to fusion of the promoter of the CYP11B1 gene that encodes 11-beta-hydroxylase with CYP11B2 the aldosterone synthase gene.

CYP11B1 cytochrome P450 family 11 subfamily B member 1, is involved in the conversion of progesterone to cortisol in the adrenal cortex. The gene maps to 8q24.3, 142872357−142879825.

CYP11B2 cytochrome P450 family 11 subfamily B member 2. The gene encodes an enzyme that has steroid 18-hydroxylase activity to synthesize aldosterone and 18-oxocortisol as well as steroid 11-beta-hydroxylase activity. The gene maps to 8q24.3, 142910559−142917843.

The fusion results in increased aldosterone levels leading to salt and water retention, mild hypokalemia metabolic alkalosis, and also low renin. The fusion then leads to hypertension in young individuals.

Adrenal hyperplasia and aldosterone producing adenomas were also noted to lead to early onset hypertension.

13.20.3 Apparent mineralocorticoid excess

This autosomal recessive condition was noted by Ahn and Gupta to result from inactivating mutations in the gene leading to decreased levels of the enzyme 11-beta hydroxysteroid dehydrogenase.

HSD11B2. The gene encodes two different isoforms. The type I isozyme has both 11-beta-dehydrogenase (cortisol to cortisone) and 11-oxoreductase (cortisone to cortisol) activities. The type II isozyme catalyzes the glucocorticoid cortisol to the inactive metabolite cortisone, thus preventing illicit activation of the mineralocorticoid receptor.

Geller syndrome was described as a mineralocorticoid excess syndrome due to activation mutations in the mineralocorticoid receptor gene, NR3C2 nuclear receptor subfamily 3 group C member 2. The protein functions as a ligand-dependent transcription factor that binds to mineralocorticoid response elements in order to transactivate target genes. Mutations in this gene cause autosomal dominant pseudohypoaldosteronism type I, a disorder characterized by urinary salt wasting. Defects in this gene are also associated with early onset hypertension with severe exacerbation in pregnancy.

Congenital adrenal hyperplasia. Different forms of this condition have been described. In CAH IV, 17 alpha hydroxylase was reported to be defective. The gene that encodes 17 alpha hydroxylase is also known as CYP17A1. Because cortisol production is inefficient, increased production of ACTH (adrenocorticotrophic hormone) results leading to adrenal hyperplasia.

Liddle syndrome was reported to be due to gain of function in SCNN1B that encodes both beta and gamma epithelial sodium channels. Gain of function mutation in SCNN1B leads to increased sodium resorption and hypertension. Patients with this syndrome are also reported to manifested hypokalemia and metabolic acidosis.

Different types of Liddle syndrome have been described. LIDS1 maps to chromosome 16p12.2 at position 23278231−23381295 and is caused by a heterozygous mutation in SCNN1B.

Liddle syndrome can also arise due to deficits in SCNN1G that maps to chromosome 16p12.2 at position 23182745−23216883.

Gordon syndrome was noted to be characterized by gain of function mutations in a serine kinase gene on chromosome 12p13.33 that encodes WNK1 leading to pseudo-hypoaldosteronism type II. Loss of function mutations in a gene on chromosome 17q21.2 that encodes WNK4, leading to pseudo-hypoaldosteronism type IIB.

Lacolley et al. (2020) noted that arterial stiffness is an independent risk factor for increased systolic hypertension and increased pulse pressure. Arterial stiffness was reported to arise due to reduced elastin/collagen ratio, elastin cross-linking, and reactive oxygen species induced inflammation. Other important factors in determining arterial stiffness included calcification and endothelial dysfunction.

In addition to mitochondrial stress, oxidative stress dyslipidemia was also reported to determine arterial stress.

13.21 Polygenic factors leading to hypertension

Ahn and Gupta (2018) noted that numerous genome-wide association studies (GWAS) have been carried out in hypertension. Multiple SNPs have been reported to be associated with hypertension, but generally have very low effect. They also discussed evidence for possible epigenetic factors that influence blood pressure.

Warren et al. (2017) analyzed association of systolic and diastolic pressure and pulse pressure in 140,886 UK biobank participants. They noted that elevated blood pressure is strongly heritable. Their study involved GWAS and exome studies. They reported validation of 107 associated loci. They further reported that results of these studies led to identification of new biological pathways involved in blood pressure determination.

Of the 107 validated loci, 102 were validated via GWAS analyses and five were identified in replicated exome studies. Twenty-four loci were reported to be associated with systolic blood pressure as the primary trait. Forty-one loci were validated as being associated with diastolic blood pressure. Forty-two loci were validated as associated with pulse pressure.

Some loci were associated with more than one of the traits. These included 24 loci correlated with both systolic and diastolic measures; 11 with systolic blood pressure and pulse pressure; 1 locus was associated with diastolic blood pressure and pulse pressure.

Four loci were associated with all three traits and these included the following:

NADK-CPSF3L NADK NAD kinase catalyzes the transfer of a phosphate group from ATP to NAD to generate NADP, which in its reduced form acts as an electron donor for biosynthetic reactions maps to 1p36.33 1751232–1780509. CPSF3L also known as INTS11 integrator complex subunit 11 associates with the C-terminal domain of RNA polymerase II large subunit (POLR2A) maps to 1p36.33 1311597–1324660.

GTFB general transcription factor IIB, one of the ubiquitous factors required for transcription initiation by RNA polymerase II. Maps to 1p22.2 88852633–88891944.

METTL21A-ACO79767.3 METTL2A methyl transferase like 2A maps to 2q33.3 207580635–207626053.

PAX2 transcription factor maps to chromosome 10 100735396–199829944.

In addition to identifying hypertension risk loci, Warren et al. reported associated variants at specific loci that represented drug targets. These included the following loci:

Product of locus	Drug
ACE angiotensin 1 converting enzyme	ACE inhibitors
CACNA2D2 calcium voltage-gated channel subunit alpha2delta 2	Calcium channel blockers
MME Membrane metallo-endopeptidase	Omapatrilat
ADRA2B adrenoceptor alpha 2B	Beta blockers
SLC14A2 solute carrier family 14 member 2	Nifedipine
PDE5A phosphodiesterase 5A	Sildenafil

It is important to note that 27 of the loci associated with hypertension were also associated with other traits, including coronary heart disease (CAD), myocardial infarction, and vascular disorders. Further studies revealed that 59 of the 107 loci acted as expression quantitative trait loci in certain tissues. Expression enrichment within DNAse1 hypersensitive sites was also identified. Specific candidate loci were shown to impact microvascular endothelium and aortic smooth muscle.

Histone markers were analyzed and activating modification, H3K4Me3, was found near promoters of involved genes in studies on endothelial cells.

Specific studies were undertaken to identify hypertension-associated variants in nonprotein coding regions and the downstream genes that they impacted. Three were identified, they included:

NOX4 NADPH oxidase 4 that functions as the catalytic subunit the NADPH oxidase complex.

KCNH4 potassium voltage-gated channel subfamily H member 4 it is a pore-forming (alpha) subunit.

LHFP2 LHFPL tetraspan subfamily member 2, tetraspan transmembrane protein.

Warren et al. noted a lack of rare variants in their analyses. It is interesting to note that specific associated gene with a number of associated variant were shown to be expressed both in arterial and venous endothelium in smooth muscle and in vascular walls.

A common variant in the NOX4 NADPH oxidase 4 was found to correlate with increased generation of reactive oxygen species.

The authors stressed that identification of increased hypertension risk early in life followed by specific lifestyle interventions, including reduced sodium intake, increased potassium intake, maintenance of optimal weight, limited alcohol intake, and regular exercise, could offset the impact of high genetic risk.

13.22 Coronary heart disease

Khera and Kathiresan (2017) emphasized that CAD is a leading cause of death worldwide. They noted that CAD, hypertension, arterial fibrillation, and stroke are all complex diseases resulting from interplay of genetic and environmental factors.

Specific monogenic causes of cardiovascular diseases were noted to include cardiomyopathies and specific cardiac rhythm disorders.

They emphasized that complex diseases are influenced by both common and rare DNA variants and studies have been carried out to include both common variant analysis and rare variant analysis. Khera and Kathiresan noted that more than 150 loci have been shown to have significant association with CAD. They noted that in addition it is necessary to establish how significantly associated variants, often in the nonprotein coding of the genome, lead to disease phenotype.

Analyses reveal several risk factors associated with CAD. These include levels of low-density lipoproteins, cholesterol, especially triglyceride-rich cholesterol fraction, levels of lipoproteins, evidence of insulin resistance, increased tendency to thrombosis, and inflammation. Other risk factors relate to vascular wall adhesion, migration across vascular endothelium, nitric oxide levels, and vascular tone.

Variants in protein coding genes that occur at altered frequencies in patients with CAD were reported to occur. These impacted:

LDLR low density lipoprotein receptor. Mutations cause the autosomal dominant disorder, familial hypercholesterolemia.

APOA5 apolipoprotein A5, plays an important role in regulating the plasma triglyceride levels, a major risk factor for coronary artery disease.

APOC3 apolipoprotein C3 protein component of triglyceride (TG)-rich lipoproteins (TRLs) including very low-density lipoproteins, high density lipoproteins (HDL), and chylomicrons.

PCSK9 proprotein convertase subtilisin/kexin type 9, plays a role in cholesterol and fatty acid metabolism. Mutations in this gene have been associated with autosomal dominant familial hypercholesterolemia.

ANGPTL3. The N-terminal chain is important for lipid metabolism, while the C-terminal chain may be involved in angiogenesis. Mutations in this gene cause familial hypobetalipoproteinemia type 2.

LPL lipoprotein lipase has the dual functions of triglyceride hydrolase and ligand/bridging factor for receptor-mediated lipoprotein uptake.

ASGR1 asialoglycoprotein receptor 1, plays a critical role in serum glycoprotein homeostasis by mediating the endocytosis and lysosomal degradation of glycoproteins.

GUCY1A3 guanylate cyclase 1 soluble subunit alpha, linked to nitric oxide signaling.

13.22.1 Heritability of coronary heart disease

Khera and Kathiresan noted that heritability of CAD was reported to be 40%–50%. There is evidence that 38% of the heritability is contributed by common variants. It is important to note that the common variants primarily occur in the nonprotein coding regions of the genome.

Specific variants in coding regions that predispose to CAD vary in frequency in different populations. In the United States, the frequency of the PCSK9 variant Y142X that impacts cholesterol levels and CAD occur with a frequency of 1%.

13.22.2 Nonmonogenic risk

The common variant associated with highest polygenic risk scores was reported to occur on chromosome 9p21. The locus was described as being within a 58-kb noncoding genomic region. There is some evidence that the locus includes sequence for a nonprotein coding antisense RNA designated ANRIL. Further studies of the 58-kb locus identified enhancers that interact with genes that encode CDKN2A, CDKN2B, MTAP, and IFNA21. However, Khera and Kathiresan noted that the CAD-related causative variant in 9p21 has not been identified.

13.23 Genetic-guided therapies

Khera and Kathiresan noted that variants in three genes present potentially target therapies. These included PCSK9 proprotein convertase subtilisin/kexin type 9, ANGPTL3. Mutations in this gene cause familial hypobetalipoproteinemia type 2 and APOC3.

Another locus linked to CAD (coronary heart disease) with significant scores is located on chromosome 1p13 and the risk allele at this locus was reported to be associated with increased risk for CAD between 15% and 20% and to be associated with increases risk between 25% and 20% for hyperlipidemia.

Aggregated polygenic risk scores also vary in different populations. This locus, designated SORT1, was described as an expression quantitative locus. However, the exact role of SORT1 relative to CAD has not been described.

Another CAD-associated locus maps to 6p24 and harbors gene that encodes PGACTR1, phosphatase and actin-related, and EDN1 endothelin, proteins that have roles in vascular biology.

Khera and Kathiresan noted that different families with increased risk for CAD may harbor different risk variants. In some families there may be blend of causal risk factors.

13.23.1 Genetics-based therapeutic targets

Khera and Kathiresan noted three targets for gene therapy in treatments of CAD-related factors. They include PCSK9, ANGPTL3, APOC3.

Musunuru and Kathiresan (2019) reported an addition GWAS locus for CAD on 15q25. This locus harbored ADAMTS7 that impacts migrations of smooth muscle cells. Mouse studies supported a role for this gene product on endothelial repair.

They noted that suitable organoids had not yet been developed for studies in CAD.

13.24 Approaches to determining polygenic risk scores

Newcombe et al. (2019) emphasized that heritability of complex traits is driven by variants across the genome and that polygenic risk score determinations that included analyses of variants across the genome had generated useful information. They noted that in determining polygenic risk scores it was important to take into account the existence of linkage disequilibrium between close loci. Specific software packages have been developed for polygenic risk score determination.

Igo et al. (2019) emphasized that overall risk score analyses should take into account other forms of information, including family history, population information, and environmental information. They also noted that a major obstacle in polygenic risk score determination is the degree of correlation between neighboring markers (linkage disequilibrium).

Different strategies are utilized to take linkage disequilibrium into account. One of these strategies is defined as pruning, noted to be a method to remove from consideration in the analyses all markers that were in linkage disequilibrium with a specific marker. Another approach is to consider all the markers jointly in a summary statistic.

Igo et al. emphasized that complex disease-associated common variants confer modest effect size.

13.25 Familial hypercholesterolemia

An American College of Cardiology report by Sturm et al. on clinical genetic testing in familial hypercholesterolemia was reviewed. Authors of this report recommended that genetic testing becomes standard of care for patients with familial hypercholesterolemia, definite or probable and that cascade testing should be considered for relatives.

They proposed that genetic testing be carried out on the genes that encode LDLR, APOB, PCSK9, with decisions for additional testing based on the phenotype.

They noted that the normal LDL cholesterol level should be below 190 mg/dl.

If an individual with hypercholesterolemia manifested no variants in the above genes, alternate molecular etiologies should be considered. These included polygenic factors, high lipoprotein levels, abnormal APOC, ABCG8 variants, LIPA (lysosomal lipase) deficiency, LDLRAP1 variants.

LIPA. This enzyme functions in the lysosome to catalyze the hydrolysis of cholesteryl esters and triglycerides.

LDLRAP1, low density lipoprotein receptor adaptor protein 1. Mutations in this gene lead to LDL receptor malfunction and cause the disorder autosomal recessive hypercholesterolemia.

Sturm et al. emphasized that the genetic test results have implication for therapeutic choices.

They noted that in homozygous familial hypercholesterolemia, LDL cholesterol levels were between 400 and 1000 mg/dl. In heterozygous familial hypercholesterolemia, LDL cholesterol levels range between 130 and 450 mg/dl. In common hypercholesterolemia level ranges between 125 and 190 mg/dl.

Very high levels of LDL cholesterol could result from homozygous LDL receptor's defects or from compound heterozygous LDLR defects. Heterozygous hypercholesterolemia could result from pathogenic variant in LDLR, PCDK9, or APOB.

Specific types of variants: Pathogenic LDLR variants were noted to include nonsense, missense, small insertion, small deletion. Genomic structural variants and rearrangements that impacted the LDLR locus have also been described. Pathogenic variants in APOB that lead to hypercholesterolemia were reported to be primarily missense variants in the region of APOB that binds to LDLR. One predominant pathogenic APOB variant was noted to occur in 10% of cases p.Arg3500Gln.

PCSK9 gain of function variants were reported to lead to familial hypercholesterolemia. The most frequent pathogenic variant was reported to be p.Arg374Tyr.

PCSK9 was noted to play a role in regulation of LDLR recycling. Gain of function variants were found to be associated with high levels of LDL cholesterol and increased CAD risk.

Specific inhibitors have been designed to treat HDL cholesterol resulting from PCSK9 gain of function variants. These inhibitors include a PCSK9 monoclonal antibody.

Loss of function PCSK9 variants were noted to lead to low levels of LDL cholesterol. The lowered levels were associated with reduced incidence of CAD.

Trinder et al. (2020) carried out analyses of UK biobank data to determine monogenic and polygenic risk for hypercholesterolemia and risk of atherosclerotic cardiovascular disease. The monogenic loci analyzed included LDLR, APOB, and PCSK9. The polygenic risk score was based on analyses of 223 single nucleotide variants. Phenotypic measures analyzed included information on coronary and carotid vascularization, myocardial infarction, history of ischemic stroke, and death.

In total, 48,741 individuals were genotyped using arrays and exome sequencing. Data analyses revealed that individuals with monogenic familial hypercholesterolemia and those with polygenic hypercholesterolemia were at increased risk for cardiovascular events.

The polygenic variants designated with rs numbers (defined single nucleotide variants, defined SNPs) occurred in introns, in intergenic regions in 3′UTR regions, and also included nonsynonymous variants in coding regions.

Variants were defined as pathogenic based on use of 5 or 6 bioinformatic tools, Meta SVM, LRT, protein effect analyzer, Mutation Taster, Polyphen2 and Soring tolerant, Soring intolerant.

Participants with a polygenic risk score higher than the 95th percentile were reported to be polygenic.

Polygenic risk scores in Europeans and East Asians overlapped while in Africans, polygenic risk scores were lower and only partially overlapped with those of Europeans and Africans.

The cohort included an additional group of individuals and within a cohort of 50,243 individuals 2379 were defined as having polygenic hypercholesterolemia and 277 had monogenic hypercholesterolemia. In 2232 individuals, hypercholesterolemia could not be accounted for by genetic risk.

Trinder therefore defined three forms of hypercholesterolemia, monogenic, polygenic, and nongenetic.

13.26 Undiagnosed diseases and application of genetic and genomic studies

Splinter et al. (2018) drew attention to the fact that in a number of patients with longstanding and often severe diseases, diagnosis was not established despite extensive medical evaluation. An undiagnosed disease network (UDN) was established by the National Institutes of Health in 2014 that included clinical sites, sequencing cores, and a coordinating core. At the time of reporting in 2018, 1519 patients had been referred and 601 were accepted for full examination. In 11% of cases diagnoses were reported to have been established by clinical investigation while in 74% of cases diagnoses were made on the basis of exome or genome sequencing. Specific databases utilized to help establish diagnoses included dbGAP database of Genotypes and Phenotypes, Phenome Centre, OMIM, l Clinvar. Other databases utilized included Exomiser and HPO, defined as a standardized vocabulary of phenotypic abnormalities, Orphanet and Decipher database of genomic variants.

It is interesting to note that the UDN network facilities include a model organism screening center with a drosophila and zebrafish core for evaluating pathogenicity of variants.

Importantly, in some cases diagnosis revealed association of a clinical phenotype with a previously undescribed chromosomal region and gene. In some cases the clinical findings in a specific patient were found to be new phenotypic variants due to defects in a known disease gene. Following molecular diagnoses, the network team considers whether the molecular findings indicate specific medical therapies or medical management and counseling.

The primary symptoms in patients evaluated in the network were neurologic in 40%, musculoskeletal in 10%, immunologic in 7%, gastrointestinal in 7%, and rheumatologic in 6%. The average age of patients was 8 ± 5 years for pediatric patients and 43 ± 16 years for adult patients.

It is important to note that the UDN includes a metabolic core and metabolic studies were carried out to support clinical diagnoses when certain sequence abnormalities were found. Examples included metabolic studies in follow-up to discovery of missense variants in the gene that encodes ATP5F1D mitochondrial ATP synthase. Metabolic studies carried out in follow-up to discovery of a pathogenic variant in NADK2 mitochondrial NAD kinase revealed elevated levels of lysine in plasma and urine.

Model organism studies were carried out to confirm pathogenicity of variants in EBF3, a transcription factor, and defective functions of this factor have ben reported to be associated with hypotonia, ataxia, and delayed development syndrome. Model organism studies were carried out to obtain additional information on the functional spectrum of CACNA1A variants.

References

Ahn SY, Gupta C: Genetic programming of hypertension, *Front Pediatr* 5:285, 2018. Available from: https://doi.org/10.3389/fped.2017.00285. eCollection 2017.PMID: 29404309.

Alkuraya FS: 2020 Curt Stern Award address: a more perfect clinical genome-how consanguineous populations contribute to the medical annotation of the human genome, *Am J Hum Genet* 108(3):395–399, 2021. Available from: https://doi.org/10.1016/j.ajhg.2020.12.009. PMID: 33667393.

Antonarakis SE: Carrier screening for recessive disorders, *Nat Rev Genet* 20(9):549–561, 2019. Available from: https://doi.org/10.1038/s41576-019-0134-2. PMID: 31142809.

Ashizawa T, Öz G, Paulson HL: Spinocerebellar ataxias: prospects and challenges for therapy development, *Nat Rev Neurol* 14(10):590–605, 2018. Available from: https://doi.org/10.1038/s41582-018-0051-6. PMID: 30131520.

Ast T, Meisel J, Patra S. et al: Hypoxia rescues frataxin loss by restoring iron sulfur cluster biogenesis, *Cell* 117(6):1507–1521, 2019. Available from: https://doi.org/10.1016/j.cell.2019.03.045. PMID: 31031004.

Balciuniene J, DeChene ET, Akgumus G, et al: Use of a dynamic genetic testing approach for childhood-onset epilepsy, *JAMA Netw Open* 2(4):e192129, 2019. Available from: https://doi.org/10.1001/jamanetworkopen.2019.2129. PMID: 30977854.

Beaudin M, Matilla-Dueñas A, Soong BW, et al: The classification of autosomal recessive cerebellar ataxias: a consensus statement from the Society for Research on the Cerebellum and Ataxias Task Force, *Cerebellum* 18(6):1098–1125, 2019. Available from: https://doi.org/10.1007/s12311-019-01052-2. PMID: 31267374.

Boutry M, Morais S, Stevanin G: Update on the genetics of spastic paraplegias, *Curr Neurol Neurosci Rep* 19(4):18, 2019. Available from: https://doi.org/10.1007/s11910-019-0930-2. PMID: 30820684.

Brennenstuhl H, Jung-Klawitter S, Assmann B, Opladen T: Inherited disorders of neurotransmitters: classification and practical approaches for diagnosis and treatment, *Neuropediatrics* 50(1):2–14, 2019. Available from: https://doi.org/10.1055/s-0038-1673630. PMID: 30372766.

Buijsen RAM, Toonen LJA, Gardiner SL, van Roon-Mom WMC: Genetics, mechanisms, and therapeutic progress in polyglutamine spinocerebellar ataxias, *Neurotherapeutics*. 16(2):263–286, 2019. Available from: https://doi.org/10.1007/s13311-018-00696-y. PMID: 30607747.

Campuzano V, Montermini L, Moltò MD, et al: Friedreich's ataxia: autosomal recessive disease caused by an intronic GAA triplet repeat expansion, *Science*. 271 (5254):1423–1427, 1996. Available from: https://doi.org/10.1126/science.271.5254.1423. PMID: 8596916.

Carey RM, Calhoun DA, Bakris, et al: Resistant hypertension: detection, evaluation, and management: a scientific statement from the American Heart Association, *Hypertension*. 72(5):e53–e90, 2018. Available from: https://doi.org/10.1161/HYP.0000000000000084. PMID: 30354828.

Castro IH, Pignataro MF, Sewell KE, et al: Frataxin structure and function, *Subcell Biochem* 93:393–438, 2019. Available from: https://doi.org/10.1007/978-3-030-28151-9_13. PMID: 31939159.

Coppola A, Cellini E, Stamberger H, et al: Diagnostic implications of genetic copy number variation in epilepsy plus, *Epilepsia* 60(4):689–706, 2019. Available from: https://doi.org/10.1111/epi.14683. Epub March 13, 2019. PMID: 30866059.

Delatycki MB, Bidichandani SI: Friedreich ataxia—pathogenesis and implications for therapies, *Neurobiol Dis* 132:104606, 2019. Available from: https://doi.org/10.1016/j.nbd.2019.104606. PMID: 31494282.

Dimmock DP, Clark MM, Gaughran M, et al: An RCT of rapid genomic sequencing among seriously ill infants results in high clinical utility, changes in management, and low perceived harm, *Am J Hum Genet* 107(5):942–952, 2020. Available from: https://doi.org/10.1016/j.ajhg.2020.10.003. PMID: 33157007.

Fahey MC, Maclennan AH, Kretzschmar D, et al: The genetic basis of cerebral palsy, *Dev Med Child Neurol* 59(5):462–469, 2017. Available from: https://doi.org/10.1111/dmcn.13363. PMID: 2804267.

Galea CA, Huq A, Lockhart PJ, et al: Compound heterozygous FXN mutations and clinical outcome in Friedreich ataxia, *Ann Neurol* 79(3):485–495, 2016. Available from: https://doi.org/10.1002/ana.24595. PMID: 26704351.

Gibson LE, Cooke RE: A test for concentration of electrolytes in sweat in cystic fibrosis of the pancreas utilizing pilocarpine by iontophoresis, *GibsPediatrics* 23(3):545–549, 1959. PMID: 13633369.

Girdea M, Dumitriu S, Fiume M, et al: PhenoTips: patient phenotyping software for clinical and research use, *Hum Mutat* 34(8):1057–1065, 2013. Available from: https://doi.org/10.1002/humu.22347.24. PMID: 23636887.

Gordon-Weeks PR: Phosphorylation of drebrin and its role in neuritogenesis, *Adv Exp Med Biol* 1006:49–60, 2017. Available from: https://doi.org/10.1007/978-4-431-56550-5_4. PMID: 28865014.

Green ED, Gunter C, Biesecker LG, et al: Strategic vision for improving human health at the forefront of genomics, *Nature* 586(7831):683–692, 2020. Available from: https://doi.org/10.1038/s41586-020-2817-4. Epub October 28, 2020. PMID: 33116284.

Hamosh A, Sobreira N, Hoover-Fong J, et al: PhenoDB: a new web-based tool for the collection, storage, and analysis of phenotypic features, *Hum Mutat* 34(4):566–571, 2013. Available from: https://doi.org/10.1002/humu.22283. Epub March 4, 2013.PMID: 23378291.

Hebbar M, Mefford HC: Recent advances in epilepsy genomics and genetic testing. F1000Res. 2020 March 12;9:F1000 Faculty Rev-185. Available from: https://doi.org/10.12688/f1000research.21366.1. eCollection 2020.PMID: 32201576

Helbig I: Genetic causes of generalized epilepsies, *Semin Neurol* 35(3):288–292, 2015. Available from: https://doi.org/10.1055/s-0035-1552922. Epub June 10, 2015. PMID: 26060908.

Igo RP Jr, Kinzy TG, Cooke, Bailey JN: Genetic risk scores, *Curr Protoc Hum Genet* 104 (1):e95, 2019. Available from: https://doi.org/10.1002/cphg.95. PMID: 3176507.

Jin SC, Lewis SA, Bakhtiari S, et al: Mutations disrupting neuritogenesis genes confer risk for cerebral palsy, *Nat Genet* 52(10):1046–1056. Available from: https://doi.org/10.1038/s41588-020-0695-1, 2020.

Kaback M, Lim-Steele J, Dabholkar D, et al: Tay – Sachs disease–carrier screening, prenatal diagnosis, and the molecular era. An international perspective, 1970 to 1993. The International TSD Data Collection Network, *JAMA* 270(19):2307–2315, 1993.

Kara E, Tucci A, Manzoni C: Genetic and phenotypic characterization of complex hereditary spastic paraplegia, *Brain*. 139(Pt 7):1904–1918, 2016. Available from: https://doi.org/10.1093/brain/aww111. PMID: 27217339.

Khera AV, Kathiresan S: Genetics of coronary artery disease: discovery, biology and clinical translation, *Nat Rev Genet* 18(6):331–344, 2017. Available from: https://doi.org/10.1038/nrg.2016.160. Epub March 13, 2017. PMID: 28286336.

Koeleman BPC: Genetics of common forms of epilepsy, *Lancet Neurol* 16(2):101–102, 2017. Available from: https://doi.org/10.1016/S1474-4422(16)30400-8. PMID: 2810214.

Köhler S, Gargano M, Matentzoglu N, et al: The Human Phenotype Ontology in 2021, *Nucleic Acids Res* 49(D1):D1207–D1217, 2021. Available from: https://doi.org/10.1093/nar/gkaa1043. PMID: 33264411.

Lacolley P, Regnault V, Laurent S: Mechanisms of arterial stiffening: from mechanotransduction to epigenetics, *Arterioscler Thromb Vasc Biol* 40(5):1055–1062, 2020. Available from: https://doi.org/10.1161/ATVBAHA.119.313129. PMID: 32075419.

Lord J, McMullan DJ, Eberhardt RY, et al: Prenatal exome sequencing analysis in fetal structural anomalies detected by ultrasonography (PAGE): a cohort study, *Lancet*. 393 (10173):747–757, 2019. Available from: https://doi.org/10.1016/S0140-6736(18)31940-8. PMID: 30712880.

Marras C, Lang A, van de Warrenburg BP: Nomenclature of genetic movement disorders: recommendations of the international Parkinson and movement disorder society task force, *Mov Disord* 31(4):436–457, 2016. Available from: https://doi.org/10.1002/mds.26527. PMID: 27079681.

Matthews AM, Blydt-Hansen I, Al-Jabri B, et al: Atypical cerebral palsy: genomics analysis enables precision medicine, *Genet Med* 21(7):1621–1628, 2019. Available from:

https://doi.org/10.1038/s41436-018-0376-y. Epub December 13, 2018. PMID: 30542205.

McMichael G, Bainbridge MN, Haan E, et al: Whole-exome sequencing points to considerable genetic heterogeneity of cerebral palsy, *Mol Psychiatry* 20(2):176–182, 2015. Available from: https://doi.org/10.1038/mp.2014.189. Epub February 10, 2015. PMID: 25666757.

Melville S, Byrd JB: Personalized medicine and the treatment of hypertension, *Curr Hypertens Rep* 21(2):13, 2019. Available from: https://doi.org/10.1007/s11906-019-0921-3. PMID: 30747306.

Musunuru K, Kathiresan S: Genetics of common, complex coronary artery disease, *Cell.* 177(1):132–145, 2019. Available from: https://doi.org/10.1016/j.cell.2019.02.015. PMID: 30901535.

Nakamori M, Panigrahi GB, et al: A slipped-CAG DNA-binding small molecule induces trinucleotide-repeat contractions in vivo, *Nat Genet* 52(2):146–159, 2020. Available from: https://doi.org/10.1038/s41588-019-0575-8. PMID: 32060489.

Newcombe PJ, Nelson CP, Samani NJ, Dudbridge F: A flexible and parallelizable approach to genome-wide polygenic risk scores, *Genet Epidemiol* 43(7):730–741, 2019. Available from: https://doi.org/10.1002/gepi.22245. PMID: 31328830.

Ng J, Heales SJ, Kurian MA: Clinical features and pharmacotherapy of childhood monoamine neurotransmitter disorders, *Paediatr Drugs* 16(4):275–291, 2014. Available from: https://doi.org/10.1007/s40272-014-0079-z. PMID: 25011953.

Oskoui M, Coutinho F, Dykeman J, et al: An update on the prevalence of cerebral palsy: a systematic review and *meta*-analysis, *J Dev Med Child Neurol* 55(6):509–519, 2013. Available from: https://doi.org/10.1111/dmcn.12080. Epub January 24, 2013. PMID: 23346889.

Panteliadis C, Panteliadis P, Vassilyadi F: Hallmarks in the history of cerebral palsy: from antiquity to mid-20th century, *Brain Dev* 35(4):285–292, 2013. Available from: https://doi.org/10.1016/j.braindev.2012.05.003. PMID: 22658818.

Pauling L, Itano HA, et al: Sickle cell anemia, a molecular disease, *Science.* 109 (2835):443, 1949. PMID: 18213804.

Perucca P, Perucca E: Identifying mutations in epilepsy genes: impact on treatment selection, *Epilepsy Res* 152:18–30, 2019. Available from: https://doi.org/10.1016/j.eplepsyres.2019.03.001. PMID: 30870728.

Pope S, Artuch R, Heales S, Rahman SJ: Cerebral folate deficiency: analytical tests and differential diagnosis, *Inherit Metab Dis* 42(4):655–672, 2019. Available from: https://doi.org/10.1002/jimd.12092. Epub May 2, 2019. PMID: 30916789.

Savatt JM, Myers SM: Genetic testing in neurodevelopmental disorders, *Front Pediatr* 9:526779, 2021. Available from: https://doi.org/10.3389/fped.2021.526779. eCollection 2021.PMID: 33681094.

Sharma S, Prasad AN: Inborn errors of metabolism and epilepsy: current understanding, diagnosis, and treatment approaches, *Int J Mol Sci* 18(7):1384, 2017. Available from: https://doi.org/10.3390/ijms18071384. PMID: 2867158.

Shribman S, Reid E, Crosby AH, et al: Hereditary spastic paraplegia: from diagnosis to emerging therapeutic approach, *Lancet Neurol* 18(12):1136–1146, 2019. Available from: https://doi.org/10.1016/S1474-4422(19)30235-2. PMID: 31377012.

Splinter K, Adams DR, Bacino CA, et al: Effect of genetic diagnosis on patients with previously undiagnosed disease, *N Engl J Med* 379(22):2131–2139, 2018. Available from: https://doi.org/10.1056/NEJMoa1714458. Epub October 10, 2018. PMID: 30304647.

Stamatoyannopoulos G: The molecular basis of hemoglobin disease, *Annu Rev Genet* 6:47−70, 1972. Available from: https://doi.org/10.1146/annurev.ge.06.120172.000403. PMID: 4581487.

Steward CA, Roovers J, Suner MM, et al: Re-annotation of 191 developmental and epileptic encephalopathy-associated genes unmasks de novo variants in *SCN1A*, *NPJ Genom Med* 4:31, 2019. Available from: https://doi.org/10.1038/s41525-019-0106-7. eCollection 2019.PMID: 31814998.

Synofzik M, Schüle R: Overcoming the divide between ataxias and spastic paraplegias: shared phenotypes, genes, and pathways, *Mov Disord* 32(3):332−345, 2017. Available from: https://doi.org/10.1002/mds.26944. Epub February 14, 2017. PMID: 28195350.

Trinder M, Francis GA, Brunham LR: Association of monogenic vs polygenic hypercholesterolemia with risk of atherosclerotic cardiovascular disease, *JAMA Cardiol* 5(4):390−399, 2020. Available from: https://doi.org/10.1001/jamacardio.2019.5954. PMID: 32049305.

van Karnebeek CDM, Sayson B, Lee JJY, et al: Metabolic evaluation of epilepsy: a diagnostic algorithm with focus on treatable conditions, *Front Neurol* 9:1016, 2018. Available from: https://doi.org/10.3389/fneur.2018.01016. eCollection 2018.PMID: 30559706.

von Stülpnagel C, van Baalen A, Borggraefe I, et al: Network for therapy in rare epilepsies (NETRE): lessons from the past 15 years, *Front Neurol* 11:622510, 2021. Available from: https://doi.org/10.3389/fneur.2020.622510. eCollection 2020.PMID: 33519703.

Waggoner D, Wain KE, Dubuc AM, et al: Yield of additional genetic testing after chromosomal microarray for diagnosis of neurodevelopmental disability and congenital anomalies: a clinical practice resource of the American College of Medical Genetics and Genomics (ACMG), *Genet Med* 20(10):1105−1113, 2018. Available from: https://doi.org/10.1038/s41436-018-0040-6. Epub June 18, 2018. PMID: 29915380.

Warren HR, Evangelou E, Cabrera CP, et al: Genome-wide association analysis identifies novel blood pressure loci and offers biological insights into cardiovascular risk, *Nat Genet* 49(3):403−415, 2017. Available from: https://doi.org/10.1038/ng.3768. Epub January 30, 2017. PMID: 28135244.

Further reading

EpiPM Consortium: A roadmap for precision medicine in the epilepsies, *Lancet Neurol* 14(12):1219−1228, 2015. Available from: https://doi.org/10.1016/S1474-4422(15)00199-4. PMID: 26416172.

Sturm AC, Knowles JW, Gidding SS, et al: Clinical genetic testing for familial hypercholesterolemia: JACC Scientific Expert Panel, *J Am Coll Cardiol* 72(6):662−680, 2018. Available from: https://doi.org/10.1016/j.jacc.2018.05.044. PMID: 30071997.

CHAPTER 14

Using insights from genomics to increase possibilities for treatment of genetic diseases

14.1 Introduction

Accurate mapping of specific genetic disease to particular gene loci and analyses of specific gene changes at mapped loci and specific changes in the gene product that alter function and that lead to disease, have direct relevance to treatment of a growing number of genetic diseases.

In considering disease treatment it is also important to understand how disease function disruptions lead to disease manifestations and altered phenotypes.

In some disorders, disease treatments are based on use of physiological, biochemical, or pharmacological substances to improve function. Treatments may in some cases involve dietary manipulations, for example, in phenylketonuria. In some cases specific proteins, cells, tissue, or organs are used to replace essential substances or functions.

Increasingly, therapies are being designed that directly impact abnormal genes, or abnormal RNAs or to add normal versions of genes to body systems.

In some disorders, treatments have regulatory approval, for example, from the Food and Drug Administration (FDA), United States or European Medicines Agency and are in active use. In other disorders treatments are at earlier stage of investigations and may be in clinical trials.

Collins et al. (2021) noted that discoveries of genes implicated in 5000 rare Mendelian diseases have had significant clinical consequences. Identification of specific gene defects serve to initiate studies on diseases mechanisms. In a growing number of cases, accurate diagnostics have had implications for therapy.

Collins et al. emphasized that diagnosis of disease had advanced but that development and validating of treatment remains challenging. Treatment approaches include development of therapies targeted to the specific molecular defects, as in cystic fibrosis and therapies designed to target specific gene defects.

In a 2021 editorial, Tremblay et al. noted that more than 7000 rare diseases in humans have been documented and that in the United States more than 30 million individuals are reported to be affected by these diseases.

In this chapter different types of therapies will be considered, those that replace missing substances or proteins or enhance functions of specific proteins or enzymes and those that involve gene therapy. It is important to note that both of these approaches may be used to treat a specific disease. For that reason it seems most fruitful to organize discussions of therapies in part on the basis of disease types.

In the following sections therapies of specific diseases will be presented.

14.2 Therapy lysosomal diseases

Bonam et al. (2019) noted that lysosomes have particularly been studied in the context of lysosomal storage diseases. However, lysosomes have also been determined to play roles in inflammatory and autoimmune diseases, in neurodegenerative diseases, in cancer, and in metabolic diseases. Lysosomes were noted to recover material for degradation via endocytosis and autophagy. There is also evidence that lysosomes play roles in nutrient sensing (Settembre and Medina, 2015).

Lysosomal storage diseases are treated in some cases with enzyme replacement therapy (ERT). Bonam et al. reviewed possibilities for targeting lysosomes with small molecules and peptides. Studies are ongoing to investigate therapies based on targeting autophagy upstream of lysosomes.

Bonam et al. reviewed roles of lysosomes in more common diseases in inflammatory and autoimmune diseases. Strategies were investigated to decrease excessive lysosomal activity. Lysosomes were noted to contain almost 60 different hydrolytic enzymes that have been grouped into categories that include nucleases, proteases, lipases, phosphatases, and sulfatases. An acid pH exists within lysosomes due to activity of acid hydrolase. Important factors in maintenance of this acidic interior include an outer glycocalyx membrane and activity of vacuolar proton ATPase that dives protons to the interior using energy derived from ATP.

Other important components to maintain lysosomal function include lysosomal-associated membrane protein LAMP1, LAMP2, and membrane GTPases, including MTOR1-associated GTPases. LAMP2 defective function was shown to lead to Danon disease, also known as X-linked vacuolar cardiomyopathy and myopathy. Important channels within lysosomes include cation ion channels including MCOLN1 mucolipin and TRP cation channel 1, member of transient receptor potential cation channel family. MCOLN1 was shown to play important roles in functions of endosomes, autophagosomes, and lysosomes.

Bonam et al. noted that autophagy and lysosomal systems play roles in foreign particle recognition. Furthermore, there is evidence that lysosomes play roles as signaling platforms. Lysosomes were noted to play important roles in cellular homeostasis.

Dysfunction of lysosomes was noted to also affect functioning of other organelles including mitochondria and peroxisomes.

Bonam et al. emphasized the key roles of lysosomes in the immune system. They noted that increased lysosomal activity was reported in several immune disease including systemic lupus erythematosus, psoriasis, Sjogren syndrome, multiple sclerosis. Lupus erythematosus was reported to be associated with altered antigen presentation. Lysosomal cathepsins were reported to be increased in rheumatoid arthritis.

Lysosomal functions were reported to be impaired in certain neurodegenerative disorders, including amyotrophic lateral sclerosis (ALS), Alzheimer disease, Parkinson's disease, Huntington disease (HD). In Alzheimer disease, lysosomal luminal balance was reported to be disturbed. In Parkinson's disease impaired glucocerebrosidase may be present. The lysosomal transport pathway was reported to be altered in HD.

Given evidence of their involvement in common diseases, Bonam et al. noted the importance of considering therapies that target lysosomes.

Bonam et al. noted that selective pharmacological molecules being designed to target lysosome include compounds related to chloroquine, drugs related to rapamycin, and drugs that target vacuolar proton ATPase. In rheumatoid arthritis, cathepsin modulators are being investigated.

14.2.1 Enzyme replacement therapy in clinical use for treatment of specific lysosomal storage diseases

Disorder	Enzyme replacement product
Gaucher	Imiglucerase, Velaglucerase alfa, Taliglucerase alfa
Fabry	Agalsidase beta, Agalsidase alfa
MPS I Hurler–Scheie	Laronidase
MPS II Hunter	Idursulfatase, Idursulfatase beta
MPS VI Maroteaux–Lamy	Galsulfase
MPS VII Sly	Vestronidase alfa
Pompe	Alglucosidase alfa
Wolman disease	Sebelipase alfa

It is important to note that Fabry disease, Hunter disease, and Danon disease are X-linked diseases.

For treatment of lysosomal storage, disease substrate reduction therapies are also used, for example, miglustat is a glucosylceramide synthase inhibitor used to treat Nieman–Pick disease and also to treat mild Gaucher disease. Other agents related to miglustat are also being produced to reduce side effects.

Genistein is a natural product used to treat Sanfilippo disease. This disorder is associated with defects in breakdown of heparan and may be due to defects in any one of four enzymes. Gene therapy trials are ongoing.

Competitive inhibitors of certain enzymes are being investigated as potential methods to facilitate substrate reduction. Studies are also being carried out to assess therapeutic impact of ion channel modulators on lysosomal storage diseases.

14.3 Neuroimmune disorders, autophagy lysosomes, and treatment

Bonam et al. (2019) noted that there is some evidence that impairment of autophagy processes may play roles in neuroimmune diseases. In addition, there is evidence for inefficiency of lysosome endosomal systems in neurodegenerative diseases that may lead to inefficient clearing of neurotoxic proteins.

They noted that certain drugs, for example, chloroquine that target lysosomes also target other organelles. Selective pharmacological molecules being designed to target lysosome include modified forms of chloroquine, drugs related to rapamycin, and drugs that impact vacuolar proton ATPase.

These investigators emphasized "although the cellular pathogenesis of LSDs is complex and still not fully understood, the approval of disease-specific therapies and the rapid emergence of novel diagnostic methods led to the implementation of extensive national newborn screening (NBS) programs in several countries."

Other forms of lysosomal disease for which defined therapies are still under investigation or in clinical trials include sphingolipidoses, neuronal ceroidlipofuscinoses (CLN1–14), glycoproteinoses, glycogen storage diseases, lipid storage diseases, posttranslational modifications disorders, mucolipidoses, multiple sulfatase deficiency, integral membrane protein disorders, Niemann–Pick NPC1, NPC2, cystinosis, mucolipidois MCOLN, sialidosis, SCARB2 deficiency (myoclonus), sialic acid storage disease, Danon disease (LAMP2), lysosomal-related organelle diseases, Hermansky–Pudlak syndrome, Griscelli syndrome, Chediak–Higashi syndrome.

Collectively, lysosomal storage diseases were reported to occur in 1 in 5000 live births. Lysosomal storage diseases and lysosome-related organelle disorders were considered.

Major disadvantages of ERT include inability of the replaced enzyme to diffuse throughout tissues, for example, in the central nervous system (CNS). Hematopoietic stem cell (HSC) therapy is also used but the mortality of the required transplantation procedures are high due to infection and graft versus host reactions. Substrate reduction therapy, for example, with miglustat is used in Gaucher disease. Another form of therapy is being explored to promote posttranslational modification and appropriate folding and trafficking of enzyme proteins Suzuki, 2014).

14.3.1 Chaperone therapies in the treatment of lysosomal storage diseases

Suzuki (2021) reviewed the development and use of chaperones to treat specific lysosomal storage diseases. He defined these chaperones as small molecules that bind to misfolded and unstable enzymes and can promote correct folding and increase stability. One chaperone developed to treat Fabry disease is migalastat. Chaperones have also been developed to treat Gaucher disease.

14.4 Mucopolysaccharidosis II (Hunter syndrome)

Mc Bride et al. (2020) noted that a specific ERT was in phase III trials in the United States. Participation in these trials was specifically approved for mucopolysaccharidosis (MPS) II in older individuals with milder forms of Hunter syndrome.

The population incidence of this syndrome was quoted as being between 1 in 60,000 and 1 in 150,000 and the incidence was stated to be higher in the Ashkenazi Jewish Population than in other populations. A series of different pathogenic changes that alter the function of iduronate sulfatase can lead to this disorder. Specific defects identified include decreased production, decreased catalytic activity, and protein misfolding.

Clinical manifestations of Hunter syndrome include skeletal alteration, joint stiffness, facial coarsening, organomegaly, increased ear infections, airway obstruction, and cognitive impairment may also occur. Cardiac defects may arise. McBride et al. noted that deposition of glycosoaminoglycans in heart valvular tissue impairs vascular function.

The clinically approved treatment trial includes intravenous administration of idursalfase (Elaprase) and was reported to lead to improvement in clinical manifestations in individuals who were not severely affected at the start of the trial.

An additional clinical trial involves use of intrathecal enzyme administration in more severely affected individuals.

McBride et al. noted the cost of ERT and remaining uncertainties of its efficacy. A specific American College of Medical Genetics and Genomics Committee was established to review MPS II therapy. Results of the study of this committee led to the conclusion that ERT should be considered in individuals with signs or symptoms of MPS II, including individuals with attenuated manifestations and individuals with severe manifestations. The committee also noted that home infusions could be considered in stable environments.

14.4.1 Precision medicine in lysosomal storage disease disorders

Pinto E Vairo et al. (2020) reviewed precision medicine in lysosomal disorders. They defined precision medicine as taking into account genetics, environment, and lifestyle. The terms precision medicine and personalized medicine were noted to be used interchangeably.

They defined different categories of lysosomal storage diseases including disorders associated with lysosomal enzyme deficiency, altered activity of lysosomal membrane components, and altered function of modifiers of lysosomal components. They reported the combined incidence of lysosomal disorders to be 1 in 4000.

Pinto E Vairo et al. focused on diagnostic technologies and clinical management of patients with lysosomal disorders. The clinical manifestations of lysosomal disorders were noted to be in part related to abnormal storage of specific substances and on the specific location of accumulated undegraded material.

Many of these disorders were noted to lead to facial coarsening, visceromegaly, skeletal changes, and to CNS impairments.

Diagnostic work-up includes biomarker detection in blood, urine, and cerebrospinal fluid (CSF) and enzyme assays.

Advances in treatment were noted including available enzyme replacement therapies. They also reviewed effectiveness of HSC transplantation. In specific conditions, HSC transplantation was noted to be associated with high rates of transplant complications including Fabry disease, lysosomal acid lipase deficiency, Pompe disease, and neuronal ceroid lipofuscinosis.

Specific conditions noted to have available ERT included: alpha mannosidosis, Fabry disease, Gaucher disease type 1, lysosomal acid lipase deficiency, MPS types I, II, IVA VI, VIII, neuronal ceroid lipofuscinosis, Pompe disease.

In addition to defined enzyme replacement therapies for lysosomal storage diseases, investigators noted that in 2020 there were ongoing clinical trials for a number of other lysosomal storage diseases.

14.4.2 Substrate reduction therapy

Pinto E Vairo et al. noted that substrate reduction therapy was approved for the therapy of specific lysosomal storage diseases including Gaucher disease and Niemann—Pick disease.

Substrate reduction therapy was noted to be in clinical trials for other lysosomal storage diseases including Fabry disease, Pompe disease, and GM2 gangliosidosis.

They also noted the importance of other aspects of treatment and management of patients with lysosomal disorders, including supportive care availability of hearing aids and orthoses, physical therapy, respiratory care, early detection and treatment of complications, diagnostic testing of lysosomal storage diseases through next generation sequencing.

Next generation sequencing was noted to have significant impact on lysosomal storage diseases and at many sites specific gene panels are used for diagnostic testing. Molecular diagnosis was noted to be particularly important in milder cases and in atypical cases.

Pinto E Vairo et al. noted that in 2020 the particular value of metabolomic studies was to assess whether or not specific therapies are effective.

In addition to defined approved therapies for lysosomal storage disorders, in 2020 there are ongoing clinical trials in a number of lysosomal disorders and also applications for new modes of therapy. Pinto E Vairo et al. documented the following:

Acid sphingomyelinase deficiency	Transplantation with donor-derived oligodendrocyte-like cells
Alpha mannosidase deficiency	Transplantation with donor-derived oligodendrocyte-like cells
Aspartylglucosaminuria	Chaperone therapy
Cystinosis	Chaperone

(Continued)

Continued

Acid sphingomyelinase deficiency	Transplantation with donor-derived oligodendrocyte-like cells
Danon disease	Gene therapy AAV liver directed
Fabry disease	Substrate reduction therapy in combination with enzyme replacement therapy
Gaucher disease	Substrate reduction therapy, gene therapy
GM2 gangliosidosis	Substrate reduction therapy
Krabbe disease	Transplantation with umbilical cord-derived oligodendroglial cells
GM1 gangliosidosis	Gene therapy

It is important to note that in clinical trials, in some cases, cell therapies or gene therapies are used in intrathecal administration to treat brain lesions.

Pinto E Vairo et al. reviewed applications of small molecules in treatment of lysosomal storage diseases. Small molecules are defined as low molecular weight synthetic compounds. Small molecules therapeutically used include molecules that enhance enzyme stability, small molecules that influence gene transcription, for example, molecules that facilitate readthrough of premature termination codon mutations.

Miglustat is a substrate synthesis inhibitor. It is a synthetic analog of D-glucose and is used in treatment of a number of different lysosomal storage diseases. Miglustat can act as a glucosylceramide synthase inhibitor; it has been used in treatment of Niemann–Pick disease.

Other small molecules under consideration for therapies include chaperones that counteract protein misfolding that can arise due to missense variants. One such molecule is migalastat that has been approved for treatment of Fabry disease.

Ambroxol is a small molecule that was reported to enhance enzyme stability and is in clinical trials for treatment of certain lysosomal storage diseases.

Small molecules that promote readthrough of premature termination codons include ataluren and certain aminoglycosides.

Premature termination codon mutations can be counteracted by nonsense mediated mRNA decay, a quality control mechanism (Finkel, 2010).

Pinto E Vairo et al. described treatments under investigation that constitute next generation ERT and included intraventricular and intrathecal therapy in this category. Strategies designed to facilitate the capacity of large molecules to cross the blood–brain barrier include the Trojan Horse approach. This was reported to involve generation of a fusion molecule where a large molecule to be transported is coupled to a molecule that binds to a receptor on membranes that form the blood–brain barrier. In some cases, the molecule to promote transfer is an antibody that targets specific membrane receptors on epithelial cells, for example, transferrin receptor or insulin receptor.

Tosi et al. (2020) reported that biocompatible and biodegradable nanoparticles with specific surface properties can facilitate delivery of drugs and certain specifically macromolecules across the blood—brain barrier through interactions with the brain capillary endothelium (Trojan Horse strategy).

14.5 Strategies designed to increase the half-life of enzymes used in enzyme replacement strategies being considered in therapy

Pinto E Vairo noted that enzymes used in therapy often have short half-lives and frequent transfusion is required. Specific strategies are being designed to address this problem including modifying the formulas, for example, using pegylated formulation of enzyme. Other strategies being considered are including enzymes in capsules that are implanted. It is, however, important that the capsules do not induce immune response.

It is important to note that antisense oligonucleotide therapies are being investigated in certain disorders classified as inborn errors of metabolism, where the underlying gene mutation impacts a splice site, for example, Pompe disease (Bellotti et al., 2020).

14.5.1 Gene therapy in lysosomal storage diseases (Fabry disease)

Khan et al. (2021) noted that enzyme replacement and chaperone therapies have been developed to treat Fabry disease but noted that frequent infusions are required and that these are intrusive. They noted that these therapies were not always particularly beneficial. They therefore undertook a clinical trial of gene therapy in five adult patients. This trial involved the use of autologous CD34 progenitor stem cells that were treated with lentivirus, which carries the normal alpha galactosidase gene. The patients in this trial did not undergo myeloablation prior to therapy.

In follow-up studies to the gene therapy, they assayed circulating levels of active alpha galactosidase in patient plasma and determined that these levels were higher than those found in the patients before therapy. They also noted that three of the five gene therapy treated patients elected to discontinue enzyme replacement infusions.

14.6 Gene-directed treatments

Collins et al. (2021) noted that a number of gene editing trials using Crispr-Cas were in place. However, long-term safety of gene editing was thought to be as yet undetermined.

Particularly relevant to therapy are gene editing approaches for single gene disorders where a specific cell type is targeted. Targeting of reagents for gene therapy or for gene editing was noted to remain a challenge. Intense efforts were being applied to identify appropriate methods for therapeutic targeting of gene mutations.

Tremblay et al. (2021) noted that initially matched HSCs were used to treat specific genetic diseases. Following preclinical studies, therapies were undertaken that included integration of a functional gene into patient HSCs. Problems that emerged in some of those studies included aberrant integration of the inserted gene into sites in the patient's genome that led in rare cases to mutagenesis.

This finding was then followed by efforts to direct the integration of the therapeutic gene into safe sites in the genome.

In cancer therapy, genetic techniques were developed to block inhibitory checkpoint molecules that inhibited destruction of tumor cells by patient's CD4 immune cells.

Although progress has been achieved, significant hurdles remain. Tremblay et al. noted that these include genotoxicity of vectors used to transfer genes or editing reagents and also difficulties in analysis and control of regulation.

Preclinical studies to investigate efficacy of gene therapy are often carried out in animal models of the specific disease. Other possibilities for preclinical investigation involve use of pluripotent stem cells and especially of cell- and tissue-specific organoids developed from pluripotent stem cells (Clevers, 2016; Kim et al., 2020a, b).

Adeno-associated viral (AAV) vectors have become the most commonly used viral vectors for gene therapy as they do not integrate into the host genome. Mendell et al. (2021) also noted that gamma-retroviral vectors have been used in some trials.

These investigators also reviewed AAV safety concerns that include hepatotoxicity risk and noted that studies suggest that patients should be tested prior to therapy to ensure that they do not have antibodies to adenoviral vectors. There is also evidence that lowest possible effective vector doses be administered.

Mendell et al. emphasized that "hundreds of researchers have dedicated their lives to the treatment of hereditary disease and goals included curative repair or minimization of symptoms." They noted that, in 2021 five gene therapy treatments have been approved for clinical use and that many gene therapy clinical trials are ongoing.

Gene therapies and gene therapy trials are often directed toward treatment of diseases diagnosed in young individuals. Increasingly, efforts are being extended to develop treatments for diseases that occur in older individuals.

Mendell et al. also noted advances in investigation of intracerebral gene transfer methods.

Byrne et al. (2021) reviewed aspects of systemic gene therapy and that the goal of this was to reach widespread body sites impacted by genetic disease. They emphasized increased use of adeno-associated viruses that remain as episomes in cells.

Byrne et al. stressed the importance of understanding the molecular pathogenesis of genetic disorders and how this has opened the way to therapy.

These authors noted that it is important to understand the tropism of vectors used in therapy and also to understand the possible immunotoxicity of vectors. Different serotypes of adenovirus are now being used in therapy. Another consideration relates to the possible development of adaptive humoral responses to the viral capsid protein. Antibodies can potentially also be produced to the protein produced by the inserted gene in cases that prior to therapy were totally deficient in that protein as a result of null mutations in the relevant gene.

Rossi and Rossi (2021) drew attention to the growing importance of RNA interference therapies in genetic diseases. RNA interference was noted to have been first described by Fire et al. (1998). They reported that 21 base-pair RNA duplexes could lead to degradation of complementary mRNAs. They described these duplexes as short inhibitory RNAs (siRNAs). There was evidence that siRNAs could be generated in the laboratory and that it was possible to deliver siRNAs to cell via lipid nanoparticles or other carriers. Studies were subsequently carried out to determine if siRNAs could be used in therapy.

Rossi and Rossi noted that in 2018 the FDA approved first siRNA for therapy, Patisiran. It was approved for treatment of hereditary transthyretin-mediated amyloidosis due to defects in the transthyretin gene, TTR gene. The Patisiran siRNA was reported to bind to mutant TTR mRNA and prevented it from producing the mutant protein that is misfolded and forms deposits.

Givosiran was noted to be the second FDA-approved siRNA and it is specifically used to treat acute hepatic porphyria due to defects in the aminolevulinate synthetase gene ALAS1. A specific conjugation modification of givosiran leads it to bind to a specific liver receptor.

The third siRNA approved for clinical use by the FDA is Lumasiran (Oxlumo) used to treat hyperoxaluria I due to deficient detoxification of glyoxylate in the liver that arise secondary to defects in the enzyme serine pyruvate aminotransferase. In this case the antisense RNA targets an enzyme glycolate oxidase that is involved in the synthesis of glyoxylate.

Rossi et al. noted the challenge of moving siRNAs beyond diseases that involve the liver. Strategies being investigated involve delivery of siRNAs into tissues with coated microneedles. Another challenge is to limit cellular destruction of inhibitory siRNAs through their uptake into endosomes.

14.7 Hemoglobinopathy treatment including gene therapy

In a 2021 review Orkin reported that advances in studies on gene structure and regulation of expression have paved the way to curative treatments in hemoglobin disorders, thalassemia, and sickle cell disease. Specific important advances include improved lentiviral vectors that facilitate gene transfer, elucidation of regulatory factors in globin expression, and improved gene engineering.

A main strategy of gene therapy in hemoglobinopathies has been directed to increase of fetal gamma globin production.

ZYNTEGLO is the first gene therapy approved for transfusion-dependent β-thalassemia (TDT), European marketing authorization for ZYNTEGLO.

Orkin and Bauer (2019) reviewed emerging gene therapy in sickle cell disease. They noted that current therapy did not rely on gene therapy.

The molecular defect in sickle cell disease was first detected by electrophoresis by Pauling et al. (1949). Eight years later Ingram (1957) identified the amino acid abnormality in sickle cell disease. Orkin drew attention to recent studies by Shriner and Rotimi (2018) that revealed that a common haplotype occurred in the beta globin gene regions of many cases of sickle cell disease. They classified haplotypes by using 27 polymorphisms in linkage disequilibrium with rs334, the sickle mutation in classified haplotypes by using 27 polymorphisms in linkage disequilibrium with rs334 in Central African Republic (CAR), Cameroon, and Arabian/Indian haplotypes. The near-exclusive presence of the original sickle haplotype was also identified in the CAR, Kenya, Uganda, and South Africa. The heterozygote advantage was determined to be 15.2%.

Orkin et al. noted that the clinical presentation, severity, and disease course in individual sickle cells patients is considered unpredictable. A major modifier of clinical course was noted to be the level of expression of gamma globin (fetal hemoglobin) beyond the postnatal period.

Therapy for sickle cell disease was noted to include administration of hydroxyurea in patients with complications. Hydroxyurea was noted to increase level of expression of fetal hemoglobin, to reduce painful crises possibly related to its capacity, to increase the size of red blood cell that may reduce sickling. However, Orkin et al. noted that hydroxyurea therapy is considered to be a "halfway measure."

Bone marrow transplantation is an important treatment for sickle cell patients. However, it requires appropriate donor cells and Orkin noted problems with the ethnic makeup of cell banks that potentially supply donor cells.

Another approach to therapy involves the use of small molecules that impact the sickle hemoglobin polymer (HBS) and agents that shift the oxygen dissociation curve to reduce formation of deoxyhemoglobin that favors sickling.

Orkin et al. reviewed genetic control of fetal hemoglobin production. The beta globin cluster includes epsilon, beta gamma, and alpha gamma. An upstream locus control region is an enhancer that loops to sequentially activate each gene in the gamma cluster. Switch to production of primarily beta globin occurs after the first few months of postnatal life.

Hereditary persistence of gamma globin occurs in individuals with specific deletions in the beta cluster and individuals with specific variants in the gamma gene promoter region. The application of Genome-wide Association Studies (GWAS) revealed that a locus on chromosome 2p BCL11A influences ongoing expression of gamma globin. A second important locus for ongoing gamma globin production was mapped to chromosome 6q. This locus is designated HBs1l-Myb and it was reported to be a quantitative trait locus for HbF production.

Orkin et al. noted that there are difficulties in gene editing in sickle cell disease due to the difficulty in obtaining sufficient HSC from sickle patient. A specific treatment used to stimulate stem cell production in sickle patients can lead to sickling crises.

14.8 Gene-directed therapies in clinical trials in hemoglobinopathies

In January 2021 two papers were published documenting results of clinical trials designed to treat beta globin hemoglobinopathies, through increased expression of gamma globin (fetal hemoglobin).

Frangoul et al. reported results of studies in two separate groups of patients, one group with blood transfusion-dependent beta thalassemia and a second group with sickle cell disease.

The patients with beta thalassemia had reduced or absent synthesis of beta globin, imbalance of alpha and beta globin levels, and ineffective hemopoiesis. These patients required frequent blood transfusion and iron chelation therapy.

The patients with sickle cell disease had episodes of hemolysis, anemia, painful episodes due to veno-occlusion, and in some cases stroke episodes and end organ damage.

In both groups, beta thalassemia patient's and sickle cell disease patient's disease manifestations were less severe in those patients whose gamma globin levels were higher.

Comprehensive studies on globin synthesis in past years revealed that the transcription factor BCL11A acts to suppress gamma globin production from late fetal life on. This suppression is in part due to expression of a specific enhancer element.

14.9 Hemoglobinopathy treatment through genetic silencing of BCL11A expression using antisense strategy

Frangoul et al. (2021) undertook studies to investigate the use of CRISPR-CAS gene editing in deleting the specific enhancer of BCL11A that suppresses gamma globin production in patient's autologous bone marrow cells. Gene editing of these cells was carried out ex-vivo. Following successful editing to eliminate BCL11A expression, genome sequencing of edited cells was carried out to ensure absence of off-target effects. Following blood cell transfusions, a brief bone marrow ablation procedure was carried out, the edited bone marrow stem cells were transfused. Patients were followed to determine if the transfused stem cells had survived and were producing appropriate hemopoietic lineage cells.

Frangoul et al. reported that following these procedures high levels of fetal hemoglobin production occurred even after 1 year. The thalassemia patients became transfusion independent and the sickle cell disease patients had no further veno-occlusive episodes.

Esrick et al. (2021) reported results of a clinical trial that involved six patients with sickle cell disease who were transplanted with CD34 autologous bone marrow stem cells that had been transfected with a short antisense RNA targeting BCL11A. Patients were followed for at least 6 months following this treatment and stable induction of high levels of gamma globin were reported.

The transfusion in these cases included between 4.9 million and 8.3 million antisense RNA transfected CD34 bone marrow stems cells.

The authors reported that neutrophil production was achieved by the transplanted bone marrow cells and gamma globin and fetal hemoglobin levels were high and remained stable.

14.10 Splice mutations and diseases

Aberrant splicing occurs in a number of different diseases. Montes et al. (2019) reviewed splicing defects and developments in animal models to help develop therapies. They noted that mistakes in splicing can lead to alteration in transcript code and also to frameshift mutations.

Accurate splicing was noted to be dependent on factors within transcript splice sites, in associated sequences, and in cofactors involved in splicing. They emphasized that transcripts of a specific gene may undergo splicing at different sites and that there is evidence that 95% of genes are alternatively spliced. Alternative splicing of a specific gene may occur at different stages of development, or in different tissue.

Montes et al. noted the importance that animal models can play in studies of splicing and impact of its alterations and that studies of such models are important for development of therapies directed at aberrant splicing.

Important elements in splicing include splice regulatory elements (SREs) that occur on primary transcripts and other important elements include trans factors, RNA binding proteins such as serine-arginine rich proteins (SR proteins), and heterogeneous ribonucleoproteins (hnRNPs).

Montes et al. noted that 10% of all mutations documented in human disease mutation databases impacted splice sites. In the Human Mutation database (Cooper et al.) mutations that impact trans splicing factors are also documented.

For analysis of splicing mutation and development of possible therapies mouse models of specific mutations have often been developed. Examples include the mdx mouse for studies of DMD mutations leading to Duchenne muscular dystrophy (DMD). A Duchenne-like phenotype in Golden Retriever dogs also serves as a model for therapeutic application. This specific mutation is reported to be an A to G nucleotide change in intron 6.

Montes et al. noted that specific splice switching oligonucleotides have been developed to bind to transcripts of SREs to prevent binding of important proteins. Splice switching oligonucleotides have also been developed to alter activity of splice enhancer or splice inhibitory factors.

In addition to splice switching oligonucleotides, small molecules have been identified that target splicing factors; examples include small molecules that impact spliceosomes, Spliceostatin. Such molecules may be directed at specific cell types.

Montes et al. reviewed splicing alteration in specific diseases including myelodysplasia syndromes due to proliferation of specific subtypes of hematopoietic cells. In myelodysplasia and in acute myeloid leukemia, mutations were reported in splice factors SRF2, U2AF, SFBP1. Splice modulators were developed to target these mutations.

Questions arise regarding the targeting of splice modulators to specific cell types, for example, splicing-directed therapies in certain forms of retinits pigmentosa.

14.11 Pluripotent stem cells in investigations of disease therapies

Neural stem cells derived from pluripotent stem cell of patients with lysosomal storage diseases have been developed to obtain further insight into stages of pathology in these disorders. Studies are also undertaken on these cells to investigate the impact of certain therapeutic agents.

Luciani et al. (2020) noted that pluripotent stem cells can also be developed into brain organoids. Neural stem cells can also be developed to radial glial cells. They specifically considered utilization of pluripotent cells differentiated to neural stem cells in the investigation of certain therapeutic agents proposed to treat lysosomal storage disorders. In Niemann–Pick disease Type 1, patient's neural cells were developed from patient pluripotent stem cells to study effects of certain pharmacologic agents including cyclodextrins, tocopherols, reported promoter autophagy, and reduced accumulation of sphingomyelins.

HSCs and stem cell-derived neurons have also been investigated to determine the effects of gene therapy.

Effects of gene therapy have been studied in neural stem cells differentiated from patients' cells in cases of mucopolysaccharide storage disease, Sandhoff disease, and Tay–Sachs disease.

14.12 Relevance to protein folding and secondary modifications

In a review in 2016, Forlino and Marini noted significant advances in determination of genetic origins and protein changes involved in causation of osteogenesis imperfecta (OI), the disorder associated with bone fragility. This was considered to be an autosomal dominant disorder due to defects in collagen. In recent years defects in a number of other proteins, enzymes, and chaperones have been shown to lead to OI and in some cases this disorder manifests autosomal recessive inheritance.

Forlino and Marini identified five different systems involved in the genesis of OI. Earlier studies identified mutation in collagens, in COL1A1 that encodes the alpha-1 chain of collagen 1 and COL1A2 that encodes the alpha2 chain of collagen A1. Collagen type 1 was reported to be the predominant protein in bone, skin, and tendons. OI could be caused by mutations that impair the structure or levels of expression of collagen chains.

Forlino and Marini also documented recessively inherited forms of OI that were primarily due to defects in proteins and enzymes involved in posttranslational modifications or folding of collagens. Recessive forms of OI have also been identified that are due to defects in collagen mineralization.

Examples of OI due to collagen processing defects include ADAMTS defects, this protein is a metalloproteinase required for generation of helical structures. It was reported to excise the N-propeptide of the fibrillar procollagens types I—III and type V. Mutations in this gene cause Ehlers—Danlos syndrome type.

The BMP/metalloprotease/Td proteinase K complex was reported to be important in collagen peptide cleavage.

Other complexes involved in posttranslational modification and folding of collagen include P3H1 prolyl 3 hydroxylase CRTAP cartilage-associated protein and CYPB also known as PPIB peptidyl-prolyl cis trans isomerase.

P3H1, CRTAP, and PPIB form a complex involved in modification of the collagen helical regions.

LH1 also known as PLOD1 procollagen-lysine,2-oxoglutarate 5-dioxygenase 1 hydroxylates proteins including collagen in the endoplasmic reticulum. Hydroxylation of lysine residues is required for cross-linking.

Cyclophilin now known as peptidyl-prolyl isomerase B, PPIB, is part of an intracellular collagen-modifying complex that 3-hydroxylates proline at position 986 (P986) in the alpha-1 chains of collagen type I.

The gene TMEM38B encodes TRICB that forms an endoplasmic reticulum cation channel.

14.12.1 Proteins that play roles in collagen folding and cross-linking

FKBP10 (FKBP65) a prolyl hydroxylase and PPIA peptidyl-prolyl isomerase A accelerate protein folding, it associates with lysyl hydroxylase LH2 (PLOD2).

SERPINH1 (HSP4) encodes a collagen chaperone. It shuttles proteins between the endoplasmic reticulum and the Golgi and was shown to bind to collagen.

14.13 Defects in ossification and mineralization

IFITM5 interferon-induced transmembrane protein 5 is thought to play a role in bone mineralization.

SERPINF1 (PEDF) inhibits angiogenesis and is reported to be a regulator of osteoid mineralization.

Osteoblast developmental defects were reported to arise due to defects in WNT1 signaling protein. CREB3L1 cAMP responsive element binding protein 3 like 1 is expressed in osteoblast.

SP7 is reported to be a bone-specific transcription factor.

OI type	Gene	Gene product	Chromosome
OI1	COL1A1	Collagen type 1 alpha 1	17q21.33
OI2	COL1A2	Collagen type 1, alpha2	7q21.3
OI2	COL1A1	Collagen type 1 alpha 1	17q21.3
OI3	COL1A2	Collagen type 1 alpha 2	7q21.3
OI4	COL1A1	Collagen type 1 alpha 1	17q21.3
OIEhlersDanlos	COL1A2	Collagen type 1 alpha 2	7q21.3
OI5	IFIT5	Interferon-induced transmembrane protein	11p15.5
OI6	SERPINF1	Angiogenesis inhibitor	17p13.3
OI7	CRTAP LEPREL3	Cartilage-associated protein	3p22.3
OI8	P3H1 LEPRE1	Leucine and proline-rich glycoprotein	1p34.2
OI9	PPIB	Peptidyl-prolyl isomerase B	15q22.31
OI10	SERPINH1	Collagen binding protein	11q13.5
OI11	FKBP10	FK506 binding protein	17q21.2
OI12	SP7 (OSX)	Osterix transcription factor	12q13.13
OI13	BMP1	Bone morphogenetic protein 1	8p21.3
OI14	TMEM38B	Trimeric intracellular cation channel protein	9q31.2
OI15	WNT1	Signaling pathway protein	12q13.12
OI16	CREB3L1	Cyclic AMP response element	11p11.2
OI17	SPARC	Osteonectin	5q33.2
OI18	TENT5A	Terminal nucleotide transferase	6q14.1
OI19	MBTPS2	Membrane-bound transcription factor	Xp22.2
OI20	MESD	Mesoderm development chaperone	15q25.1

OI types 1, 2, 3, 4, and 5 are inherited as autosomal dominant characteristics.

OI types 6–18 and OI20 are autosomal recessive traits.

Type 19 is inherited as an X-linked characteristic. All other described types above are inherited as autosomal recessive disorders.

14.14 Osteogenesis imperfecta treatment

Rossi et al. (2019) reported that significant genetic heterogeneity occurs in OI but that different forms overlap in clinical features. They noted that treatment goals include decreasing fracture risk, while maximizing activity and growth.

Rossi et al. noted that bisphosphonate therapy has proven useful. Bisphosphonates inhibit osteoclasts. Other therapeutics under consideration include inhibitors of transforming growth factor (TGF) beta and inhibitors of sclerostin. Sclerostin is a glycoprotein secreted by osteophytes. It is reported to be a negative regulator of bone mass. Delgado-Calle et al. (2017) reported that sclerostin has become a target for medication to treat osteoporosis and other skeletal diseases. A specific antibody has been developed as a sclerostin inhibitor.

Rossi et al. reported that an additional treatment under consideration is an osteoanabolic therapy to increase osteoblast activity. These treatments include teriparatide, a recombinant form of parathyroid hormone.

They noted that there is evidence that TGF beta signaling leads to lowering of bone mass and that clinical trials were ongoing to investigate whether administration of inhibitors of TGF beta signaling are particularly useful in treatment of OI.

Denosumab is a monoclonal antibody that targets RANKL, a cytokine that is involved in osteoclast generation and survival. RANKL suppression is proposed to suppress bone resorption. Side effects of denosumab have been reported.

Rossi et al. reported that in OI low bone mass and bone fragility and low levels of trauma lead to fractures, pain, immobility, skeletal deformities, and impaired growth. They also drew attention to extra-skeletal defects that occur in OI. These include dental abnormalities, blue sclera, and hearing loss reported in some cases.

Rossi et al. reported that Sillence et al. (1979) distinguished four clinical types of OI: Type 1 mild, nondeforming, Type II perinatal lethal, Type III nonlethal with progressive deformities, Type IV moderately severe. Rossi et al. noted that 19 different genetic types of OI are now distinguished. However, they fit clinically into the four Sillence types. It is interesting to note that deleterious mutations in COL1A1 and COL1A2 occur in OI with different degrees of severity, mild, moderate, severe, and lethal. CO1A1 and COL1A mutations are predominantly associated with autosomal dominant AI.

Maioli et al. (2019) reported results of studies on 364 Italian patients with OI. They carried out detailed phenotype analyses and sequencing of COL1A1 and COL1A2 genes. The whole range of OI phenotypes was reported to occur in patients with COL1A1 or COL1A2 mutations, and mutations occurred at different positions within the genes. They reported that mutations that altered codons for glycine in the gene were particularly associated with lethal forms of OI.

14.15 Ongoing clinical trials related to cell and gene therapy

Condition	Gene/product	Details
X-linked SCID	IL2RG Interleukin receptor gamma	Lentiviral vector CD34 + cells
Wiskott–Aldrich syndrome	WAS (actin nucleation promoting factor)	Lentiviral vector CD34 + cells
X-chronic granulomatous disease	CYBB cytochrome b245beta (PHOX)	Lentiviral vector CD34 + cells
Leukocyte adhesion deficiency	CD18 ITGB2integrin subunit beta	Lentiviral vector CD34 + cells
SCID immunodeficiency	DCLRE1C DNA cross-link repair	Lentiviral vector CD34 + cells
Malignant T/B cells	DNA nucleotidylexotransferase (TDT)	Gene editing
Fanconi anemia	FANCA	FANCA lentiviral vector T-cells
Nonspherocytic hemolytic anemia	PKLR pyruvate kinase	Lentiviral vector CD34 cells
Severe combined immunodeficiency	ADA deficiency	Strimvelis CD34 cells with ADA
Metachromatic leukodystrophy ARSA	Arylsulfatase A	Lentiviral vector CD34 cells
X-linked adrenoleukodystrophy	ABCD1 ATP-binding cassette D	Lentiviral vector CD34 cells
Mucopolysaccharidosis MPS I	IDUA alpha-L-iduronidase	Stem cells and other cells
Sanfilippo A	SGSH N-sulfoglucosamine sulfohydrolase	Stem cells and other cells
Fabry	Galactosidase alpha	Stem cells and other cells
Cystinosis	CTNS cystinosin lysosomal cystine transporter	Stem cells and other cells

14.15.1 Hematopoietic stem cells and progenitors

Gene therapies using HSCs and progenitors have been used in the treatment of a number of genetic diseases. Questions have arisen regarding the potential use of nonhematopoietic cells.

Ferrari et al. (2020) reviewed aspects of gene therapy in HSCs. They reported that hematopoietic cell transplants have been used to treat specific genetic diseases for more than 50 years since early reports of treatment of patients with specific types of immunotherapy using HSC transplants.

Ferrari et al. noted that improvements have been made in donor matching, as incomplete matching of donor and recipient resulted in graft versus host reactions in some cases.

Autologous HSC therapy together with gene transfer has been used for treatment of a number of genetic disorders. In successful cases, the treated autologous stem cells were reported to undergo further division so that the benefits of therapy extended over time. Improvements have been made in gene therapy, and viral vectors such as gamma-retroviral vectors that presented problems are no longer used. In addition to their use in gene therapy, Ferrari et al. noted that HSCs can potentially be used as delivery vehicles for therapeutic proteins.

14.15.2 Collection of hematopoietic stem cells for therapy

This was reported to require multiple collections of bone marrow samples from iliac crest. In some cases this collection is followed by leukapheresis to specifically collect CD34+ cells that represent cells that include HSCs and their progenitors.

In some situations, specific agents are administered to patients to promote hematopoiesis and if stimulation is significant it is sometimes possible to collect adequate number of stem cells by leukapheresis of peripheral blood. One pretreatment agent used is plexifor that promotes release of stem cells into peripheral blood. However, plexifor cannot be used in all patients.

Ferrari et al. noted that lentiviral vectors were often used for gene transfer. Lentiviral vectors were reported to more frequently integrate into gene bodies than into gene promoters. In addition, modifications to vectors were made to inactivate viral promoter regions to limit transcription of viruses following introduction and to produce replication defective viruses.

Ferrari et al. also reviewed gene editing approaches using Crispr-Cas, or TALEN effector nucleases of zinc finger nucleases. They emphasized that targeted gene editing could ensure that the targeted gene remained under control of endogenous regulatory elements. They noted that gene editing proof of concept studies had been carried out in hemoglobinopathies, in severe combined immunodeficiency, and in Wiskott–Aldrich syndrome.

It is known that double-stranded DNA breaks induced by gene editing nucleases can be repaired by homologous recombination if donor homologous sequence is available. In some cases, donor sequence elements need to be introduced. There is evidence that introduction of donor template sequence using adeno-associated virus has yielded encouraging results.

DNA double-stranded breaks can also be repaired by nonhomologous end joining. This mechanism could be useful to induce knockout of damaging sequences.

Ferrari et al. noted that chemical modification in CRISPR-CAS editing reagents was being investigated. These include chemical modifications of the single-strand guide RNA and adaptation of the CAS9 ribonucleoproteins.

Modifications in agents used to transfer sequences were also being investigated such as use of adenovirus type 6.

Evaluation of the proportion of treated cells correctly modified prior to transfer were necessary. They noted that high doses of modified HSCs were often required.

In some cases, preconditioning of patients was required to reduce the number of cells within the bone marrow so that the transplanted cells could gain a foothold. This preconditioning required immunosuppression and was not recommended in some patients.

Ferrari et al. explored evidence of utility of HSCs as delivery vehicles for certain enzymes or proteins. Questions arise regarding the efficacy of modified HSCs in treating conditions that impacted brain, bone, or other organs.

There is some evidence that enzyme proteins released from bone marrow stem cells could be advantageous in reducing enzyme deficiency. There is evidence that HSCs give rise to monocytes and microglia that pass through capillaries and paravascular channels in the interior of the brain.

14.16 Coagulation disorders

14.16.1 Hemophilia

In a 2019 review, Nathwani noted that gene therapy offered a viable solution to the treatment of hemophilia types A and B. Increments of 5% in factor A or B levels were sufficient to reduce bleeding episodes.

The worldwide incidence of Hemophilia A, factor 8 deficiency, was reported to be between 1 in 4000 and 1in 5000 individuals. The incidence of Hemophilia B, factor 9 deficiency, was reported to be 1 in 2000.

A report by Iorio et al. (2017) of data from the World Federation of Hemophilia included information on the number of hemophilia cases in 10 different countries.

Nathwani reported that patients with these disorders often have F8 or F9 levels less than 1% of normal. Manifestations of the disorder include episodes of severe bleeding particularly intraarticular bleeding leading to arthropathy and joint destruction.

In mild hemophilia, bleeding can follow relative minor trauma. Nathwani particularly mentioned the danger of intracerebral bleeding.

In factor 8 deficiency, prophylactic administration of this factor is administered but administration has to be frequent. Nathwani noted that synthetic forms of factor 8 and pegylated forms have longer half-life. One form of recombinant factor 8 was reported to be fused with domain 3 of Von Willebrand protein and to longevity of factor 8. Other forms of recombinant factor 8 are also available.

A significant problem that arises in treatment of Hemophilia A or be with the coagulation factors F8 or F9 is the development of neutralizing antibodies,

Factor 9 was reported to encode a vitamin K-dependent coagulation factor that is present in an inactive form. Cleavage of a specific peptide from the inactive form generates a heavy chain and a light chain. This cleavage is carried out by a protein encoded by factor 11. The activated form of factor 9 serves to activate factor 10 in the presence of calcium phospholipids and factor 8.

Gene	Location	Proteins encoded
F8	Xq28	Factor 8
F9	Xq27.1	Factor 9
F10	13q34	Factor 10
F11	4q35,2	Factor 11

An antibody emicizumab was developed that binds to factor 9a and to factor X and facilitates activation of factor 10 that binds to factor 8 and facilitates hemostasis. However, this antibody must be used in combination with antithrombin to avoid thrombosis,

Nathwani noted that another therapeutic approach involves reducing circulating levels of endogenous anticoagulant including antithrombin AT3 with antisense molecules or antibodies. They reviewed gene therapy studies in the hemophilia and particularly emphasized later trials that utilized AAV serotype 8 vectors that were reported to have liver tropism that can lead to efficient transduction of hepatocytes. They also noted that there is low prevalence of antibodies to AAV serotype 8 in humans.

A number of different clinical trials were reported in which therapy was used for factor 9 deficiency. The AAV vectors were noted to have limited packaging capacity and this represented a problem in factor 8 deficiency given the large size of the factor 8 gene.

Factor 8 gene location and size:
F8 Xq28 155,022,723–154,835,792 gene size 1,166,931.
F9 Xq27.1 139,530,720–139,563,459 gene size 967,261.

The F8 encoding gene is partitioned into six domains A1, A2, B (the largest domain), A3, C1, and C2.

However, studies have revealed that the factor 8 B domain can be removed without diminishing anticoagulant efficiency. The studies used in that particular trial involved AAV5.

It is important to note that more than 250 unique factor 8 mutations have been reported that lead to hemophilia, The CHAMP database classifies mutation according to the degree of severity of associated hemophilia.

Donadon et al. (2018) reported that a specific exon 19 missense mutation is highly prevalent: F8 c6046 C > T/ p R2016W. They demonstrated that this mutation impacts activity of the protein and also level of secretion. Using an antisense RNA to mask sequence in the exon 19 region they established that a splice regulator occurs in this region. Nathwani et al. emphasized that durability of the transgene needed to be carefully monitored. It is also important to determine prior to therapy whether or not the patient has antibodies to AAV.

14.17 Von Willebrand factor and disease

Sharma and Flood (2017) reported that Von Willebrand disease (VWD) is the most common inherited bleeding disorder. Both qualitative and quantitative defects occur. VWD 1 is a partial deficiency, VWD 3 is complete deficiency. Qualitative defects occur in VWD type 2A and were reported to be associated with impaired subunit multimerization.

Clinical manifestations include positive family history of the disorder characterized by easy bruising, frequent nose bleeds, gum bleeding, excessive menstrual bleeding; gastro-intestinal bleeding is reported in some cases.

14.17.1 Laboratory tests

Sharma and Flood documented test for Von Willebrand factor, VWF, using antigen and also test based on interaction of VWD with risocetin cofactor. The latter test was noted to reflect the degree of binding of VWF to platelet binding and specific binding of the VWFa1 domain to the platelet glycoprotein Glycoprotein Ib (GPIb), also known as CD42.

In addition to binding to platelets, VWF was noted to also bind to exposed collagen at wound sites.

Hassan et al. (2012) reviewed structure and function of VWF. The VWF encoding gene was reported to have 52 exons. VWF is synthesized in endothelial cells and in megakaryocytes. They reported that VWF mediates interaction between platelets and vascular endothelium. VWF is subsequently cleared by proteolysis that involves activity of the metalloproteinase ADAMTS13.

The VWF polypeptide was reported to have five domains A, B C, D, E, and subdomain. These domains are sometimes referred as TIL domains. Different functions were reported for different domains. Dagil et al. (2019) reported strong interactions between the TIL E domain of VWF and coagulation factor VIII that is important in hemostasis. This interaction was reported to prolong the half-life of factor VIII.

Sharma and Flood reported that many variants occur in the VWF gene and that some of these variants previously reported as pathogenic occur in asymptomatic individuals.

Clinically three different types of VWD are recognized VWD 1, 2, and 3. All three are due to defects in the VWF encoding gene on chromosome 12p13.31. Type 2 VWD was reported to be due to qualitative variants in VWF and two subtypes are recognized VWF2A, 2B and 2M based on different alterations in function.

14.18 Platelet receptors for Von Willebrand factor

It is important to note that coagulation disorders can also result from defects on receptors on platelets that bind VWF. Bernard–Soulier syndrome arises due defects in VWF receptors. Three different receptors are potentially defective in this syndrome.

The VWF factor receptor was described to be a heterodimer composed of two different glycoprotein subunits, GP1BA mapped by a gene on chromosome 17p13.2 and GP1BB encoded by a gene on chromosome 22Q11.21. Other important platelet glycoproteins that play important roles in coagulation include GP5 encoded on 3q29 and GP9 (also known as CD42) encoded by a gene on 3q21.3.

Bernard–Soulier syndrome is sometimes referred to as Von Willebrand receptor disorder. It is characterized by low platelet count, very large platelets, and bleeding disorder.

A second inherited disorder of platelets is referred to as Glanzmann thrombasthenia. It is caused by defects in components of the plasma membrane, ITGA2B and ITGA3.

ITGA2B integrin subunit alpha 2b encoded on 17q21.31.

ITGA3 integrin subunit alpha encoded on 17q21.33.

Löf et al. (2018) noted that when there is no vascular injury VWF has low affinity for platelets. Altered dynamics in blood flow caused by vessel injury are thought to play a key role in altering VWF to have affinity with platelets and to bind to receptors on platelets. The altered hydrodynamics that occur following vessel injury were reported to lead to conversion of the A1 domain of VWF from low affinity state to a high affinity state.

Dagil et al. (2019) noted that VWF and coagulation factor VIII together play important roles in hemostasis. They also noted that VWF binding was reported to prolong the half-life of factor VIII in plasma.

14.19 Understanding mechanisms of rare diseases that may lead to therapy

McEneaney and Tee (2019) in a review of tuberous sclerosis emphasized how understanding of pathology in disorders due to genetic mutations facilitates diagnosis, improves management, and can perhaps even pave the way to definitive treatment.

Important work following identification of the causative gene involves delineating the normal function of the gene and determining how normal function is disrupted by gene mutations.

They noted the extent to which the burden of disease can vary, even in individuals with defects in the same gene and even in individuals with the same mutation in the same gene.

They noted that 60%–70% of cases of tuberous sclerosis were noted to arise de novo. Skin lesions were reported to occur in 70% of patients. They include hypomelanotic macules, angiofibromas, ungual fibromas, confetti skin lesions, and shagreen patch.

Brain lesions include subependymal nodules, giant cell astrocytoma, and cortical dysplasias. Renal lesions were reported to occur in 80% of patients and to include cysts, renal angiofibromas, TSCPKD (Tuberous sclerosis-Polycystic kidney) contiguous syndrome occurs in some patients. Seizures were reported to occur in a high percentage of patients.

Lung lesions include pulmonary lymphangioleiomatosis.

Cardiac rhabdomyomas were reported in 50%.

Mutations in TSC2 gene were reported to often occur in the GAP domain. This domain is noted to impact RHEB (Ras homolog, mTORC1 binding) small protein. TSC1 and TSC2 gene products were reported to form a complex to inactivate RHEB and to therefore inhibit MTOR (mammalian target of Rapamycin) gene activity.

The TSC2 gap domain was reported to interact with calmodulin and to impact calcium signaling.

Impaired function of the tuberous sclerosis complex leads to hyperactive mTOR that increases mRNA and protein production. This has led to implementation of rapamycin treatment in tuberous sclerosis, as rapamycin suppresses mTOR. Questions arise as to how long rapamycin treatment should continue. Termination of rapamycin treatment has been shown to lead to rebound growth of tuberous sclerosis lesions.

Tuberous sclerosis is also noted to be associated with increased endoplasmic reticulum stress (Johnson et al., 2015).

14.20 Trinucleotide repeat disorders: progress toward therapy

In 2015 the Huntington disease (HD) consortium published results of a GWAS to identify loci in the human genome that impacted the age of onset of symptoms in HD. They identified a locus on chromosome 15 that was associated with onset of disease symptoms. The chromosome 15 locus contained two high priority genes MTMR10 and FAN1 [Fanconi anemia (FA) FANC1, FANCD2-associated nuclease]. This study also yielded some evidence of significant loci on chromosomes 8 and 3 that were associated with age of onset of HD.

Bettencourt et al. (2016) carried out a GWAS in individuals with CAG repeat expansions leading to HD, or spinocerebellar ataxia. They analyzed DNA repair genes in 1462 patients with these disorders. Results revealed that most significant correlation of age of onset of disease manifestation and SNP (single nucleotide polymorphism) markers occurred with rs3512 in FAN1 and rs1805323 in PMS2 mismatch repair system component on chromosome 7p22.1 (also known as MLH4).

They concluded that DNA repair gene variants significantly modified age of onset in HD and spinocerebellar ataxia and proposed that somatic expansion of trinucleotide repeats were significantly impacted by DNA repair gene variants.

They noted that earlier studies had revealed that the number of trinucleotide repeats in the HTT (Huntington) gene on chromosome 4 influenced pathology and that further expansion of the repeat number occurred during aging and influenced the extent of pathology in HD.

Earlier studies had revealed that the number of trinucleotide repeats in the HTT gene on chromosome 4 further expanded in number during aging and were associated with the extent of pathology in HD.

More recent studies have indicated that factors elsewhere in the genome influence repeat expansion. These studies included GWAS studies in individuals with varying degrees of disease severity. Studies from the HD consortium in 2015 identified three loci elsewhere in the genome that impacted HD severity. One such locus was mapped to chromosome 15q13.2–13.3. This locus was noted to include the gene that encodes FAN1, a nuclease that functions in conjunction with FANCD2 and FANC1 in the DNA damage repair pathway. Evidence therefore suggested that FAN1 could be part of the DNA damage repair pathway. They also noted that FAN1 nuclease complexed with MLH1 and PMS2 that play roles in DNA mismatch repair.

In comprehensive in vitro studies, Goold et al. (2019) demonstrated that increased expression of FAN1 suppressed expansion of the HTT repeat and that knockdown of FAN1 enhanced expansion of the HTT repeat.

Additional studies included transcription-wide association studies with analyses of levels of expression in the dorsolateral prefrontal cortex samples and correlation with information on HTT disease progression. These studies revealed that decreased expression of FAN1 was associated with earlier onset and more pathology progression in HD patients.

Goold et al. investigated impact of FAN1 expression in an in vitro system in which HTT genes with different lengths of trinucleotide repeats were introduced. Over time, degrees of expansion of the trinucleotide repeat were followed in this system. They were also able to demonstrate that increased expression of FAN1 slowed repeat expansion.

Studies were also carried out to determine the roles of different domains of FAN1. These studies revealed that the nuclease domain of FAN1 was not involved in suppression of report expansion.

Importantly FAN1 was shown by Zhao et al. to protect against repeat expansion in the Fragile X gene in mice.

Goold et al. noted that instability of trinucleotide repeats probably arises as a results of strand slippage and formation of unusual loops and quadruplexes during DNA replication and transcription.

The exact mechanism through which FAN1 was proposed to act was proposed to possibly be related to its capacity to act as a scaffold to which other repair proteins bind.

Information obtained indicated that a FAN1 interactome exists and that this includes MLH1. MLH3, PMS2, and PCNA (proliferating cell nuclear antigen) mismatch repair proteins. FAN1 was therefore predicted to delay pathology and decrease age of onset of symptom manifestation in HD.

MLH1. The protein encoded by this gene can heterodimerize with mismatch repair endonuclease PMS2 to form MutL alpha, part of the DNA mismatch repair system.

MLH3. MLH1 is also involved in DNA damage signaling and can heterodimerize with DNA mismatch repair protein MLH3.

PMS2 forms heterodimers with MLH1, can correct DNA mismatches, and small insertions and deletions that can occur during DNA replication.

PCNA. The encoded protein acts as a homotrimer and helps increase the processivity of leading strand synthesis during DNA replication.

Wang et al. (2014) described how DNA interstrand cross-links (ICLs) that are toxic lesions in neurodegenerative disease can be repaired by FAN1.

DNA ICLs were reported to be highly toxic lesions found in cancer and degenerative diseases. Repair of ICLs can be carried out by components of the FA pathway and through FA-independent processes reported to involve the FAN1 nuclease.

Kim et al. (2020b) carried out studies on the functional effects of FAN1 in modifying DNA maintenance. They noted that specific variants in FAN1 impacted its DNA binding activity. However, the major effect of FAN1 relative to HD was related to the level of expression of FAN1 mRNA. Knockout of FAN1 in HD1 pluripotent stem cells was shown to be associated with increased CAG trinucleotide expression. They also noted that individual variation in the FAN1 haplotype could influence level of response to FAN1 expression.

Kim et al. published data on genetic and functional analysis of the FAN1 locus in modifying HD. They defined the chromosomal location of the HD modifier locus as 15q13.2−15q13.3 and as the segment that included nucleotides 30.900.00−31,500.000 and they identified 13,294 single nucleotide variants in this region. On the basis of variants in this region they defined four different haplotypes, 15AM1, 15AM2, 15AM3, and 15AM4. The 15AM1 and 15AM3 haplotypes demonstrated reduced binding of FAN1 to the HD repeat. A variant in the 15AM2 haplotype that led to delay in onset of HD pathology was found to be associated with increased expression of FAN1.

This onset delaying capacity of AM2 haplotype in the FAN1 region was considered to likely have therapeutic relevance. Therapeutic procedures to be considered included either introduction of exogenous FAN1 or stimulation of upregulation of FAN1 expression.

Lahue (2020) noted that age of onset of clinical manifestations of HD was traditionally attributed to the length of the CAG trinucleotide HTT expansion. However, more recent studies indicate that somatic instabilities in the trinucleotide repeat expansion region, partly due to aberrant DNA repair mechanisms, influence the age of onset of clinical manifestations.

Scahill et al. (2020) carried out a study in 54 young adults, average age 29 years determined to be at risk for HD due to CAG repeat expansion in HTT and estimated to be 24 years younger than expected age of onset for clinical manifestations. They carried out a battery of neuropsychological assessments, magnetic resonance imaging (MRI) studies, and studies of particular markers in blood and CSF.

Results of their studies indicated that no brain functional abnormalities were present. The only difference found in the HD at risk individuals compared with controls was that levels of neurofilament light protein in CSF and blood were elevated indicating some degree of neurodegeneration. The authors noted that these findings could provide guidance as to when to commence therapy in this disorder.

14.21 Toward Huntington disease therapy

Tabrizi et al. (2019) reviewed strategies for HD modifications. Strategies included methods to lower the levels of mutant HTT to reduce pathologies in this dominantly inherited disorder. They reviewed pathologic mechanisms in HD. And noted that normal HTT likely acts as a scaffold protein that interacts with other protein. There is also evidence that normal HTT regulates transcription of certain genes including the brain-derived neurotrophic factor gene.

Pathologic consequences of the mutant HTT gene were reported to include transcriptional dysregulation and mutant HTT was reported to lead to synaptic dysfunction, impaired nuclear pore function, impaired mitochondrial function, and impaired proteastasis.

Given the important functions of normal HTT, it is important that one normal HTT gene and its protein product remain functional.

Therapeutic strategies in HD that are in early-stage investigations in model organisms include gene editing. Zinc finger nuclease, Talens, or CRISPR-Cas editing could potentially remove the mutant genome segment with the CAG repeat expansion. It is important to note that gene editing approaches have been carried out in model systems including cells and animal models.

RNA targeting with antisense oligonucleotides or with inhibitory RNAs have been carried out in preclinical studies. Intrathecal administration of antisense RNA to target mutant RNA are reported to be in phase 2a clinical trials. Tabrizi noted that one possible risk of this therapy is that it could potentially reduce synthesis of CAG containing segments of normal length and CAG repeats in locations other than the HTT region.

RNA inhibition studies and small molecule splice modulatory agents were reported to be in clinical trials in 2019.

In considering gene editing approaches, Tabrizi et al. noted that these target the mutant HTT gene specifically and have less impact on other genes. They documented such therapies as including DNA binding elements linked to nuclease effectors that act to cleave DNA. Repair of cleaved DNA by homologous recombination was preferred. However, this requires that the correct DNA sequence also be included in the system that delivers the gene editing agents.

Tabrizi et al. noted that major factors to be considered in gene editing include aspects of delivery of the therapeutic agents to the CNS.

14.21.1 RNA targeting and RNA interference approaches in gene therapy in Huntington disease

Antisense oligonucleotides (ASO) and small molecule RNA modulators potentially lead to translation suppression or enhance degradation of mutant HTT RNA and reduced production of mutant HTT protein.

In a model organism, a transgenic mouse model of HD cholesterol conjugation of siRNAs anti-HTT siRNA delivered specifically to the striatum was shown to delay onset of pathological manifestations of HD.

Tabrizi et al. noted that these studies raise questions as to whether early delivery of therapy will be necessary to avoid disease. They also noted that it will be essential to determine that the siRNAs do not have any off-target effects and that they do not induce immune responses.

In human studies, administration of allele-specific antisense oligonucleotide ASO IONIS HTT_{RX} RG6042 was reported to lower levels of mutant HTT in patient's CSF. Reportedly, that RNA−DNA hybrid molecule that results following administration of RG6042 is thought to be degraded by RNAseH1.

The antisense binding close to the translation initiation site was also thought to lead to translation arrest. There is also some evidence that the antisense oligonucleotide binding impaired downstream splicing of the mutant HTT transcript.

Tabrizi et al. (2019) reported that clinical trials of ASO RG6042 administered via lumbar puncture were in phase 3 clinical trials. They noted that over the course of time additional modifications of this antisense molecule had occurred and that unintended binding of this ASO to other sites is reduced by careful selection of the sequences to be targeted so that the sequences unique in the genome are targeted.

14.22 Protein clearance

Additional therapeutic strategies being considered involve clearance of the mutant HTT protein. Possibilities exist to selectively target specific proteins for proteolytic degradation. An approach referred to as PROTAC aims to specifically target mutant HTT for proteolytic degradation.

Tabrizi et al. emphasized that it is critical that production and levels of normal HTT proteins be retained. They also emphasized that therapeutic agents need to target the CNS, and this represents a specific challenge.

With respect to delivery agents, they noted that capsid serotypes of AAV vectors differ in their cellular tropisms and this can be used to advantage. However, immune responses to vector components must be considered. They noted that direct infusion of inhibitory RNAs can potentially by-pass the use of vectors.

Tabrizi et al. also noted that the FAN1 nuclease that promotes DNA ICL repair was shown to stabilize somatic CAG expansion. MSH3 mismatch repair protein recognizes insertion deletion loops that result from somatic repeat expansion and instability.

DNA repair proteins are being considered in the therapy of HD.

14.23 Polyglutamine cerebellar ataxias

These disorders were reviewed in 2019 by Buijsen et al. They noted that there are six known forms of autosomal dominant spinocerebellar ataxias due to expansion of CAG repeats in the gene coding regions. Buijsen et al. noted that increasing knowledge of pathological mechanisms involved in these diseases are promoting investigations into therapeutic measures to slow disease progression. Key features of these disorders include progressive ataxia. Other manifestations are dysarthria and oculomotor defects. In addition, one-third of families with a clinical diagnosis of autosomal dominant cerebellar ataxia (ADCA) did not have a genetic diagnosis. Population frequency of ADCA was reported to be approximately 5 per 10,000 in several countries including Portugal, Norway, and Japan. The prevalence of different forms of ADCA differed. Importantly with increasing length of the CAG repeats, onset of clinical manifestations occurred earlier.

Several disease mechanisms have been proposed including disrupted gene function and toxicity of the polyglutamine derived from CAG repeats.

Different genes that manifest the CAG repeat expansions that lead to ADCA include the following:

SCA1 ATXN1 reported to bind chromatin and acts as a transcriptional repressor.

SCA2 ATXN2 reported to control distinct steps in posttranscriptional gene expression.

SCA3 ATXN3 suggested to play role in autophagy and proteastasis.

SCA6 CACNA1A subunit of calcium channel predominantly expressed in neuronal tissues.

SCA7 ATXN7 was reported to associate with microtubules and to stabilize the cytoskeleton.

SCA17 TBP (TATA box binding protein) serves as the scaffold for assembly of transcription complex.

Therapeutic approaches being considered included promotion of proteastasis through induction of autophagy in some forms of the disorder. Gene therapies including RNA interference therapy and exon skipping approaches were being investigated in some preclinical models.

The repeat expansions and polyglutamine expansions were reported to impair autophagy and activity of the ubiquitin proteosome system. Expanded polyglutamine sequences led to accumulation of large insoluble protein aggregates. These aggregates were shown to be present, particularly in neurons in the cerebellum. The aggregates likely disrupt cellular function.

In a review in 2018 Ashizawa et al. reported that more than 40 autosomal dominant forms of spinocerebellar ataxias have been identified. They noted that analyses of the molecular pathology in a number of forms of spinocerebellar ataxia have led to development of interesting treatment strategies. This applies to the polyglutamine repeat forms of these disorders but also to other spinocerebellar ataxia due to abnormalities in RNA or to generation of toxic proteins.

They noted further that MRI and analysis of magnetic resonance spectroscopic biomarkers facilitate assessment of disease activity.

Ashizawa documented different forms of spinocerebellar ataxia due to nucleotide repeat expansions. It is important to note that CAG repeat expansion and CAA repeat expansions represent polyglutamine expansion in the coding region.

Gene region	Repeat expansion	Type of spinocerebellar ataxia
5'UTR	CAG	SCA12
Exon	CAG	SCA1, SCA2, SCA3, SCA6, SCA7, SCA17 DRPLA
Intron	ATTCT	SCA10
	TGGAA	SCA31
	ATTC	SCA37
	GGCCTG	SCA36
3'UTR	CTG	SCA8

Ashizawa et al. documented pathophysiological mechanisms in these disorders noting that they included development of aggregates, impairment of protein function, perturbation of neural circuit, impairment of bioenergetics, and alterations in DNA repair.

They noted that therapeutic strategies could include gene editing, transcription inhibition with targeting of disease-specific transcripts, modification of splicing, blocking of translation of mutant mRNA, clearance of aberrant protein, and promotion of protein folding.

Specific treatments were also aimed at the pathophysiological effects of the mutant gene and gene products.

Ashizawa et al. noted that it was important to consider the CAG repeat expansions in coding regions, the polyglutamine expansions as a distinct category. These included the gene products listed below. It is important to note that the functions of ataxins (ATXN) are listed as unknown NCBI genes 2021.

SCA1 ATXN1.
SCA2 ATXN2.

SCA3 ATXN3.
SCA6 CACNA1A Calcium ion channel.
SCA7 ATXN17.
SCA17 TBP TATA box binding protein important in initiation of transcription.
DRPLA ATN1 atrophin this protein accumulates to abnormal concentrations in dentatorubral-pallidoluysian atrophy.

Important new information relevant to therapy of polyglutamine expansion diseases has emerged. Nakatani et al. (2015) demonstrated that a specific small molecule naphthyridine azaquinolone (NAT) binds to expanded CAG repeats. This binding was shown to prevent repeat expansions and to in fact enhance repeat contraction.

Expansion of trinucleotide repeats over time is thought to play important roles in polyglutamine diseases. The binding of NAT appears to be specific for expanded CAG repeats. There is evidence that expanded repeats form unusual slipouts during replication and certain DNA repair proteins act to eliminate these slipouts and the NAT was shown to prevent repair of these slipouts (Nakamori et al., 2020).

In 1991 Lupski et al. determined that the Charcot–Marie–Tooth CMT1A was associated with a duplication on chromosome 17p. In 1993 Roa and Lupski reported that in the majority of cases of the autosomal dominant disorders, CMT1A a 1.5-megabase duplication occurred in17p11-p12 and that the PMP22 gene occurred in the duplicated region.

Pantera et al. (2020) defined pes cavus as very frequent in CMT1a and that distal symmetrical muscle wasting may be more severe in legs than in arms. Sensory manifestation occurred in a stocking glove distribution. Reflexes were noted to be depressed or absent; importantly they noted that pain may be a manifestation. They reiterated that elevated levels of PMP22 mRNA and PMP2 protein represent the key problem.

PMP22 protein is defined as a 160-aminoacid transmembrane glycoprotein mainly expressed in Schwann cells. Following translation, the PMP22 is glycosylated and is localized in the endoplasmic reticulum with a chaperone protein calnexin. Control of levels of PMP22 synthesis seems critically important. Boutary et al. (2021) reported that small changes in the levels of PMP22 influence myelination and sensory functions.

It is important to note that alteration in PMP22 have been reported in a number of different disorders associated with neuropathy: CMT1A, CMT2E, Dejerine–Sottas syndrome, Roussy–Levey syndrome.

Boutary et al. (2021) reviewed Charcot–Marie–Tooth disease and current and emerging treatment options. This review was initiated with a description of Schwann cells that are key to efficient conduction of nerve impulses. Schwann cells were noted to separate axon segments. They noted that CMT1 is the most prevalent inherited disease of the peripheral nervous system; it is an autosomal dominant disorder reported to occur in 50%–80% of CMT cases. Boutary et al. noted that in CMT1 nerve conduction velocities were reduced to less than 38 milliseconds. In addition, muscle stretch reflexes were reduced. Biological studies revealed the occurrence of structures referred to as onion bulbs. These are concentric layers of Schwann cell processes and collagen that surround an axon.

The PMT22 gene was noted to have two promoters leading to generation of two different transcripts, 1A and 1B, and the two transcripts have different patterns of tissue expression. Transcript 1A generated from promoter 1 leads to myelination of Schwann cell. Transcript 1B generated from the second promoter was noted to occur in nonmyelin producing tissues. PMP22 synthesis was noted to be under tight regulation.

PMP22 protein was reported to constitute 2%–5% of the content of the myelin sheath.

The normal function of the myelin sheath was noted to be dependent upon proper protein folding of PMP22 and its proper function that includes roles in cholesterol homeostasis in Schwann cells. P<P22 interacts with the cholesterol efflux protein ABCA1 (ATP-binding cassette subfamily A member 1) in maintenance of cholesterol homeostasis in the myelin sheath.

Unequal crossing over into the 17p11.2 region leads to duplication or deletion of a genomic segment. Duplication of an 1.5-Mb segment containing the *PMP2* gene leads to CMT1A. Deletion of this genomic segment leads to liability of hereditary neuropathy with liability to pressure palsy.

Boutary et al. reported that CMT, a subtype of Charcot–Marie–Tooth disease, arises due to mutations in PMP22. Some of these mutations lead to gain of function, others lead to loss of function. Specific point mutations in PMP22 were noted to impair proper folding of the protein. Misfolded PMP22 protein accumulated in the endoplasmic reticulum or could accumulate in cells and lead to cellular stress.

CMT1A due to duplication was reported to be associated with overexpression of PMP22 leading to neurological impairment.

Boutary et al. noted that a number of pharmacological therapies have been tried to treat CMT1A, though none of these have been successful. Treatments under investigation include use of neurotrophin 3. Other medications were reported to be in early clinical trials. They noted that treatment possibilities being explored included use of small interfering RNAs. Antisense mRNAs that target and degrade mRNA are being investigated.

Special procedure for targeting these inhibitory RNAs or antisense oligonucleotides is being explored. Lipid nanoparticles can be used in siRNA delivery. Phosphothiolate modifications are being investigated to promote stability of antisense oligonucleotides.

A small protein monocyte chemoattractant protein and heat shock proteins are being investigated to target aggregates of PMP22.

14.24 Duchenne muscular dystrophy

DMD was described in the mid-19th century by the French neurologist Guillaume-Benjamin-Amand Duchenne in 1861. The clinical features of this disorder include progressive muscle weakness, loss of muscle mass, frequent falls, and an unusual strategy on rising referred to as the Gower's sign; when rising from a squatting position, they use their hands on their legs for support, indicating weakness of the leg muscles. In the later stages of the disease cardiomyopathy and respiratory failure occur. This sign is named after William Richard Gowers, neurologist 1845–115. In the later stages of the disease cardiomyopathy and respiratory failure occur.

One-third of individuals are reported to have cognitive impairment. The gene defective in this disorder, the DMD gene, was discovered by Kunkel et al. (1986).

The DMD gene maps to Xp21.2-p21.1. It is approximately 2.2 megabases in size and has 79 exons and encodes a protein referred to as dystrophin. Transcripts can potentially be derived from any one of seven promotors and at least seven different splice forms can be derived (Kunkel et al., 1989).

Female carriers of dystrophin deletions or pathogenic mutations can manifest symptoms. These can include asymmetric bilateral leg weakness myalgia, cramps, fatigue, enlarged calf muscles (pseudohypertrophy), and dilated cardiomyopathy with elevated serum creatine kinase (CK) levels.

Aartsma-Rus et al. (2016) reviewed aspects of genetic diagnosis in DMD. They noted that a milder form of the disorder, Becker muscular dystrophy, arises due to specific defects in this same gene. They emphasized that as certain mutation-specific therapies had become available, it became particularly important to ensure specific genomic diagnoses in patients with DMD defects.

Aartsma-Rus et al. described the function of the DMD encoded protein. It was noted to function as a "shock absorber" that links the action of the muscle contractile apparatus to the connective tissue that surrounds the muscle fiber. Loss of the shock absorber causes repeated muscle contraction to lead to muscle damage, inflammation, and degeneration. Eventually muscle fibers are replaced by fat and fibrous tissue.

The mutations leading to Becker muscular dystrophy were noted to be less disruptive of muscle function and to be less damaging.

Aartsma-Rus et al. (2016) reported that one in every three patients diagnosed with DMD defects represents a new mutation. In addition, a large number of disease-causing mutations occur in this gene and the range of different mutations is large. Included are nucleotide point mutations, splice site mutations, deletions, and duplications.

The deletions or duplications were reported to occur particularly between exons 45 and 55 and two exons could be deleted. In some cases, 10 exons were duplicated. Particularly important are deletions or duplications that disrupt the reading frame of the gene. If the reading frame is not disrupted, transcription of the gene can occur and transcripts will be shorter or longer than usual, and can retain some degree of function. However, if deletions or duplications disrupt the reading frame, an aberrant transcript can result or many stop codons may be generated leading to transcription termination.

Aartsma-Rus et al. noted that small deletions or insertions can also disrupt the reading frame leading to transcription impairment or to transcription termination.

Damaging mutation also includes disruption of splice donor or splice acceptor sites. Specific deep intronic mutations were also reported to lead to aberrant transcript processing.

They noted that DMD missense mutations were rare in patients. However, in some locations they were noted to be disruptive. One such region includes the cysteine-rich domain of DMD. Binding of dystrophin to dystroglycan was noted to be dependent on intact cysteines.

Genomic rearrangements involving the X chromosome and leading to translocation of the X chromosome to an autosome were found to lead to DMD in some patients, as the normal X chromosome was preferentially inactivated.

In-frame deletions that do not result in functional DMD transcripts, include deletion of exons 64–70 since these exons encode protein segments that interact with the extracellular matrix. In-frame deletions of exons 32–45 or deletions of exon 2–10 also lead to nonfunctional products as those exons encode segments that bind to actin.

Skipping of exon 44 was shown to be associated with a milder form of DMD. Milder forms of muscular dystrophy were also associated with in-frame deletion between exons 45 and 55 and with deletions between exons 10 and 40.

Bladen et al. (2015) analyzed data compiled in a database of 7149 patients with DMD. In this dataset 68% of patients had large deletions, 11% had large duplications. Small changes were defined as changes that encompass a single exon.

14.24.1 Laboratory diagnosis of DMD

Aartsma-Rus et al. emphasized that important clinical chemistry signs of the disorder includes elevation of serum levels of muscle enzymes, particularly CK and also aspartate and alanine aminotransferase.

Genomic studies should follow. These include deletion and duplication analyses. If these are not found individual DMD exon sequencing is carried out. Additional studies can include muscle biopsy with analysis of dystrophin.

14.25 DMD therapy molecular approaches

Chamberlain and Chamberlain (2017) noted that the finding that some patients with large in-frame deletions in the dystrophin gene had milder forms of muscular dystrophy have spurred design to determine if gene therapy with a small version of the DMD gene could be helpful.

A key problem in gene therapy relates to the size of the gene that must be inserted into a vector. They noted that AAV vectors have a carrying capacity of 5 kb. Duchenne minigenes to be utilized for therapy can be 4 kb in size as gene delivery cassettes need to be refined with elements related to muscle delivery and must include regulatory cassettes. AAV and AAV9 vectors were noted to have minimal immunologic activity.

Chamberlain and Chamberlain noted that systemic protocols for gene delivery had been established in canine models of DMD. The clinical trials were considered to be in the Phase 1 stage. An important assessment that needs to be made is to establish that systemically administered gene therapy vector reaches muscles throughout the body. Initially this assessment required muscle biopsy. However, other potential investigations of vector distribution include development of appropriate assessment of muscle biomarkers in serum and MRI of muscle. Ultimately assessment of muscle function will be necessary.

An important fact that has emerged in gene therapy investigations is that high dose administration of vector is to be avoided as this tends to lead to the development of neutralizing antibodies.

Chamberlain and Chamberlain noted that other forms of gene therapy under investigation in DMD involve efforts to increase expression of utrophin, alpha-7 integrin, and GALNT2 (BGALNT2). Overexpression of these gene products was shown to reduce muscle pathology in the mdx mouse model of DMD. Studies are also in progress to investigate utrophin gene therapy in DMD.

Chamberlain and Chamberlain also discussed possibilities for gene editing. They noted that it was not clear whether editing AAV vector with gene editing reagents would require access to muscle stem cells for this to be successful.

Ramos et al. (2019) reviewed development of microdystrophins for therapeutic use in patients with DMD. They particularly explored effects of modification of the central rod domain of dystrophin to determine if specific modification could improve functionality. The impacts of specific modifications were tested in the mouse model of DMD.

In a 2019 review of molecular therapy in DMD, Davies and Guiraud (2019) noted that there was no effective therapy for this disorder. However, specific clinical therapies were available in a limited number of cases; these include exon skipping therapies and stop codon readthrough applications.

They emphasized that dystrophin serves as an important link between F actin in the cytoskeleton, the dystrophin-associated complex in the sarcolemma and laminin.

Davies and Guiraud documented four domains of dystrophin: the N-terminal domain with a binding site for F actin; the central rod domain with 24 spectrin repeats and interspersed hinges; a cysteine-rich domain that binds to the dystroglycan complex; and a C-terminal domain.

Mendell et al. (2021) noted that truncated versions of the DMD gene have been developed for therapy by three different companies.

Neuronal nitric oxide synthase was reported to bind to the N-terminal and rod domains.

They noted that some patients with mild muscular dystrophy had a gene that was 46% the size of the normal gene and had lost portions of the gene within the central rod domain. This finding inspired the development of dystrophin minigenes for therapy. Import of the minigenes on function in the DMD mouse model was further improved when the nNOS binding sites were included in the minigene.

14.26 Utrophin

Loro et al. (2020) reported on the therapeutic relevance of posttranscriptional upregulation of utrophin in treatment of DMD.

They noted that newer steroid medications that have fewer side effects can potentially beneficially be used in these patients; these include vamorolone and deflazacort. They also noted ongoing clinical trials of gene therapy in DMD involving the use of shortened versions of the dystrophin gene, and the use of therapies to target specific mutation.

Another proposed therapy involved increasing expression of utrophin, a dystrophin-related protein that is primarily expressed in fetal life. Utrophin is reported to have functional domain similar to dystrophin and to interact with the glycoprotein complex.

Studies on the mdx mouse reported that utrophin activation alleviated manifestation of muscular dystrophy and led to studies of potential utrophin activation.

Loro et al. reported that utrophin has a broad tissue distribution and that distinct promoters drive expression of different isozymes in different tissues and specific regulatory factors exist. They also noted that utrophin is subject to posttranscriptional regulation. Specific elements present in the 5′ and 3′ untranslated regions and RNA binding factors were reported to impact the ultimate levels of utrophin product.

They carried out studies to identify small molecules that impact posttranscriptional regulation and potentially impact enhanced tissue levels of utrophin. To identify such factors, they established a library of 3127 compounds with established biological and pharmacological activities. Studies were carried out to determine if specific compounds influenced utrophin levels in a dose-dependent manner.

Loro et al. identified 27 promising substances and 10 of these were then analyzed to determine if they led to increased utrophin levels and functional improvements in the mdx mice. Trichostatin was the top scoring compound, and additional other small molecules with positive effects were also identified.

Other preclinical studies revealed that trichostatin had pro-myelogenic effects on C2C12 stem cells and also the trichostatin improves muscle structure in mdx mice.

Trichostatin was shown to bind to the utrophin gene promoter. There is also evidence that trichostatin inhibits histone deacetylase activity.

Spinazzola and Kunkel (2016) noted that the population incidence of DMD is 1 in 3500 males and that diagnosis most commonly occurs between the second and third years. They reported pharmacological therapies for downstream therapies in DMD. The dystrophin-associated complex was noted to be a multimeric protein complex associated with sarcolemma of skeletal and smooth muscle. Documented treatment approaches under investigation were directed at gene repair or delivery of mini-dystrophin, promotion of readthrough of aberrant stop codons, and promotion of exon skipping, to ensure appropriate transcription reading frames.

Spinazzola and Kunkel noted that only a few patients could currently benefit from these therapies. It was important to continue to search for beneficial pharmacological interventions. Long-term treatment with antiinflammatory glucocorticoids has been shown to have prolonged independent mobility and to delay cardiomyopathy. However, this treatment has significant side effects including significant weight gain and compression fractures of the vertebrae. Novel glucocorticoid analogs have been developed and applied to therapy of Duchenne patients.

They noted that the NFkappa B transcription factor pathway acts to moderate inflammation and cytokines. Specific inhibitors of NFkappa B are being investigated as treatment options. The Nemo peptide acts to inhibit NFkappa B.

Other potential treatment strategies involved investigation of substances to promote muscle growth. One possible avenue along these lines involves investigation of inhibitors of myostatin as it negatively regulates skeletal muscle cell differentiation and proliferation. Insulin-like growth factor is known to promote myoblast differentiation and is being investigated in treatment of DMD patients.

TGF beta has numerous actions including promotion of fibrosis. Losartan was reported to be an inhibitor of TGF beta. Losartan is widely used as an antihypertensive. A clinical trial of Losartan therapy in DMD is underway.

Other mechanisms known to be impaired in DMD include calcium conduction and mitochondrial function. Spinazzola and Kunkel noted that these mechanisms are also under consideration for therapeutic intervention in DMD. Possible interventions to counteract oxidative stress include CoQ10 administration of the synthetic derivative of coenzyme Q10 idebenone.

Sun et al. (2020) noted evidence that DMD is a disease that impacts muscle stem cell as satellite cells and that there is evidence that dysfunctional dystrophin leads to abnormal polarity and abnormal division of muscle stem cells. They focused on use of human pluripotent stem cells and their potential in treating DMD. The pluripotent stem cells can be differentiated to a myogenic lineage, and then be used for autologous transplantation.

14.27 Spinal muscular atrophy (autosomal recessive proximal muscular atrophy)

Arnold and Fischbeck (2018) reviewed autosomal recessive proximal muscular atrophy [spinal muscular atrophy (SMA)]. The original description of severe SMA occurred in two reports published in 1891, one by Werdnig and another by

Hoffmann. In the 1950s, a milder form of the disorder was described by Kugelberg and Welander.

The key clinical findings of SMA include symmetric muscle weakness primarily in the upper body including upper extremities and impacting intercostal and trunk muscles, neck muscles, and bulbar musculature.

The most abundant tissue expression is in spinal cord, brain, and muscle. More recent studies indicate a wider tissue distribution for SMN1 gene expression.

Both forms of autosomal recessive SMA were mapped to chromosome 5q13 and the gene responsible for the disorders of SMN1 was identified by Lefebvre in 1995. A highly homologous gene designated SMN2 was mapped centromeric to SMN1.

Earlier studies reported that SMA1 occurred in 1 in 6000 to 1 in 10,000 births.

SMN1 location in 5q13 70924941−70953015.

SMN2 location in 5q13 70049523−70077595.

Four types of SMA are sometimes distinguished:

Type I Severe infantile (Werdnig−Hoffmann) onset birth to 6 months.

Type II SMA intermediate infantile onset before 18 months.

Type III Juvenile Kugelberg−Welander (SMA3) milder form onset between 18 months and 30 years.

Type IV Adult onset, onset later than 30 years.

Detailed genomic studies revealed that the 5q13 demonstrates a high degree of genomic instability. The SMN1 and SMN2 genes are highly homologous. One important difference is a single nucleotide in a splice enhancer in the terminal intron that leads transcripts of SMN2 to be missing exon 7. Four other nucleotide differences occur between the SMN1 and SMN2 genes, but these differences have no functional consequences.

Arnold and Fischbeck (2018) reported that 96% of SMA patients were found to be homozygous for deletions involving exon 7 or exons 7 and 8. Remaining patients were compound heterozygotes with an SMN1 deletion in one chromosome and a pathogenic SMN1 mutation on the other chromosome. These mutations can include small deletions or insertions or nucleotide substitutions. The most common mutations noted were p.Tyr272Cys exon 6, c.399−402 del AGAG, c.770−780 dup11. pThr274Ile.

SMA patients were reported to lack functioning SMN1 but may have some functioning SMN2. Interestingly, some individuals have more than one SMN2 gene and sense SMN can produce protein with some function patients with more copies of SMNs can be more mildly affected. SMN2 copy number is a modifier of the SMA phenotype.

The SMN protein was reported to localize to nuclei and to play roles in mRNA splicing it localizes to components of the spliceosome, and it was reported to influence RNA metabolism in the cytoplasm.

Deficiency of functional SMN proteins therefore impacts transcripts of other genes.

In 2016, a specific antisense nucleotide nusinersen (spinraza) was approved for therapy of SMA. Nusinersen is an antisense RNA that blocks the enhancer inhibitory sequence in the terminal intron of SMN2. It is important to note that other therapeutic approaches to treat SMA are also being investigated.

Arnold and Fischbeck reviewed recommendations for carrier test in SMA. The standard test was reported to be specific for deletions in SMN1. It is, however, important to note that other SMN1 pathogenic mutations involved in PCR assays have been developed to detect homozygous or heterozygous deletion in SMN1, SMA. Studies that investigate the number of copies of SMN2 are also useful.

Glascock et al. (2018) described a treatment algorithm diagnosed with SMA following NBS. They recommended immediate initiation of treatment for affected newborns who were found to have two or three copies of SMN2.

14.28 Antisense oligonucleotides in neurodegenerative diseases

Bennett et al. (2019) reviewed use of antisense nucleotide under investigation for therapies in specific neurodegenerative diseases, including HD, Alzheimer disease, and amyotrophic lateral sclerosis. They emphasized that beneficial medical therapies were not available for many neurodegenerative disorders including those cited above.

Bennett et al. described antisense oligonucleotides as oligonucleotide analogs, 12–30 nucleotides in length, that bind to RNAs including messenger RNA, noncoding RNAs, or microRNAs. Following their synthesis antisense oligonucleotides are sometimes modified to increase their cellular uptake, enhance tissue distribution, and increase binding efficiency.

There are several mechanisms identified through which antisense oligonucleotides impact function of the RNAs to which they bind. They can promote degradation of mRNA though RNAse H activity. Such antisense RNAs must also harbor nucleotides that promote RNAse H binding.

In some cases, binding of antisense RNA can modulate RNA splicing or modulate splice site selection. Antisense RNA can also impact translation or can impact microRNA binding.

Mechanisms though which antisense RNAs function must initially be tested in cell culture systems.

Modifications of antisense RNAs were noted to impact their pharmacological properties, tissue distribution, and clearance. Specific modifications include morpholino modifications and phosphorothioate modifications.

It is important to note that after antisense oligonucleotides intrathecal administration, a high percentage of the compound (up to 80%) appeared in the circulation. Studies on the half-life of administered antisense oligonucleotides indicate that this may be 6 weeks–6 months following injection.

Bennett et al. reported that studies have also been carried out on the spinal cord following lumbar intrathecal administration of antisense oligonucleotides. Concentrations of compound were found to be highest in the lower spine and in the brain there is a concentration gradient with highest levels in the cortex hippocampus and Purkinje cells.

They reviewed clinical trials in place and studies on neurodegenerative diseases. The most advanced antisense oligonucleotide studies involve nusinersen (spinraza) designed to treat autosomal recessive SMA, specifically by modifying a splice inhibitor in the SMN2 gene and thereby increasing expressing of SMN2.

The SMN2 gene does not normally express functional SMN protein because a C to T mutation in a splice enhancer within the terminal intron leads to exon 7 being skipped. Antisense RNA that blocks this inhibitory mutation leads functional protein to be expressed.

Bennett et al. noted that new therapies are being investigated in treatment of amyotrophic lateral sclerosis. One therapy is edavarone described as a power antioxidant that reduces motor neuron death.

Other compounds being studied include an antisense RNA that binds to the SOD1 mutation in ALS that is reported to have toxic gain of function. A specific compound was reported to be in Phase 1 and Phase 2 clinical trials. Antisense oligonucleotides directed to bind C9ORF72 repeat expansion were reported to be under development to treat amyotrophic lateral sclerosis.

14.29 Dynamic mutability of microsatellite repeats

Chatterjee et al. (2015) investigated the dynamic mutability of microsatellite repeats. They demonstrated that cold, heat, hypoxic stress, and oxidative stress impacted the mutability of CAG repeats in human cells.

A specific replication origin licensing factor CDT1, chromatin licensing and DNA replication factor 1, was reported to play an important role in CAG repeat expansion.

Schmidt and Pearson (2016) reported that the mammalian mismatch repair (MMR) system constituted a major driving force in CAG and CTG repeat expansions. Important components of the MMR system are listed below along with the chromosome assignments of the encoding loci.

MSH2 mutS homolog 2, 2p21-p16.3.
MSH3 mutS homolog 3, 5q14.1.
MSH6 mutS homolog 6 2p16.3.
MLH1 mutL homolog 1, 3p22.2.
MLH3 mutL homolog 3.
PMS2 PMS1 homolog 2, mismatch repair system component 7p22.1.

The formation of unusual DNA and R-loop structures was reported to play key roles in repeat instability.

Kadyrova et al. (2020) reported that a specific MLH1–MLH3 heterodimer acts as an endonuclease that promotes triplet repeat expansion. This heterodimer referred to as MutLγ (MLH1–MLH3 heterodimer) was shown to act as an endonuclease that nicks DNA. This endonuclease was shown to cleave the strand opposite the DNA loop and this strand break was shown to facilitate DNA repeat expansion.

14.29.1 Genomic tandem repeats in autism

In a 2021 article, Hannan noted that approximately half of the human genome is composed of repetitive DNA sequences. These also include tandem repeat sequences. Furthermore, at least 50 different human single gene disorders were noted to be due to expansion of tandem repeat sequences.

Hannan noted that the role of tandem repeat expansion in polygenic disorders was not well-studied. One exception was evident for tandem repeat expansions in autism spectrum disorders.

These disorders are known to arise due to single gene defects and can also potentially be polygenic or oligogenic in origin based on evidence of increased liability in families.

Evidence for genomic tandem repeat expansion in autism was published by Trost et al. (2020). They investigated expansion of repeats between 2 bp and 20 bp in a study of 17,231 genomes in families with autism and they also analyzed sequences from control populations.

Overall their study revealed that tandem repeat expansion had a prevalence of 23.3% in children with autism and a frequency of 20.7% in children without autism.

Mitra et al. (2021) developed a new bioinformatic method MONSTR for analyzing tandem repeats in DNA sequencing data generated from parents and offspring. The samples analyzed were from the Simons Simplex data on autism families. Results of their analyses revealed that genome-wide de novo autosomal tandem repeat expansions occurred at significantly higher frequency in autism probands than in nonautistic siblings with a mean frequency of 54.65 in autism cases and a frequency of 53.05 in nonautistic siblings.

Mitra et al. then set out to determine potential deleterious repeat expansions. They identified 35 deleterious expansions and 25 of these occurred in autism probands. These included deleterious expansions in genes that had previously been implicated in autism on the basis of point mutations. The implicated genes included:

PDCD1 programmed cell death 1, that encodes an immune-inhibitory receptor.

KCNB1 potassium voltage-gated channel subfamily B member 1.

AGO1 argonaute RISC component 1, involved in RNA inhibition.

CACNA2D3 calcium voltage-gated channel auxiliary subunit alpha2delta 3.

FOXP1 forkhead box P1, transcription factor.

RFX3 regulatory factor X3 transcription factor.
MED13L mediator complex subunit 13L.

14.30 Ocular gene therapy

In a 2018 review, Rodrigues et al. reported that subretinal injection of AAV vectors containing the RPE65 gene was reported to improve vision in patients with Leber's congenital amaurosis type 2 in studies carried out by Jacobson et al. (2012). This led to FDA approval for use of LUXTURNA in treatment of this disorders.

Leber's congenital amaurosis is known to be genetically heterogeneous. Kumaran et al. (2017) reviewed Leber's congenital amaurosis as an early onset retinal diseases (LCA/EOSRD) They grouped gene mutations causing these disorders into five different classes based on their function. These included genes involved in guanine synthesis, photoreceptor morphogenesis, ciliary transport, phototransduction, and retinoid cycle.

The first gene therapy approved treatment for LCA/EOSRD disorders involved RPE65 that functions in the retinoid cycle. Rodrigues et al. reported that in 2018 12 clinical trials were in place for LCA/EOSRD.

In considering the phenotype of these disorders, Rodrigues noted that Theodor Leber in 1869 described early onset of severe visual impairment with roving eye movements and eye poking. A variety of different fundal images developed over time that include abnormal fundus disk pallor and vessel shrinkage.

In a 2020 review Garafalo et al. emphasized that improved phenotyping and improved genetic characterization had led to great progress in diagnosis and treatment of retinal diseases. In this review, genes determined to be implicated in inherited retinal diseases in patients seen in Philadelphia and Toronto were documented.

The most commonly implicated gene identified in patients with inherited retinal diseases in Philadelphia were reported to be:

ABCA4 ATP-binding cassette subfamily A member 4, member of the superfamily of ATP-binding cassette (ABC) transporters.

RHO rhodopsin protein is found in rod cells in the back of the eye and is essential for vision in low-light conditions.

RPGR retinitis pigmentosa GTPase regulator protein localizes to the outer segment of rod photoreceptors and is essential for their viability.

PRPH2 peripherin 2, cell surface glycoprotein found in the outer segment of both rod and cone photoreceptor cells. It may function as an adhesion molecule involved in stabilization.

RPR65 retinoid isomerohydrolase performs the essential enzymatic isomerization step in the synthesis of 11-cis retinal.

The most commonly implicated genes with identified in patients with inherited retinal diseases in Toronto were reported to be:

ABCA4 ATP-binding cassette (ABC) subfamily A member 4, member of the superfamily of ABC transporters.

USH2A usherin protein is found in the basement membrane, and may be important in development and homeostasis of the inner ear and retina.

VMD2 also known as BEST encodes bestrophin1. Bestrophins are generally believed to form calcium-activated chloride-ion channels in epithelial cells but they have also been shown to be highly permeable to bicarbonate ion transport in retinal tissue.

CHM Rab escort protein defects lead progressive dystrophy of the choroid, retinal pigment epithelium, and retina.

CNGA3 cyclic nucleotide-gated channel subunit alpha 3 cyclic nucleotide-gated cation channel protein family, which is required for normal vision and olfactory signal transduction.

However, considering all patients with early onset inherited eye disease seen in both centers, 58 different genes were implicated. Across different centers, between 36% and 92% of cases had autosomal recessive disorders, X-linked disorders occurred in 10%−25% of cases, and autosomal dominant disorders occurred in 5%−39% of cases. With respect to X-linked inheritance, heterozygous females were reported to manifest patchy regional pigmentary retinopathy.

Additional gene therapies for early onset retinal diseases developed since the RPE65 gene therapy treatment. These included MYO7A gene therapy in Usher syndrome 1B (Lopes and Williams, 2015) and ABCA4 treatment of Stargardt disease (Han et al., 2014).

Garafalo et al. noted that studies in models of autosomal recessive retinitis pigmentosa due to defects in PDE6B, a rod cyclic GMP phosphodiesterase, were being carried out using subretinal AAV transferred gene therapy.

Several clinical trials were noted to be in place that involved subretinal therapy in retinal disease due to CHM (choroidemia) defects.

Garafalo et al. noted that in addition to subretinal gene augmentation in treatment of incurable retinal disease, studies were ongoing to investigate use of vitreal injections.

Kumaran et al. (2017) reviewed Leber congenital amaurosis (LCA) referred to as severe childhood retinal disease and a milder form of the disorder EOSRD (early onset severe retinal disease). They particularly emphasized roles in Leber amaurosis of mutations in the following genes:

GUCY2D, guanylate cyclase 2D, retinal, retina-specific guanylate cyclase, which is a member of the membrane guanylyl cyclase family.

NMNAT1 nicotinamide nucleotide adenylyltransferase 1 m catalyzes a key step in the biosynthesis of nicotinamide adenine dinucleotide.

CEP290 centrosomal protein 290, protein is localized to the centrosome and cilia and has sites for N-glycosylation, tyrosine sulfation, phosphorylation, N-myristoylation, and amidation.

AIPL1 aryl hydrocarbon receptor interacting protein like 1. Mutations in this gene may cause approximately 20% of recessive LCA.

In the milder disease EOSRD mutations in others including:

RPE65 retinoid isomerohydrolase RPE65, protein encoded by this gene is a component of the vitamin A visual cycle of the retina, which supplies the 11-cis retinal chromophore of the photoreceptor opsin visual pigments.

LRAT lecithin retinol acyltransferase catalyzes the esterification of all-trans-retinol into all-trans-retinyl ester. This reaction is an important step in vitamin A metabolism in the visual system.

RDH12 retinol dehydrogenase 12, an NADPH-dependent retinal reductase whose highest activity is toward 9-cis and all-trans-retinol.

In 2020 Cehajic-Kapetanovic et al. reported results of a first in human clinical gene therapy trial to treat retinitis pigmentosa due to mutations in RP GTPase regulator (RPGR) gene. Visual improvements were documented.

References

Aartsma-Rus A, Ginjaar IB, Bushby KJ: The importance of genetic diagnosis for Duchenne muscular dystrophy, *Med Genet* 53(3):145–151, 2016. Available from: https://doi.org/10.1136/jmedgenet-2015-103387.

Arnold ES, Fischbeck KH: Spinal muscular atrophy, *Handb Clin Neurol* 148:591–601, 2018. Available from: https://doi.org/10.1016/B978-0-444-64076-5.00038-7. PMID: 29478602.

Ashizawa T, Öz G, Paulson HL: Spinocerebellar ataxias: prospects and challenges for therapy development, *Nat Rev Neurol* 14(10):590–605, 2018. Available from: https://doi.org/10.1038/s41582-018-0051-6. PMID: 30131520.

Bellotti AS, Andreoli L, Ronchi D, Bresolin N, Comi GP, Corti S: Molecular approaches for the treatment of Pompe disease, *Mol Neurobiol* 57(2):1259–1280, 2020. Available from: https://doi.org/10.1007/s12035-019-01820-5.

Bennett CF, Krainer AR, Cleveland DW: Antisense oligonucleotide therapies for neurodegenerative diseases, *Annu Rev Neurosci* 42:385–406, 2019. Available from: https://doi.org/10.1146/annurev-neuro-070918-050501. PMID: 31283897.

Bettencourt C, Hensman-Moss D, Flower M, et al: DNA repair pathways underlie a common genetic mechanism modulating onset in polyglutamine diseases, *Ann Neurol* 79(6):983–990, 2016. Available from: https://doi.org/10.1002/ana.24656. PMID: 27044000.

Bladen CL, Salgado D, Monges S, et al: The TREAT-NMD DMD Global Database: analysis of more than 7,000 Duchenne muscular dystrophy mutations, *Hum Mutat* 36(4):395–402, 2015. Available from: https://doi.org/10.1002/humu.22758. PMID: 25604253.

Bonam SR, Wang F, Muller S: Lysosomes as a therapeutic target, *Nat Rev Drug Discov* 18(12):923–948, 2019. Available from: https://doi.org/10.1038/s41573-019-0036-1. PMID: 31477883.

Boutary S, Echaniz-Laguna A, Adams D, et al: Treating PMP22 gene duplication-related Charcot–Marie–Tooth disease: the past, the present and the future, *Transl Res* 227:100–111, 2021. Available from: https://doi.org/10.1016/j.trsl.2020.07.006.

Buijsen RAM, Toonen LJA, Gardiner SL, van Roon-Mom WMC: Genetics, mechanisms, and therapeutic progress in polyglutamine spinocerebellar ataxias, *Neurotherapeutics* 16(2):263–286, 2019. Available from: https://doi.org/10.1007/s13311-018-00696-y. PMID: 30607747.

Byrne BJ, Corti M, Muntoni F: Considerations for systemic use of gene therapy, *Mol Ther* 29 (2):422−423, 2021. Available from: https://doi.org/10.1016/j.ymthe.2021.01.016. PMID: 33485465 Review.

Cehajic-Kapetanovic J, Xue K, Martinez-Fernandez de la Camara C, et al: Initial results from a first-in-human gene therapy trial on X-linked retinitis pigmentosa caused by mutations in RPGR, *Nat Med* 26(3):354−359, 2020. Available from: https://doi.org/10.1038/s41591-020-0763-1. PMID: 32094925.

Chamberlain JR, Chamberlain JS: Progress toward gene therapy for Duchenne Muscular Dystrophy, *Mol Ther* 25(5):1125−1131, 2017. Available from: https://doi.org/10.1016/j.ymthe.2017.02.019. PMID: 28416280.

Chatterjee N, Lin Y, Santillan BA, Yotnda P, Wilson JH: Environmental stress induces trinucleotide repeat mutagenesis in human cells, *Proc Natl Acad Sci U S A*. 112 (12):3764−3769, 2015. Available from: https://doi.org/10.1073/pnas.1421917112.

Clevers H: Modeling development and disease with organoids, *Cell* 165(7):1586−1597, 2016. Available from: https://doi.org/10.1016/j.cell.2016.05.082. PMID: 27315476.

Collins FS, Doudna JA, Lander ES, Rotimi CN: Human molecular genetics and genomics—important advances and exciting possibilities, *N Engl J Med* 384(1):1−4, 2021. Available from: https://doi.org/10.1056/NEJMp2030694.

Dagil L, Troelsen KS, Bolt G, et al: Interaction between the a3 region of Factor VIII and the TIL'E' domains of the von Willebrand factor, *Biophys J* 117(3):479−489, 2019. Available from: https://doi.org/10.1016/j.bpj.2019.07.007. PMID: 31349985.

Davies KE, Guiraud S: Micro-dystrophin genes bring hope of an effective therapy for Duchenne Muscular Dystrophy, *Mol Ther* 27(3):486−488, 2019. Available from: https://doi.org/10.1016/j.ymthe.2019.01.019. PMID: 30765324.

Delgado-Calle J, Sato AY, Bellido T: Role and mechanism of action of sclerostin in bone, *Bone* 96:29−37, 2017. Available from: https://doi.org/10.1016/j.bone.2016.10.007. Epub October 12, 2016. PMID: 27742498.

Donadon I, McVey JH, Garagiola I: rt clustered *F8* missense mutations cause hemophilia A by combined alteration of splicing and protein biosynthesis and activity, *Haematologica* 103(2):344−350, 2018. Available from: https://doi.org/10.3324/haematol.2017.178327.

Esrick EB, Lehmann LE, Biffi A, et al: Post-transcriptional genetic silencing of *BCL11A* to treat sickle cell disease, *Engl J Med* 384(3):205−215, 2021. Available from: https://doi.org/10.1056/NEJMoa2029392. MID: 33283990.

Ferrari S, Jacob A, Beretta S, et al: Efficient gene editing of human long-term hematopoietic stem cells validated by clonal tracking, *Nat Biotechnol* 38(11):1298−1308, 2020. Available from: https://doi.org/10.1038/s41587-020-0551-y. Epub June 29, 2020. PMID: 32601433.

Finkel RS: Read-through strategies for suppression of nonsense mutations in Duchenne/Becker muscular dystrophy: aminoglycosides and ataluren (PTC124), *J Child Neurol* 25(9):1158−1164, 2010. Available from: https://doi.org/10.1177/0883073810371129. Epub June 2, 2010.

Fire A, Xu S, Montgomery MK, Kostas SA, Driver SE, Mello CC: Potent and specific genetic interference by double-stranded RNA in *Caenorhabditis elegans*, *Nature*. 391 (6669):806−811, 1998. Available from: https://doi.org/10.1038/35888. PMID: 9486653.

Forlino A, Marini JC: Osteogenesis imperfecta, *Lancet*. 387(10028):1657−1671, 2016. Available from: https://doi.org/10.1016/S0140-6736(15)00728-X.

Frangoul H, Altshuler D, Cappellini MD, et al: CRISPR-Cas9 gene editing for sickle cell disease and β-thalassemia, *N Engl J Med* 384(3):252–260, 2021. Available from: https://doi.org/10.1056/NEJMoa2031054. Epub 2020 Dec 5. PMID: 33283989.

Garafalo AV, Cideciyan AV, Héon E, et al: Progress in treating inherited retinal diseases: early subretinal gene therapy clinical trials and candidates for future initiatives, *Prog Retin Eye Res* 77:100827, 2020. Available from: https://doi.org/10.1016/j.preteyeres.2019.100827. Epub December 30, 2019. PMID: 31899291.

Glascock J, Sampson J, Haidet-Phillips A, et al: Treatment algorithm for infants diagnosed with spinal muscular atrophy through newborn screening, *J Neuromuscul Dis* 5(2):145–158, 2018. Available from: https://doi.org/10.3233/JND-180304. PMID: 29614695.

Goold R, Flower M, Moss DH, et al: FAN1 modifies Huntington's disease progression by stabilizing the expanded HTT CAG repeat, *Hum Mol Genet* 28(4):650–661, 2019. Available from: https://doi.org/10.1093/hmg/ddy375. PMID: 30358836.

Han Z, Conley SM, Naash MI: Gene therapy for Stargardt disease associated with ABCA4 gene, *Adv Exp Med Biol* 801:719–724, 2014. Available from: https://doi.org/10.1007/978-1-4614-3209-8_90. PMID: 24664763.

Hannan AJ: Repeat DNA expands our understanding of autism spectrum disorder, *Nature* 589(7841):200–202, 2021. Available from: https://doi.org/10.1038/d41586-020-03658-7. PMID: 33442037.

Hassan MI, Saxena A, Ahmad F: Structure and function of von Willebrand factor, *Blood Coagul Fibrinolysis* 23(1):11–22, 2012. Available from: https://doi.org/10.1097/MBC.0b013e32834cb35d. PMID: 22089939.

Ingram VM: The sulphydryl groups of sickle-cell haemoglobin, *Biochem J* 65(4):760–763, 1957. Available from: https://doi.org/10.1042/bj0650760. PMID: 13426098.

Iorio A, Stonebraker JS, Brooker M, Soucie JM: Data and Demographics Committee of the World Federation of Hemophilia. Measuring the quality of haemophilia care across different settings: a set of performance indicators derived from demographics data, *Haemophilia.* 23(1):e1–e7, 2017. Available from: https://doi.org/10.1111/hae.13127.

Jacobson SG, Cideciyan AV, Ratnakaram R, et al: Gene therapy for leber congenital amaurosis caused by RPE65 mutations: safety and efficacy in 15 children and adults followed up to 3 years, *Arch Ophthalmol* 130(1):9–24, 2012. Available from: https://doi.org/10.1001/archophthalmol.2011.298.

Johnson CE, Hunt DK, Wiltshire M, et al: Endoplasmic reticulum stress and cell death in mTORC1-overactive cells is induced by nelfinavir and enhanced by chloroquine, *Mol Oncol* 9(3):675–688, 2015. Available from: https://doi.org/10.1016/j.molonc.2014.11.005.

Kadyrova LY, Gujar V, Burdett V, Modrich PL, Kadyrov FA: Human MutLγ, the MLH1-MLH3 heterodimer, is an endonuclease that promotes DNA expansion, *Proc Natl Acad Sci U S A* 117(7):3535–3542, 2020. Available from: https://doi.org/10.1073/pnas.1914718117. PMID: 32015124.

Khan A, Barber DL, Huang J, et al: Lentivirus-mediated gene therapy for Fabry disease, *Nat Commun* 12(1):1178, 2021. Available from: https://doi.org/10.1038/s41467-021-21371-5. PMID: 33633114.

Kim J, Koo BK, Knoblich JA: Human organoids: model systems for human biology and medicine, *Nat Rev Mol Cell Biol* 21(10):571–584, 2020a. Available from: https://doi.org/10.1038/s41580-020-0259-3. PMID: 32636524.

Kim KH, Hong EP, Shin JW, et al: Genetic and functional analyses point to FAN1 as the source of multiple Huntington disease modifier effects, *Am J Hum Genet* 107 (1):96–110, 2020b. Available from: https://doi.org/10.1016/j.ajhg.2020.05.012. Epub June 25, 2020. PMID: 32589923.

Kumaran N, Moore AT, Weleber RG, Michaelides M: Leber congenital amaurosis/early-onset severe retinal dystrophy: clinical features, molecular genetics and therapeutic interventions, *Br J Ophthalmol* 101(9):1147–1154, 2017. Available from: https://doi.org/10.1136/bjophthalmol-2016-309975. PMID: 28689169.

Kunkel LM, Beggs AH, Hoffman EP: Molecular genetics of Duchenne and Becker muscular dystrophy: emphasis on improved diagnosis, *Clin Chem* 35(7):B21–B24, 1989.

Kunkel LM, Hejtmancik JF, Caskey CT, et al: Analysis of deletions in DNA from patients with Becker and Duchenne muscular dystrophy, *Nature*. 322(6074):73–77, 1986. Available from: https://doi.org/10.1038/322073a0. PMID: 3014348.

Lahue RS: New developments in Huntington's disease and other triplet repeat diseases: DNA repair turns to the dark side, *Neuronal Signal* 4(4):NS20200010, 2020. Available from: https://doi.org/10.1042/NS20200010. eCollection December, 2020. PMID: 33224521.

Löf A, Müller JP, Brehm MA: A biophysical view on von Willebrand factor activation, *J Cell Physiol* 233(2):799–810, 2018. Available from: https://doi.org/10.1002/jcp.25887.

Lopes VS, Williams DS: Gene therapy for the retinal degeneration of Usher syndrome caused by mutations in MYO7A, *Cold Spring Harb Perspect Med* 5(6):a017319, 2015. Available from: https://doi.org/10.1101/cshperspect.a017319. PMID: 25605753.

Loro E, Sengupta K, Bogdanovich S, et al: High-throughput identification of post-transcriptional utrophin up-regulators for Duchenne muscle dystrophy (DMD) therapy, *Sci Rep* 10(1):2132, 2020. Available from: https://doi.org/10.1038/s41598-020-58737-6. PMID: 32034254.

Luciani M, Gritti A, Meneghini V: Human iPSC-based models for the development of therapeutics targeting neurodegenerative lysosomal storage diseases, *Front Mol Biosci* 7:224, 2020. Available from: https://doi.org/10.3389/fmolb.2020.00224. eCollection 2020. PMID: 33062642.

Lupski JR, de Oca-Luna RM, Slaugenhaupt S, et al: DNA duplication associated with Charcot–Marie–Tooth disease type 1A, *Cell* 66(2):219–232, 1991. Available from: https://doi.org/10.1016/0092-8674(91)90613-4.

Maioli M, Gnoli M, Boarini M, et al: Genotype-phenotype correlation study in 364 osteogenesis imperfecta Italian patients, *Eur J Hum Genet* 27(7):1090–1100, 2019. Available from: https://doi.org/10.1038/s41431-019-0373-x.

McBride KL, Berry S, Braverman N, et al: ACMG Therapeutics Committee: Treatment of mucopolysaccharidosis type II (Hunter syndrome): a Delphi derived practice resource of the American College of Medical Genetics and Genomics (ACMG), *Genet Med.* 22(11):1735–1742, 2020. Available from: https://doi.org/10.1038/s41436-020-0909-z.

McEneaney LJ, Tee AR: Finding a cure for tuberous sclerosis complex: from genetics through to targeted drug therapies, *Adv Genet* 103:91–118, 2019. Available from: https://doi.org/10.1016/bs.adgen.2018.11.003.

Mendell JR, Al-Zaidy SA, Rodino-Klapac LR, et al: Current clinical applications of in vivo gene therapy with AAVs, *Mol Ther* 29(2):464–488, 2021. Available from: https://doi.org/10.1016/j.ymthe.2020.12.007.

Mitra I, Huang B, Mousavi N, et al: Patterns of de novo tandem repeat mutations and their role in autism, *Nature*. 589(7841):246–250, 2021. Available from: https://doi.org/10.1038/s41586-020-03078-7. PMID: 33442040.

Montes M, Sanford BL, Comiskey DF, Chandler DS: RNA splicing and disease: animal models to therapies, *Trends Genet* 35(1):68–87, 2019. Available from: https://doi.org/10.1016/j.tig.2018.10.002.

Nakamori M, Panigrahi GB, Lanni S, et al: A slipped-CAG DNA-binding small molecule induces trinucleotide-repeat contractions in vivo, *Nat Genet* 52(2):146–159, 2020. Available from: https://doi.org/10.1038/s41588-019-0575-8.

Nakatani R, Nakamori M, Fujimura H, Mochizuki H, Takahashi MP: Large expansion of CTG•CAG repeats is exacerbated by MutSβ in human cells, *Sci Rep* 5:11020, 2015. Available from: https://doi.org/10.1038/srep11020. PMID: 26047474.

Nathwani AC: Gene therapy for hemophilia, *Hematol Am Soc Hematol Educ Program* 2019(1):1–8, 2019. Available from: https://doi.org/10.1182/hematology.2019000007. PMID: 31808868.

Orkin SH, Bauer DE: Emerging genetic therapy for sickle cell disease, *Annu Rev Med* 70:257–271, 2019. Available from: https://doi.org/10.1146/annurev-med-041817-125507. Epub October 24, 2018. PMID: 30355263.

Orkin SH: MOLECULAR MEDICINE: Found in Translation, *Med (NY)* 2(2):122–136, 2021. Available from: https://doi.org/10.1016/j.medj.2020.12.011. PMID: 33688634.

Pantera H, Shy ME, Svaren J: Regulating PMP22 expression as a dosage sensitive neuropathy gene, *Brain Res* 1726:146491, 2020. Available from: https://doi.org/10.1016/j.brainres.2019.146491.

Pauling L, Itano HA, et al: Sickle cell anemia a molecular disease, *Science*. 110(2865):543–548, 1949. Available from: https://doi.org/10.1126/science.110.2865.543. PMID: 15395398.

Pinto E. Vairo F, Rojas Málaga D, et al: Precision medicine for lysosomal disorders, *Biomolecules* 10(8):1110, 2020. Available from: https://doi.org/10.3390/biom10081110. PMID: 32722587.

Ramos JN, Hollinger K, Bengtsson NE, et al: Development of novel micro-dystrophins with enhanced functionality, *Mol Ther* 27(3):623–635, 2019. Available from: https://doi.org/10.1016/j.ymthe.2019.01.002. PMID: 30718090.

Rodrigues GA, Shalaev E, Karami TK, et al: Pharmaceutical development of AAV-based gene therapy products for the eye, *Pharm Res* 36(2):29, 2018. Available from: https://doi.org/10.1007/s11095-018-2554-7. PMID: 30591984.

Rossi JJ, Rossi DJ: siRNA drugs: here to stay, *Mol Ther* 29(2):431–432, 2021. Available from: https://doi.org/10.1016/j.ymthe.2021.01.015.

Rossi V, Lee B, Marom R: Osteogenesis imperfecta: advancements in genetics and treatment, *Curr Opin Pediatr* 31(6):708–715, 2019. Available from: https://doi.org/10.1097/MOP.0000000000000813. PMID: 31693577.

Scahill RI, Zeun P, Osborne-Crowley K, et al: Biological and clinical characteristics of gene carriers far from predicted onset in the Huntington's disease Young Adult Study (HD-YAS): a cross-sectional analysis, *Lancet Neurol* 19(6):502–512, 2020. Available from: https://doi.org/10.1016/S1474-4422(20)30143-5.

Schmidt MHM, Pearson CE: Disease-associated repeat instability and mismatch repair, *DNA Repair (Amst.)* 38:117–126, 2016. Available from: https://doi.org/10.1016/j.dnarep.2015.11.008.

Settembre C, Medina DL: TFEB and the CLEAR network, *Methods Cell Biol* 126:45–62, 2015. Available from: https://doi.org/10.1016/bs.mcb.2014.11.011.

Sharma R, Flood VH: Advances in the diagnosis and treatment of Von Willebrand disease, *Blood*. 130(22):2386–2391, 2017. Available from: https://doi.org/10.1182/blood-2017-05-782029. PMID: 29187375.

Shriner D, Rotimi CN: Whole-genome-sequence-based haplotypes reveal single origin of the sickle allele during the holocene wet phase, *Am J Hum Genet* 102(4):547–556, 2018. Available from: https://doi.org/10.1016/j.ajhg.2018.02.003. Epub March 8, 2018. PMID: 29526279.

Spinazzola JM, Kunkel LM: Pharmacological therapeutics targeting the secondary defects and downstream pathology of Duchenne muscular dystrophy, *Expert Opin Orphan Drugs* 4(11):1179–1194, 2016. Available from: https://doi.org/10.1080/21678707.2016.1240613. PMID: 28670506.

Sun C, Serra C, Lee G, Wagner KR: Stem cell-based therapies for Duchenne muscular dystrophy, *Exp Neurol* 323:113086, 2020. Available from: https://doi.org/10.1016/j.expneurol.2019.113086.

Suzuki Y: Chaperone therapy for molecular pathology in lysosomal diseases, *Brain Dev*. 43(1):45–54, 2021. Available from: https://doi.org/10.1016/j.braindev.2020.06.015. PMID: 32736903.

Suzuki Y: Emerging novel concept of chaperone therapies for protein misfolding diseases, *Proc Jpn Acad Ser B Phys Biol Sci*. 90(5):145–162, 2014. Available from: https://doi.org/10.2183/pjab.90.145. PMID: 24814990.

Tabrizi SJ, Ghosh R, Leavitt BR: Huntingtin lowering strategies for disease modification in Huntington's disease, *Neuron* 101(5):801–819, 2019. Available from: https://doi.org/10.1016/j.neuron.2019.01.039. PMID: 30844400.

Tosi G, Duskey JT, Kreuter J: Nanoparticles as carriers for drug delivery of macromolecules across the blood–brain barrier, *Expert Opin Drug Deliv* 17(1):23–32, 2020. Available from: https://doi.org/10.1080/17425247.2020.1698544.

Tremblay JP, Annoni A, Suzuki M: Three decades of clinical gene therapy: from experimental technologies to viable treatments, *Mol Ther* 29(2):411–412, 2021. Available from: https://doi.org/10.1016/j.ymthe.2021.01.013. PMID: 33472032.

Trost B, Engchuan W, Nguyen CM, et al: Genome-wide detection of tandem DNA repeats that are expanded in autism, *Nature*. 586(7827):80–86, 2020. Available from: https://doi.org/10.1038/s41586-020-2579-z.

Wang R, Persky NS, Yoo B, et al: DNA repair. Mechanism of DNA interstrand cross-link processing by repair nuclease FAN1, *Science*. 346(6213):1127–1130, 2014. Available from: https://doi.org/10.1126/science.1258973. PMID: 2543077.

Further reading

Genetic Modifiers of Huntington's Disease (GeM-HD) Consortium: CAG repeat not polyglutamine length determines timing of Huntington's disease onset, *Cell* 178(4):887–900.e14, 2019. Available from: https://doi.org/10.1016/j.cell.2019.06.036. PMID: 31398342.

Genetic Modifiers of Huntington's Disease (GeM-HD) Consortium: Identification of genetic factors that modify clinical onset of Huntington's disease, *Cell* 162

(3):516–526, 2015. Available from: https://doi.org/10.1016/j.cell.2015.07.003. PMID: 26232222.

Kim J, Hu C, Moufawad El Achkar C, et al: Patient-customized oligonucleotide therapy for a rare genetic disease, *N Engl J Med* 381(17):1644–1652, 2019. Available from: https://doi.org/10.1056/NEJMoa1813279. PMID: 31597037.

Kohn DB, Hershfield MS, Puck JM, et al: Consensus approach for the management of severe combined immune deficiency caused by adenosine deaminase deficiency, *J Allergy Clin Immunol* 143(3):852–863, 2019. Available from: https://doi.org/10.1016/j.jaci.2018.08.024.

Roa BB, Lupski JR: Molecular basis of Charcot–Marie–Tooth disease type 1A: gene dosage as a novel mechanism for a common autosomal dominant condition, *Am J Med Sci* 306(3):177–184, 1993. Available from: https://doi.org/10.1097/00000441-199309000-00010. PMID: 8128981.

Stirnadel-Farrant H, Kudari M, Garman N, et al: Gene therapy in rare diseases: the benefits and challenges of developing a patient-centric registry for Strimvelis in ADA-SCID, *Orphanet J Rare Dis* 13(1):49, 2018. Available from: https://doi.org/10.1186/s13023-018-0791-9. PMID: 29625577.

Index

Note: Page numbers followed by "*f*" refer to figures.

A

ABCA4, 172
Aberrant splicing, 158, 321
ABO, 33
   information on, 33–34
   mapping of ABO locus to chromosome, 35
Accurate splicing, 158–159, 321
Achondroplasia (ACH), 18–19
   mutations reported as pathogenic in achondroplasia multiple submitters, 144
Acute intermittent porphyria, 216–217
Acute lymphoblastic leukemia (ALL), 242
Acute myeloid leukemia (AML), 248
Adeno-associated viral vectors (AAV vectors), 163, 317
Adenosine deaminase RNA specific (ADAR), 116
   ADAR1, 171
   ADAR2, 171
Adenosine deaminase tRNA specific (ADAT), 116
Adenosine deaminase-related immunodeficiency (ADA-related immunodeficiency), 162
Adenosine triphosphatases (ATPases), 69
Adenylate kinase 1 (AK1), 35
ADH1A, 29–30
ADH1C, 30
ADH3. *See* ADH1C
Adrenoleukodystrophy, 224
Adult cancers, 248–250
AICDA. *See* AID
AID, 117
AKT serine threonine kinases, 12–13
Alpha thalassemia, 181
α-N-acetylglucosaminidase (NAGLU), 224
Alternate polyadenylation, 109–110
Alternate splicing of transcripts, 110
Ambroxol, 315
American College of Medical Genetics (ACMG), 45, 185
   recommendations, 84
Amino acids, 27, 30
Aminoacidopathies, 226
Aminoacidurias, 185
Aminoacyl tRNA synthases, 113–114
   noncanonical functions, 114
Angelman syndrome (AS), 70, 74–75
ANO1, 175
Antisense oligonucleotides (ASO), 155
   in neurodegenerative diseases, 347–348

Antisense therapies, 159–160
Apical ectodermal ridge (AER), 17
Apolipoprotein A5 (APOA5), 95
Apolipoprotein L1 (APOL1), 197
Apparent mineralocorticoid excess (AME), 295
Aromatic L-amino-acid decarboxylase deficiency (AADCD), 286
Array-based CFH, 9
Aryl hydrocarbon receptor interacting protein like 1 (AIPL1), 351
Aspartylglucosaminuria, 195–196
Assay for transposase accessible chromatin (ATAC), 54
   sequencing, 103
AT-rich interaction domain 5B (ARID5B), 108
Ataxins (ATXN), 293, 338
ATP-binding cassette subfamily A member 1 (ABCA1), 340
ATP-binding cassette subfamily A member 4 (ABCA4), 350–351
ATP-binding cassette transporter (ABC transporter), 172
ATPase Na+/K+ transporting subunit alpha 3, 69
ATRX syndrome, 101
Autophagy lysosomes, 312
Autosomal dominant (AD), 193–194
   disorder, 143
   gene mutations, 246
   genes with mutations leading to autosomal dominant cerebral palsy, 283–284
Autosomal dominant cerebellar ataxia (ADCA), 291, 337
Autosomal dominant cerebellar ataxia with deafness and narcolepsy (ADCADN), 78
Autosomal recessive (AR), 148–149, 193–194, 284
   ataxias, 291–292
   compound heterozygous, 284
   proximal muscular atrophy, 345–347
Azathioprine, 214

B

B-cell receptor, 252
Bardet Biedl syndrome (BBS), 103
Base editing, 169
Beckwith–Wiedemann syndrome (BWS), 57, 70
Bernard–Soulier syndrome, 331
Bestrophins, 351

11-beta hydroxysteroid dehydrogenase, 296
BH4. *See* Tetrahydrobiopterin (BH4)
BHLHA9 gene, 17
Biallelic recessive mutation in LMNA, 49
Biochemistry, 27–28
Biotransformation
   of medicinal compounds, 210
   other factors and processes involved in biotransformation of drugs, 211–212
Bisulfite sequencing, 56
Blackfan – Diamond anemia, 113
Blood groups, 33
Blue sclerotics, 19–20
Bone morphogenetic proteins (BMPs), 145
BRCA1, 79–80
Breast cancer risk genes, 258–259
British antilewisite (BAL), 227
Bruton tyrosine kinase (BTK), 189

C

C9ORF 72 mutations, 160
Cancer
   cancer-inducing mechanisms, 255
   immunotherapy, 262–264
   therapy, 260–262
      new approaches to, 229–231
      RAS signaling pathway, 231
      synthetic lethality in cancer treatment, 230
Carney complex, 53
Carrier screening, 273–274
Cartilage hair hypoplasia syndrome, 113
CASC5, 49
Catalog of Somatic Mutations in Cancer (COSMIC), 258
CD42. *See* Glycoprotein Ib (GPIb)
Cell-free DNA (cfDNA), 256
   analyses in testing for tumors, 256
Cell-free fetal DNA (cff DNA), 51–52
Cell-free RNA (cfRNA), 256
Cell-free studies including transcriptome analyses, 257–258
Central nervous system (CNS), 312
Centrosomal protein 290 (CEP290), 351
Cerebral Autosomal Dominant Arteriopathy with Subcortical Infarcts and Leukoencephalopathy (CADASIL), 67
Cerebral palsy, 281–285
   and genetic factors, 281
Cerebrospinal fluid (CSF), 256, 314
Ceroid lipofuscinoses, 224–226
   clinical manifestations, 225–226
Chaperone therapies in treatment of lysosomal storage diseases, 312

Charcot–Marie–Tooth disease, 22, 339–340
CHARGE syndrome, 136
Checkpoint inhibitors (CPI), 263
Chemical analyses and metabolism, 37–38
Childhood ALL, 243
Chimeric antigen receptor T-cells (CAR-T cells), 263–265
   therapy and cytokine release syndrome, 265
Chloroquine, 312
Chromatin, 5–6
   conformation capture, 55
   modeling genes, 280
   remodeling, 98–99
   structure, 53–55
Chromatographic techniques, 28
Chromodomain helicase (CHD)
   CHD7, 136
   CHD8, 105
Chromogenic staining methods, 30
Chromosomal map positions, gene products involved in, 139–140
Chromosomal microarray, 271
Chromosomal mosaicism, 12
Chromosomes, 3, 41
   applications of studies of, 6–9
   genomes and sequence, 82
   microarray analyses, 8–9
Chromothripsis, 248
Cis acting DNA elements, 54
Classic type `9q34. 3 COL5A AD collagen type V alpha 1 chain, 150
Cleavage stimulation factor subunit 2 (CSTF2), 109–110
Clinical findings, 23
Clinical genetics, 133–134, 138
Clinical genome database (CGD), 138
Clinical Genome Resource (ClinGen), 45, 83
Clinical syndromes, 134
Clinical trials, 166–168
   related to cell and gene therapy, 326–328
ClinVar, 45, 275
Coagulation disorders, 328–330
   hemophilia, 328–330
Coffin – Siris syndrome (CSS), 78
Cohesin complex, 101
Cohesinopathies, 101–102
COL1A1, 323
COL1A2, 323
COL3A, 149
Colchicine, 6–7
Color blindness, 19
Column chromatography, 28
Common epilepsies, 280–281
Comparative genomic hybridization (CGH), 8

Comparative phenomics, 133
Congenital adrenal hyperplasia (CAH), 295–296
Congenital malformations and syndromes, 134–136
    accounting for phenotypic differences in individuals with same genetic defect, 135–136
    inborn errors of development, 134–135
    twin studies and analysis of gene effects on phenotype, 135
Connective tissue disorders, 148–151
Conserved noncoding elements (CNEs), 122
Contractural arachnodactyly (CCA1), 147
Coproporphyria, 217
Copy number change assessment, 84–86
Copy number variants (CNVs), 67
Cornelia de Lange syndrome, 101–102
Coronary heart disease (CAD), 298–300
Creatine kinase (CK), 341
CRISPR-Cas
    CRISPR-CAS9 systems, 166, 169
    studies, 255
    theta, 171
CTLA4, 263
CXORF5 gene, 76
Cyclic nucleotide-gated channel subunit alpha 3 (CNGA3), 351
Cyclophilin, 323
CYP2B6, 211
CYP2C19, 211
CYP2D6, 211
Cystic fibrosis (CF), 183–184, 227
    problems, 176
    therapy, 175–176
Cystic fibrosis transmembrane regulator (CFTR), 175, 183–184
Cytochrome P450 monooxygenase system, 210
Cytosine guanine nucleotides (CpG), 6

D

Database of Genotypes and Phenotypes (dbGaP), 275
Databases, 137–138
    of importance in searching for gene, genotype phenotype correlations, 275–276
ddATP, 42
ddCTP, 42
ddGTP, 42
ddTTP, 42
Deafness, 18
DECIPHER, 275
Decipher database, 138
Denosumab, 325
Desert hedgehog (DHH), 141

Developmental Disorders Study, 87–88
Developmental epileptic encephalopathies, genes involved in, 279–280
Developmental origins of pediatric cancers, 246–247
Diagnostic testing for inborn errors of metabolism leading to seizures, 277–278
Dideoxynucleotides, 42
Differentially methylated regions (DMR), 56
Diffuse large B-cell lymphoma (DLBCL), 252
Digenic inheritance, 48
    additional evidence for, 49–50
Disease treatments, 309
DLK1 locus, 77
DLX5 gene, 17
DLX6 gene, 17
DNA, 5–6
    consequences of determination of DNA structure, 6
    damage and repair, 250–251
    methylation, 55–56, 98–99
        episignatures and phenotypic correlations, 151–152
    modifications of DNA sequences, 6
    sequencing, 41–44, 72
        application of, 214–215
    structure, 5
DNMT3A, 56
DNMT3B, 56
DNMT3L, 56
Documentation
    of genotype phenotype correlations databases, 137–138
    of inherited disorders
        ACH, 18–19
        blue sclerotics and fragility of bone, 19–20
        color blindness, 19
        deafness, 18
        Duchenne muscular dystrophy, 22–23
        ectrodactyly, 17
        genetic causes of specific diseases and family studies, 23
        hemophilia, 18
        hereditary optic atrophy, 20
        Huntington's chorea, 20–22
        Treasury of Human Inheritance, The, 17
Dosage sensitivity, 83–84
Dravet syndrome, 280
DRD2 dopamine receptor, 209
Driver gene mutations, 248–250
Drugs
    and binding to receptors, 209–210
    drug-induced hemolytic anemias, 216
    responses, 212–215

Duchenne muscular dystrophy (DMD), 22–23, 321, 340–342
  laboratory diagnosis, 342
  therapy molecular approaches, 342–344
DUX1, 48
Dwarfism, 18–19
Dynamic mutability of microsatellite repeats, 348–350
  genomic tandem repeats in autism, 349–350
Dynamin 2 (DNM2), 160
Dysmorphology syndromes, 139–140
Dystrophic myotonica, 22
DYT1, 68–69
DYT11. See SGCE myoclonus syndrome

E

Ectrodactyly, 17
Editing, 162
Ehlers – Danlos syndrome (EDS), 148–151
Electrophoresis, 29–30
Electrophoretic techniques, 28
EnaC. See SCNN1A sodium channel (EnaC)
Encyclopedia of DNA Elements Project (ENCODE Project), 135–136
Endocrine disorders, 185
Endogenous antisense RNAs, 157–158
Enhancer, 98, 104
Environmental factors, 212–215
Enzyme replacement therapy (ERT), 223–224, 310
  in clinical use for treatment of specific lysosomal storage diseases, 311
Enzymes, 28
Enzymology, 28, 30
Epidermal growth factor receptors (EGFRs), 209
Epigenetic factors relevant to gene expression, 98–102
  cohesinopathies, 101–102
  disorders of epigenetic machinery leading to neurodevelopmental disorders, 101
Epilepsy, 276–277
  genetics, genomics, and relevance to therapy, 279–280
  genetics, mechanisms, and therapy, 278–279
Epimap, 102–103
  combinations of variants in different genes and impact of phenotype, 103
Epivariations, 79–80
Epstein – Barr Virus (EBV), 253
Eteplirsen, 156
Exome sequencing, 67, 272–273
Exomiser, 276
Exon splicing, 110f

Exosome component 3 (EXOSC3), 115–116
Expanded carrier screening, 190–191
Expression quantitative trait loci (eQTL), 95
Eye diseases, 171–172

F

F508 deletion (F508del), 184–185
Fabry disease, 316
Factor interacting with PAP (polyA polymerase), 109–110
Factor V Leiden, 196–197
Familial hypercholesterolemia (FH), 197–198, 301–303
Familial Mediterranean fever (FMF), 196
Family studies, 23
FANCD2, 21
FANCI-associated nuclease 1 (FAN1), 21
Fanconi anemia (FA), 332
Fatty acid oxidation defects, 186
FBN1 mutations, 148
  in Marfan syndrome, 147–148
FGF8 gene, 17
FGFR1 gene, 17
Fibrillin2 (FBN2), 147
Fibroblast growth factor pathway (FGF pathway), 134
Fibroblast growth factor receptor (FGFR)
  FGFR3, 18–19
  receptor defects, 143–144
  signaling pathway, 142–143
FIRRE LNC RNA, 97
FKBP10, 323
Fluorescent staining methods, 30
Fluorochromes, 28
Fluorometer, 28
Fluorophores, 28
5-fluorouracil, 214
Fluorpyrimidine, 214
FMR1, 78
Fomivirsen, 156
Food and Drug Administration (FDA), 309
Fragility of bone, 19–20
Fragment-based drug discovery, 219
Frataxin (FXN), 21
Friedreich ataxia (FRDA), 290
Functional genome analysis, 54

G

Galactosemia, 186–187
Galactosylceramidase (GALC), 224
GalNac. See N-acetyl galactosamine (GalNac)
Gastrointestinal stromal tumors (GIST), 241
GATA binding protein 4 (GATA4), 107

Geller syndrome, 296
Gene therapy, 155, 162–164
  by adding genes, 162
  early gene therapy applications, 162–164
  in lysosomal storage diseases, 316
  in specific diseases, 171–172
    eye diseases and retinal degeneration, 171–172
Gene(s), 4–5
  applications of studies of, 6–9
  editing, 165–166
    delivery of agents for, 167–168
    delivery of reagents for editing, 166
    early discoveries, 165–166
  environment interactions, 200
  expression, 96–98
    applications of studies of expression to clinical medicine, 6–9
    gene-directed therapies in clinical trials in hemoglobinopathies, 320
    gene-directed treatments, 316–318
    gene-related therapies, 221
    involved in developmental epileptic encephalopathies, 279–280
    with more than one promoter, 105–106
    products with germline mutations, 237–240
  transfer, 164
  variants, 215–216
    drug-induced hemolytic anemias, 216
    G6PD deficiency, 215–216
Genetic disorders, 57–58
  approaches to target identification and therapeutic design in, 220–221
  with high frequency in certain populations, 191–195
    mutation heterogeneity in Tay – Sachs HEXA mutations, 194–195
    predominant manifestations of disorders, 193–194
    Tay – Sachs disease due to hexosaminidase mutations, 191–193
  with increased frequency in other specific populations, 195–196
    aspartylglucosaminuria, 195–196
    familial Mediterranean fever, 196
Genetic(s), 4–5
  alterations in cancers in children, adolescents, and young adults, 247–248
  diseases, 186
  and environmental factors and additional aspects of population screening, 200–201
    iodine deficiency, hypothyroidism, 201
    severe visual impairment in children, 200–201

factors, 212–215
genetic-guided therapies, 300–301
genetics-based therapeutic targets, 300–301
and genomic studies in congenital anomalies and/or neurodevelopmental anomalies, 271–273
  exome sequencing, 272–273
  microarray analyses, 271–272
heterogeneity, 133
mosaicism in inborn errors of immunity, 13
polymorphisms in CYP450 system, 210–211
Genistein, 311
Genome-wide association studies (GWAS), 95, 297, 319
Genome-wide methylation analyses, 78
Genomes
  applications of studies of, 6–9
  sequencing
    in cancer, 255–256
    in pediatric developmental defects, 69–70
Genomic copy number variants and epilepsies, 281
Genomic data leading to therapeutics, 160–161
  additional RNA modifications to improve use in therapy, 161
Genomic function, 53–55
Genomic medicine, 271
  in common diseases in adults, 295–297
  hypertension, 295
Genomic studies to guide diagnosis and therapy in epilepsy, 276–277
  metabolic pathways and epilepsy, 276–277
Genomic tandem repeats in autism, 349–350
Genomics databases, 138
Genotype phenotype axis, 103–104
Germline mutations, 13
  and developmental origins of cancer, 244–245
    medulloblastoma, 244–245
  genes with, 237
Germline succinate dehydrogenase gene mutations and cancer predisposition, 241–242
Givosiran, 318
GLI1 mutations, 142
GLI3 mutations, 142
Globoid cell leukodystrophy. *See* Krabbe disease
Glucocorticoid remediable aldosteronism (GRA), 295–296
Glucose 6 phosphate dehydrogenase deficiency (G6PD deficiency), 190, 215–216
Glycogen storage diseases, 222–223
Glycoprotein Ib (GPIb), 330
GNAS locus, 77–79
GnomAD database, 138
Golodirsen, 156
Gonadal mosaicism, 11

Gordon syndrome, 296–297
GPC3 gene, 76
GTP cyclohydrolase 1 (GCH1), 286
Guanylate cyclase 2D (GUCY2D), 351

H

Hair cell regeneration, efforts to promote, 174–175
Half-life of enzymes, strategies designed to increase, 316
Haploinsufficiency, 83–84
Haplotype, 42
  analysis, 51–53
  phasing, 10, 51
    noninvasive prenatal screening, 51
HapMap consortium, 42
Hearing impairment, 18
Hematopoietic stem cells (HSCs), 161, 312
  collection of hematopoietic stem cells for therapy, 164–165, 327–328
  and progenitors, 326–327
  and therapies, 164
Hemochromatosis, 199–200
Hemoglobinopathies, 181–183, 186
  treatment
    including gene therapy, 318–320
    through genetic silencing of BCL11A expression using antisense strategy, 320–321
Hemolytic disease, 35–36
Hemophilia, 18, 328–330
  hemophilia A, 18
  hemophilia B, 18
Hereditary breast and ovarian cancer (HBOC), 197–198
Hereditary gastrointestinal cancers, 242–244
  pediatric cancers, 242–244
Hereditary optic atrophy, 20
Hereditary particles, 3
Heritability of coronary heart disease, 300
Heterogeneous ribonucleoproteins (hnRNPs), 321
Hexosaminidase gene (HEXA), 191
HGMD database, 138
High-pressure liquid chromatography, 181
Histone methylation, 100f
Histone modification, 98–99
HLA typing, 214–215
Homeobox D3 (HOXD3), 108
HTRA1, 67
Human disorders associated with impaired ribosomal biogenesis or function, 112–113
Human genetic variation, 200
Human Genome Project, 271
Human Phenotype Ontology database (HPO database), 137, 274–275
Hunter syndrome, 313–316
Huntington disease therapy, 335–336
  RNA targeting and RNA interference approaches, 336
Huntington's chorea, 20–22
Hydrogen bonds, 5
Hypermobile Ehlers–Danlos syndrome, 150–151
Hypertension, 295
  apparent mineralocorticoid excess, 296–297
  glucocorticoid remediable aldosteronism, 295–296
  personalized medicine and treatment of hypertension, 295
  polygenic factors leading to, 297–298

I

Immunodeficiencies, 186
  newborn screening for, 188
Imprinting, 72–73
  disorders, 73
  imprinted genomic regions, 56–57
In situ hybridization, 8
Inborn errors
  of development, 134–135
  genetic mosaicism in inborn errors of immunity, 13
  of metabolism, 222–223
Incomplete penetrance, 68–69, 125–126
Indian hedgehog (IHH), 141
Ingenious staining methods, 30
Inheritance, 4
Inotersen, 156
Insulin-like growth factors (IGFs), 67
  IGF2, 75
Interferon-induced transmembrane protein 5 (IFITM5), 324
Interstrand crosslinks (ICLs), 334
Iodine deficiency, hypothyroidism, 201
Isoelectric focusing, 181
Ivacaftor, 184

K

Kabuki syndrome, 57
Kagami–Ogata syndrome, 80
KCNQ1OT1, 75
KDM5C mutations, 57
KHDC3L, 81
KIF7, 141
KNL1 kinetochore scaffold 1 protein. *See* CASC5
Krabbe disease, 224
KRAS G12C, 260
KRAS pathway, 238f

L

L-3,4-dihydroxyphenylalanine (DOPA), 286
Landsteiner Wiener glycoprotein (LW glycoprotein), 36
Laws of Mendel, 3–4
Leber's congenital amaurosis, 350
Leber's congenital amaurosis as an early onset retinal diseases (LCA/EOSRD), 350
Leber's disease, 20
Lecithin retinol acyltransferase (LRAT), 352
Lentiviral vectors, 163
Lethal Congenital Contracture syndrome (LCCS1), 116
Leukemia, 251–253
LH1, 323
Liddle syndrome, 296
Linkage maps, 42
Loeys – Dietz syndromes
  phenotypic features, 146
  type 4 TGFB2 mutations, 146
  type 5 TGFB3 mutations, 147
Long noncoding RNAs (lncRNAs), 76–77, 96–97, 157–158
Long-range sequencing
  and identification of structural genomic variants leading to disease, 53
  relevance to diagnosis of rare disorders, 48–49
Long-read sequencing (LRS), 10
  applications of, 44–45
  for detection of genomic variants, 9–10
Loss of heterozygosity (LOH), 259
Low density lipoprotein receptor gene (LDLR gene), 79–80, 98
Lumasiran, 318
Lymphocyte depleted Hodgkin disease (LDCHL), 253
Lymphomas, 251–253
Lynch syndrome, 197–198
Lysosomal storage diseases, 186, 223–224, 310
  chaperone therapies in treatment of, 312
  enzyme replacement therapy in clinical use for treatment of specific, 311
  gene therapy in, 316
  precision medicine in lysosomal storage disease disorders, 313–314
Lysosomes, 310

M

Maccaca mulatta, 35
Maccacca rhesus, 35
Magnetic resonance imaging (MRI), 335
MALAT1, 97
Marfan syndrome
  15q21.1 FBN1, 147–148
  FBN1 mutations in, 147–148
  multiple submitters, 148
Marrams, 20–21
Massively parallel sequencing. *See* Second generation sequencing
MATH1 transcription factor, 174
Maxam – Gilbert sequencing method, 41
MBNL1, 108
McCune – Albright syndrome, 77
MED12, 105
MED13, 105
MED13L, 105
Mediator mutations, 105
Medulloblastoma, 244–245
MEFV gene, 196
MEG3, 78
Meiosis, 3
Mendelian genetics, 31
Mercury vapor lamp, 28
Metabolic Basis of Inherited Diseases, The, 30–31
Metabolic pathways, 276–277
5-methyl cytosine, 170
Methyl-malonyl-CoA-mutase (mut), 226
Methylation, 70
  analyses, 55–56
  and cancer, 57–58
  dietary components for, 99*f*
  methylation-specific antibodies, 54–55
Methylmalonic acidemia, 226
Microarrays, 9
  analyses, 271–272
Microexons, 111
MicroRNAs, 112
  as mRNA inhibitors, 157
  in therapeutic use, 121
Microsatellite instability, 264*f*
Miglustat, 315
Migroms, 20–21
Mineralization, defects in, 324–325
Mipomersen, 156
miR17, 157
mir21, 157
Mitochondria, 226
Mitochondrial diseases and explorations of treatments, 228–229
  search for treatment of specific mitochondrial disorders, 228–229
Mitochondrial fatty acid oxidation defects and carnitine shuttle disorders, 222
Modified penetrance, 122–123

Molecular analyses and therapies relevant to hearing loss, 172–175
  efforts to promote hair cell regeneration, 174–175
  RNA-based therapies in deafness, 174
Molecular profiling, 260–262
Molecular-based therapeutics, 184–185
Monoamine neurotransmitter disorders, 285–288
Monoclonal antibodies as therapeutic agents, 219–220
Monozygotic twins, 135
Mosaicism, 11–13
  chromosomal mosaicism, 12
  detection, 12
  and genetic diseases, 12–13
Mucopolysaccharidosis II (MPS II), 313–316
Multilocus imprinting disorders, 80–82
Multiple endocrine neoplasia and associated gene defects, 240
Muscular dystrophies, 22
Muscular skeletal diseases, 186
Mutation, 159
Mutation heterogeneity in Tay – Sachs HEXA mutations, 194–195
MYD88, 252
Myeloid leukemia, 253–255
Myotonic dystrophy, 22

N

N-acetyl galactosamine (GalNac), 158
N-acetyl transferases, 211
  NAT1, 212
  NAT2, 212
Na+/K+-ATPases, 69
Naphthyridine azaquinolone (NAT), 294, 339
NBseq, 187
Netrospin, 70
Neural stem cells, 161
Neurexin 1 (NRXN1), 105
Neurofibromatosis type 1 (NF1), 220–221
Neurofibromin, 220–221
Neuroimmune disorders, 312
Newborn, 35–36
Newborn screening (NBS), 181–183, 185–187, 312
  aminoacidurias, 185
  endocrine disorders, 185
  fatty acid oxidation defects, 186
  galactosemia, 186–187
  for immunodeficiencies, 188
  in other parts of world, 190
Next generation sequencing (NGS), 9–10, 44, 66
Next generation sequencing in clinical neurology, 66–67

Nicotinamide adenine dinucleotide phosphate (NADP), 215
Nicotinamide nucleotide adenylyltransferase 1 (NMNAT1), 351
NIH (National Institutes of Health) somatic cell gene editing program, 168–169
NIPBL, 101–102
NLRP2 mutation, 81
NLRP5 mutation, 81
NLRP7 mutation, 81
Noncoding variants, 95
Noninvasive prenatal screening, 51
Noninvasive prenatal testing (NIPT), 52, 88–90
Nonmonogenic risk, 300
Nonprotein coding genomic regions, variants in, 50–51
Nonrepeat cerebellar ataxias, 294
Nonsense mutations and human disease, 119
Nonsense-mediated decay (NMD), 115, 117–119
NORAD, 97
NOVA1, 108
Nuclear paraspeckle assembly transcript 1 (NEAT1), 97
Nuclei, 3
Nucleic acids, 5
Nucleotides, therapies designed to block, 155
Nusinersen, 156

O

OAT1B1, 214
OCT7. See POU class 3 homeobox 2 protein (POU3F2)
Ocular gene therapy, 350–352
Oligodendrocyte transcription factor 1 (OLIG1), 107
Oligonucleotide therapies, 155–158
  delivery challenges in, 158
  long noncoding RNAs, small RNAs, endogenous antisense RNAs, 157–158
  microRNAs as mRNA inhibitors, 157
  RNA inhibition in therapies, 156–157
  steric block oligonucleotides, 156
"One phenotype many genes", 68–69
  incomplete penetrance, 68–69
Online Mendelian Inheritance in Man (OMIM), 66, 137–138
OOEP, 81
OPN1MW, 19
OPSIN1LW, 19
Optic atrophy 1 (Opa1), 20
Optical DNA mapping (ODM), 70
  in human genome studies, 70
Organic acidemias, 186–187, 226

Organic acids, 30
Ornithine transcarbamylase (OTC), 106
  gene promoters and enhancers, 106—107
Orphanet, 275
Ossification, defects in, 324—325
Osteogenesis imperfecta (OI), 20, 322—323
  treatment, 325
Osteosarcoma, 245—247
  autosomal-dominant gene mutations, 246
  developmental origins of pediatric cancers, 246—247
  syndromic genes associated with osteosarcoma, 246
Oxford Nanopore sequencing, 10

P

p57(KIP2), 75
Pacific biosystems (PacBio), 10
  sequencing, 10
Pallister — Hall syndrome (PH syndrome), 142
Passenger gene mutations, 248—250
Patched (PTCH), 141
  PTCH1 mutations, 142
Pathogen sensitivity, 200
Pathogenic/likely pathogenic mutations, 149
Patisiran, 157, 318
Pediatric cancers, 242—244
Pedigrees, 21
Penetrance
  in inherited eye diseases, 124—125
  of mutations, 122—123
Peptidyl-prolyl isomerase B (PPIB).
  See Cyclophilin
Peripheral T-cell lymphomas, 253
Peripherin 2 (PRPH2), 48, 350
Peroneal muscular atrophy, 22
Personalized medicine and treatment of hypertension, 295
Pfeiffer syndrome, 143
Pharmacodynamics, 209—210
Pharmacogenes, 213—214
Pharmacokinetics, 209—210
PhenCards, 138
PhenoCB, 275
PhenoDB, 275
Phenolyzer, 138
Phenome-wide association studies, 139
PhenomeCentral, 275
Phenotype, 96—98, 133—134
  descriptions, 133
Phenotypic abnormalities
  diagnoses and management in disorders with, 274—275

HPO database, 274—275
  resources, 275
Phenotypic defects due to defects in sonic hedgehog signaling pathway, 140—142
Phenotypic differences in individuals with same genetic defect, 135—136
Phenotypic heterogeneity, 68
Phosphoinositide 3-kinase (PI3K), 12—13
Physical maps, 42
Placental mosaicism, 11
PLAG1, 72—73
Plasmodium vivax, 200
Platelet receptors for Von Willebrand factor, 331
Pleiotropism, 133
Pleiotropy, 68
Pluripotent stem cells
  for investigation of disease manifestations and effects of therapies, 161
  in investigations of disease therapies, 322
PMP22 protein, 339
POLII, 108
POLR1C gene, 113
POLR1D gene, 113
Polyadenylation, 109
Polygenic factors leading to hypertension, 297—298
Polygenic risk scores, approaches to determining, 301
Polyglutamine cerebellar ataxias, 291, 337—340
Polymorphic DNA markers, 42
Polypyrimidine tract binding protein 1 (PTBP1), 108
Population analyses, 83
Population-wide screening of adults, 197—199
  promoting diverse population screening, 199
Porphyrias, 196, 216—217
Posttranscriptional control, 116
Posttranscriptional regulation, 112
Postzygotic mosaicism, 11
Postzygotic mutations, 13
POU class 3 homeobox 2 protein (POU3F2), 107
POU3F4, 50
Prader — Willi syndrome (PWS), 57, 74—75
Precision medicine in lysosomal storage disease disorders, 313—314
Preclinical trials, 166—168
Preinitiation complex (PIC), 113
Prenatal exome sequence analysis, 86—87
Primary immunodeficiency, 125—126
Primary microcephaly, 49
Prime editing, 170
Prime editing RNA (peg RNA), 170
PRKAR1A gene, 53

Procollagen-lysine, 2-oxoglutarate 5-dioxygenase 1 (PLOD1). See LH1
Programmable base editing, 169–170
Promoters, 104
Properties of enzymes, 29
Protein(s), 27
　clearance, 336–337
　in collagen folding and cross-linking, 323
　folding, 28
　　and secondary modifications, 322–323
Proteus syndrome, 12
Protoporphyrinogen oxidase (PPO), 216

R

Rare disease medicine, 273
RAS signaling pathway, 231
RAS/MAP signal transduction pathway, gene products involved in, 139–140
RBFOX1, 108
RBFOX2, 108
RBFOX3, 108
Recessive disorders carrier screening in specific populations, 181
Recurrent miscarriage, investigations of causes of, 88
Regulatory elements, 97–98
Regulatory genome, 96–98
Retinal degeneration, 171–172
Retinitis pigmentosa GTPase regulator protein (RPGR), 350, 352
Retinoblastoma and RB1, 240–241
Retinoid isomerohydrolase (RPE65), 172, 352
Retinol dehydrogenase 12 (RDH12), 352
Rh blood group system, 35–36
RHCc gene, 36
RHD genotyping, 36
RHDd gene, 36
RHEe gene, 36
Rhodopsin protein (RHO protein), 350
Ribosomes biogenesis, functions, and defects, 112–113
Ribosomopathies, 113
Right ventricle (RV), 190
RNA
　binding proteins, 116
　editing, 171
　inhibition in therapies, 156–157
　modifications and regulation of gene expression, 116–117
　polymerase II, 108
　RNA-based therapies in deafness, 174
　sequencing in diagnosis of genetic diseases, 121–122
　surveillance, 115–117
　targeted therapeutics, 119–120
　targeting, 336
　therapies designed to block RNA derived from specific gene, 155
RNA inducing silencing complex (RISC), 156–157
RNA interference (RNAi), 174
　approaches, 336
RNAse H1 enzyme, 155

S

S-Adenosyl methionine, 100f
"Safe harbor" sites (SHS), 163
Sanfillipo syndrome, 224
Sanger method, 41
Sclerostin, 325
SCNN1A sodium channel (EnaC), 175
Screening for X-linked disorders, 189–190
Second generation sequencing, 44
Secretor status, 34–35
Separation methods, 27
Sequence variant interpretation, 45–49
　long-range sequencing relevance to diagnosis of rare disorders, 48–49
　reinterpretation of data and secondary findings, 47–48
Sequencing. See also Whole genome sequencing
　in prenatal diagnosis, 88–90
　reactions, 42
Serine-arginine rich proteins (SR proteins), 321
SERPINF1, 324
SERPINH1, 323
Severe visual impairment in children, 200–201
SGCE myoclonus syndrome, 68–69
Short inhibitory RNAs (siRNAs), 120–121, 156–157, 318
Short stature homeobox gene (SHOX), 95–96
Sickle cell disease, 181
Sickle hemoglobin polymer (HBS), 319
Signaling pathways, 134
Silver–Russell disorder, 70
Silver–Russell syndrome (SRS), 75–77
　SRS1 and SRS3, 75
Simpson–Golabi–Behmel syndrome 1, 76
Simpson–Golabi–Behmel syndrome 2, 76
Single nucleotide polymorphic markers (SNPs), 271
　microarrays, 9
Single nucleotide variants (SNVs), 70
Small RNAs, 157–158
SMCHD1, 48
Smoothened (SMO), 141

Solute carrier 26A9 (SLC26A9), 175
Somatic mosaicism, 11–12
Somatic mutations in cancer, 258
Sonic hedgehog signaling, 134
Sonic hedgehog signaling pathway, phenotypic defects due to defects in, 140–142
SORT1, 98
SP7, 324
Spastic paraplegias and ataxias, 288–290
　clinical features, 289–290
Spectrophotometry, 28
Spinal muscular atrophy (SMA), 188, 345–347
Spinocerebellar ataxias, 293–294
　nonrepeat cerebellar ataxias, 294
Splice mutations and diseases, 158–159, 321–322
Splice regulatory elements (SREs), 159, 321
Splice switching oligonucleotides, 159
Spliceosome, 111
Spliceostatin, 322
SRY-box transcription factor 2 (SOX2), 107
Standardized phenotype documentation, 137–138
Stem cells. *See also* Pluripotent stem cells; Hematopoietic stem cells (HSCs)
　collection of hematopoietic stem cells for therapy, 164–165
　hematopoietic stem cells and therapies, 164
　and importance in gene therapy, 164–165
Steric block oligonucleotides, 156
Stevens–Johnson syndrome, 214
Structural genomic variants, 10–11, 82–84
　ACMG recommendations, 84
　clinical significance, 11
　dosage sensitivity and haploinsufficiency, 83–84
　population analyses, 83
Substrate reduction therapy, 314–316
Succinate dehydrogenases, 241
SUFU, 141
Superoxide dismutase (SOD1), 121
Suppression of nonsense mutations, 118–119
SUPT5H, 108
Syndromic genes associated with osteosarcoma, 246
Synthetic lethality, 262
　in cancer treatment, 230

T

T-box transcription factor 20 (TBX20), 107
T-cell, 262
T-cell receptor excision circles (TRECs), 188
TALENs (plant transcription factors), 165
Tandem mass spectrometry, 187
Tay–Sachs disease due to hexosaminidase mutations, 191–193
TCOF1 encoding gene, 112–113
TERT promoter mutations, 98
Tetrahydrobiopterin (BH4), 286
Therapeutics, 213–214
　targets and developing therapies, 217–218
Therapy lysosomal diseases, 310–311
Therapy-related genetic and genomic information, 260–262
Thiopurines-mercaptotransferase activity (TPMT), 213–214
Third generation sequencing, 44
TLE6, 81
TMEM38B gene, 323
Toll like receptor 9 (TLR9), 252
Topologically associated domains (TADs), 101
TOR1A encoding gene, 68–69
TP53, 17, 259
Transcription, 162
　elongation, 108
　factors, 101, 107–108
　　transcription elongation and RNA polymerase II, 108
　initiation and promoters, 105–107
　　genes with more than one promoter, 105–106
　　ornithine transcarbamylase gene promoters and enhancers, 106–107
　termination, 109
Transcriptome sequencing, 70–72
Transforming growth factor (TGF), 325
　beta, 134, 144–147
　TGFB1, 175
Transfusion, 33, 35
Transfusion-dependent β-thalassemia (TDT), 319
Translation, 112–113, 117
　of biomedical observations to treatments and health improvements, 218–219
　of mRNA to proteins and associated defects leading to disease, 113–114
Transporter defects, 226–227
　cystic fibrosis, 227
　Wilson's disease, 227
Treacher Collins syndrome, 112–113
Treasury of Human Inheritance, The, 17, 21
Trichostatin, 344
Trinucleotide repeat disorders, 332–335
tRNAs, 114–115
TSC2 gene, 332
Tuberous sclerosis, 331
Twin studies and analysis of gene effects on phenotype, 135
Tyrosine kinase activity, 210

U

UBE3A, 73–74
UDP glucuronyl transferase enzymes and medication biotransformation, 212
Undiagnosed disease network (UDN), 303
Undiagnosed diseases and application of genetic and genomic studies, 303–304
Uniparental disomy (UPD), 49
Urea cycle disorders, 187, 187f
US Food and Drug Agency (FDA), 260
Usherin protein (USH2A), 171, 351
Utrophin, 344–345

V

Variability, 96–98, 133
Variable expressivity, 69
Variable genomic abnormalities in individuals with same phenotype, 136–137
Variable penetrance of disease due to polymorphisms in regulatory factors, 123–124
Variable phenotypes associated with specific mitochondrial mutations, 136
Variant Curation Expert Panels, 46
Variants, 212–215
  in nonprotein coding genomic regions, 50–51
Variegate porphyria, 216
Vascular EDS, 149
Vectors, 163
Vitamins, 27
VMD2, 351

Von Willebrand disease (VWD), 330
  laboratory tests, 330

W

Whole genome sequencing. *See also* Long-read sequencing (LRS)
  measurement toolkit for assessing clinical utility of, 65–67
  next generation sequencing in clinical neurology, 66–67
  of metastatic solid tumors, 259–260
Wilson's disease, 227
WNT signaling, 134
WNT10B gene, 17

X

X-linked dominant, 285
X-linked recessive, 285
X-ray crystallography, 28
Xanthine oxidase, 213–214
XIST, 97

Y

Yoyo1, 70

Z

ZAC1. *See* PLAG1
Zinc finger nuclease (ZFN), 165–166
Zinc fingers, 165
ZYNTEGLO, 319